Building Foundation Analysis: Soil–Structure Interaction

Building Foundation Analysis: Soil–Structure Interaction

Guest Editors

Yuxin Ban
Qiang Xie
Xiang Fu

MDPI

Basel • Beijing • Wuhan • Barcelona • Belgrade • Novi Sad • Cluj • Manchester

Guest Editors

Yuxin Ban
School of Civil and Hydraulic Engineering
Chongqing University of Science and Technology
Chongqing
China

Qiang Xie
School of Civil Engineering
Chongqing University
Chongqing
China

Xiang Fu
School of River and Ocean
Chongqing Jiaotong Unversity
Chongqing
China

Editorial Office

MDPI AG
Grosspeteranlage 5
4052 Basel, Switzerland

This is a reprint of the Special Issue, published open access by the journal *Buildings* (ISSN 2075-5309), freely accessible at: https://www.mdpi.com/journal/buildings/special_issues/5N9343N849.

For citation purposes, cite each article independently as indicated on the article page online and as indicated below:

Lastname, A.A.; Lastname, B.B. Article Title. *Journal Name* **Year**, *Volume Number*, Page Range.

ISBN 978-3-7258-4244-5 (Hbk)
ISBN 978-3-7258-4243-8 (PDF)
https://doi.org/10.3390/books978-3-7258-4243-8

Contents

About the Editors

Yuxin Ban

Professor Ban Yuxin is a faculty member at the School of Civil Engineering, Chongqing University of Science and Technology. She holds a PhD in Civil Engineering from Chongqing University, specializing in research related to geotechnical engineering and rock mechanics. She has hosted one National Natural Science Foundation Youth Fund project, one Chongqing Science and Technology Commission general project, and one Chongqing Education Commission research project; published over 20 high-level journal articles and obtained three authorized patents; and guided graduate students to undertake one Master's Innovation Program project and publish one SCI paper. She serves as a member of the Working Committee of Chongqing Society of Rock Mechanics and Engineering and a reviewer for a well-known journal in the industry.

Qiang Xie

Prof. Qiang Xie is a full professor and doctoral supervisor at Chongqing University's School of Civil Engineering, specializing in geotechnical engineering with a focus on rock and soil mechanics, geological hazards, and environmental geotechnics. Prof. Xie teaches core courses like Soil Mechanics and Foundation Engineering, leads over 40 research projects, and holds leadership roles in academic societies such as CSRME. His work bridges theoretical research and practical applications in civil engineering, contributing significantly to geotechnical testing technologies and hazard mitigation strategies.

Xiang Fu

Prof. Xiang Fu is an associate professor and master's supervisor at Chongqing Jiaotong University, specializing in geotechnical engineering and BIM technology. He holds a Ph.D. and completed postdoctoral research at the Changjiang River Scientific Research Institute of the Changjiang Water Resources Commission3. Appointed as a Deputy Director of a research institute, he serves as an expert member of the Chongqing Rail Transit Construction Expert Committee (Geotechnical/BIM) and the Chongqing Municipal Bureau of Planning and Natural Resources' Natural Disaster Preliminary Response Expert Group3. His interdisciplinary expertise bridges geotechnical engineering, smart construction technologies, and disaster management.

buildings

MDPI

Article

Influence of Stress Level on the Determination of In Situ Stress by DRA Method

Yuxin Ban [1], Qiang Xie [2,3,4,*], Jiahui Tang [2], Jun Duan [2], Xiang Fu [5] and Xiaoqiang Xue [2]

1 School of Civil and Hydraulic Engineering, Chongqing University of Science and Technology, Chongqing 401331, China; banyuxin@cqust.edu.cn
2 School of Civil Engineering, Chongqing University, Chongqing 400045, China
3 Key Laboratory of New Technology for Construction of Cities in Mountain of Education, Chongqing University, Chongqing 400045, China
4 National Joint Engineering Research Center of Geohazards Prevention in the Reservoir Areas (Chongqing), Chongqing 400045, China
5 School of River and Ocean, Chongqing Jiaotong University, Chongqing 400074, China
* Correspondence: xieqiang2000@cqu.edu.cn

Abstract: Obtaining reliable in situ stress field data is the precondition of the design and excavation of underground engineering. The deformation rate analysis (DRA) method based on the rock deformation memory effect is an effective method for testing in situ stress. Cyclic loading tests were carried out on dense granite to research the influence of stress level on the identification accuracy of the DRA method. The DRA curves of horizontal and longitudinal strains are compared to find the difference between the accuracy in different directions. The results showed that in the range of 15%-60% uniaxial compression strength, the rock deformation memory effect is obvious, which indicates that the DRA method is applicable in low-stress regions, while the measuring accuracy is not as good as that in high-stress areas. When the specimen is subjected to low pre-stress, the accuracy is higher when using the horizontal strain than the longitudinal strain. The accuracy of both horizontal and longitudinal DRA methods increases with the increase in pre-stress. The measuring accuracy is similar when the pre-stress is higher than 45%UCS. What is more, the discrete element numerical model is calculated to verify the test results and reveal the microscopic mechanism of the DRA method under the control of crack evolution. The characteristics of rock memory are related to the evolution of crack closure, cracking, and expansion. The research results can provide an experimental basis for promoting the theoretical development and field application of the testing in situ stress with the DRA method.

Keywords: rock deformation memory; deformation rate analysis (DRA); in situ stress; stress level; PFC

1. Introduction

In situ rock stress is one of the important indexes to determine the instability possibilities for various kinds of underground engineering [1]. Mastering a reliable three-dimensional in situ stress field at a low cost is necessary for underground engineering design, excavation, and stability evaluation [2,3]. Acquiring accurate three-dimensional spatial distributions and variations of in situ stress fields aids in predicting and controlling deformation and instability risks in subsurface structures and provides a scientific basis for optimizing design and support strategies.

The in situ stress measuring methods can be divided into two main categories, including in situ testing methods and laboratory methods. The hole wall caving method, the core stress relief method, and the hydraulic fracturing method are typical in situ test methods [4–6]. While these techniques have achieved notable success in practical engineering applications, each has inherent limitations. For instance, the borehole caving method relies on borehole shape data, the core stress relief method presumes the rock mass to be an ideal linear elastic body, and the hydraulic fracturing method assumes alignment of the borehole axis with a principal stress direction—assumptions difficult to fulfill in complex geological conditions. Consequently, laboratory methods, including the Acoustic Emission and Deformation Memory Method, are widely accepted to estimate in situ stress due to low cost, ease of operation, and minimal test environmental requirements. The Acoustic Emission method is proposed based on the Kaiser effect, which is proven to be affected by many factors, such as size, water content, and timeliness [7,8]. These challenges underscore the limited applicability of conventional methods in complex environments, especially in scenarios requiring high-precision measurements and long-term monitoring, where significant research gaps persist.

The deformation rate analysis (DRA) method, a novel in situ stress measurement technique based on the deformation memory effect (DME), introduces a new technological approach to stress measurement [9]. The DRA method involves cyclic loading of core samples from previous excavation projects in the laboratory, enabling in situ stress testing [10]. This technique is distinguished by its simplicity, short testing cycle, low cost, minimal interference, and independence from the physical and mechanical properties of the samples. Compared to traditional methods, the DRA method allows for large-scale testing and statistical analysis, thereby enhancing measurement precision and reliability [11].

Although it has been studied for more than 40 years, the systematic research of the experiment and theory still needs to be improved as some controversial conclusions were made when determining the suitable range of the pre-stress level in DRA methods. Within this suitable range, a clear inflection corresponding to the maximum pre-stress can be detected to ensure test accuracy [12]. In early studies, it is considered that the DRA method is only useful during high-stress zones, i.e., 30–60% of the uniaxial compressive strength (UCS), as irreversible plastic strain is generated in the rock [13,14]. However, some scholars believe that the recognition of pre-stress in the elastic region is also satisfactory. For example, Yamamoto et al. [15] found that the DRA method can measure the pre-stress of granite at 1–6 MPa. Hunt et al. [16], Seto et al. [17], and Park et al. [18] also confirmed by experiments that the DRA method can successfully measure the pre-stress of rock even if the stress level is lower than 20% UCS. Aghill [19] reported that the lowest pre-stress is seen in sandstone specimens in the elastic deformation stage. Consequently, the research on the influence of pre-stress level on the DRA-recognition accuracy method still needs to be strengthened.

Some studies turn to numerical simulation methods to explain the microscopic mechanism explanation of DME, and a microcrack initiation and development model is proposed. When the loading stress value exceeds the peak value of the previous loading stress, microcracks will occur. The generation of new microcracks will cause an increase in inelastic strain, resulting in a downward inflection point in the DRA curve. Hunt et al. [16] confirmed that the Kaiser effect is related to the generation and propagation of microcracks and concluded that when the initial loading stress is smaller than the initial stress value of microcracks, the rock deformation memory effect does not exist. That is, the stress value of the first loading needs to be in the inelastic stress region or above the initial crack threshold stress range to form the rock deformation memory effect. Obviously, this viewpoint is inconsistent with the fact that early stress can be identified in the elastic zone. Tang et al. [12] simulate the deformation memory effect of rock in low-stress regions with the contact model in Particle Flow Code (PFC). A viscoelastic friction model composed of elastic element, viscous element, and Saint Venant body is further proposed. Wang et al. [20] suggested that the time interval between rock coring and the DRA test should be strictly controlled to prevent the occurrence of memory weakening. Consequently, more work should be conducted to explain the mechanism of the DRA method from microcrack behaviors.

In this paper, pre-stress levels fall into four different levels, from 15% to 60%. UCS is recognized with the DRA method to validate its effectiveness, and the micro-mechanism is revealed from the microcrack development in simulation calculation. In Section 2, cyclic uniaxial compression tests are conducted on granite specimens to test the deformation. In

Section 3, the sensitivity of the DRA method with horizontal and longitudinal strains is compared. In Section 4, a discrete element numerical model is established to reveal the microscopic mechanism of rock memory characteristics under different pre-stress levels.

2. Experimental Methods

2.1. Principle of DRA Method

Both reversible and irreversible strains are generated when rock is subjected to stress exceeding the elastic limit. When the stress level exceeds the maximum pre-stress value, new microcracks initiate and gradually expand inside the rock specimen, and new irreversible strain appears inside the specimen. During the unload period, the stress is smaller than the pre-stress value, there is no new crack initiation, the original crack does not expand, and the strain is reversible strain. With the help of the strain difference $\Delta\varepsilon_{ij}$ between different cycles, the gradient change of irreversible strain is reflected. $\Delta\varepsilon_{ij}$ is defined as follows [9]:

$$\Delta\varepsilon_{ij} = \varepsilon_j(\sigma) - \varepsilon_i(\sigma)(j > i) \tag{1}$$

where $\varepsilon_i(\sigma)$ and $\varepsilon_j(\sigma)$ are the strains corresponding to the *i*-th and *j*-th cyclic loading when the stress is σ.

In Equation (1), the reversible strain part between two cycles is eliminated, and irreversible strain difference is obtained. When the load is the same, the strain difference always exists in different loading stages, indicating that the inflection point of the stress–strain difference curve (DRA curve) always exists. Consequently, the rock deformation memory effect can be monitored through cyclic loading and unloading tests of rock specimens [21]. The stress value corresponding to the inflection point on the DRA curve is the pre-exist maximum stress for the rock.

The parameter Felicity ratio (*FR*) is widely used to identify the Kaiser effect [7]. Felicity ratio (*FR*), defined as the ratio of the subsequent recognized stress σ_{DRA} to the previous peak stress σ_{max}, is usually adopted as a quantitative parameter to evaluate the stress and deformation memory capacity [8]. The pre-stress identified results can be divided into three categories: accurate ($0.90 \leq FR \leq 1.10$), good ($0.85 \leq FR < 0.9$ or $1.10 < FR \leq 1.10$), and inaccurate ($FR < 0.85$ or $1.15 < FR$) according to the corresponding *FR* value ranges [22,23].

2.2. Sample Preparation

The cyclic loading test is carried out on dense granite rock samples, which are collected from an outcrop quarry in Wutian Mountain, Fujian Province. The initial ground stress is small enough to be ignored. The average density of granite specimens is approximately 2.8 g/cm^3. The specimen is processed in a $\Phi 50 \times 100$ mm cylinder as recommended by ISRM (Figure 1). The angle error between the end face and the axis of the specimen is controlled within $\pm 0.25°$, and the unevenness is no more than 0.05 mm.

2.3. Test Equipment

The test system consists of a rigid loading machine, a load control unit, and a strain acquisition unit, as shown in Figure 1. A servo control testing machine is adopted, whose maximum load is 2000 kN. The force control mode was adopted in the test, and the loading rate was 1 kN/s. A four-channel 1/4 bridges strain acquisition equipment is used to collect strain, and four channels share a compensation plate. The sampling rate (continuous sampling) is set to 2 Hz.

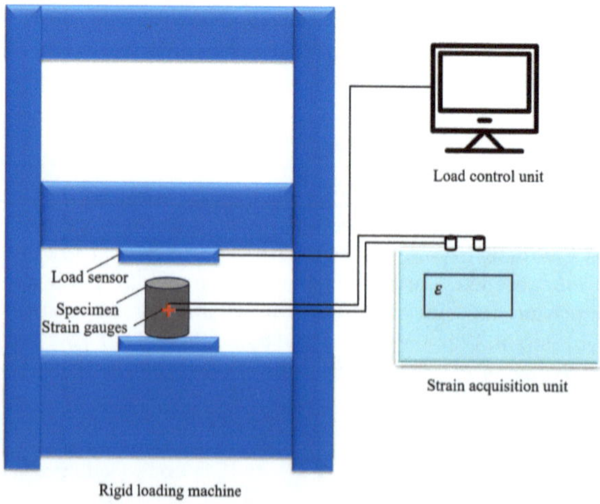

Figure 1. The test system and the granite specimens.

2.4. Test Scheme

The experimental groups are shown in Table 1. Group M specimens are subjected to a single uniaxial compression test to determine the uniaxial compressive strength (UCS) of rock samples. The average UCS is 101.87 MPa. Cyclic uniaxial compression tests are carried out on groups G1, G2, G3, and G4 with different pre-applied load levels. The pre-load levels are 30 kN, 60 kN, 90 kN, and 120 kN, respectively, and the corresponding pre-stresses are 15% UCS (15.28 MPa), 30% UCS (30.56 MPa), 45% UCS (45.84 MPa), and 60% UCS (61.12 MPa), respectively.

Table 1. Specimens grouping.

Groups	Pre-Load *B*(T)/kN	Cyclic Peak Load *C*(T)/kN	Pre-Stress/MPa
M1/M2	/	/	/
G1-1/G1-2	30	60	15.28
G2-1/G2-2	60	90	30.56
G3-1/G3-2	90	120	45.84
G4-1/G4-2	120	160	61.12

Five graded cyclic loadings are applied, as shown in Figure 2. T = 120 s is the holding time, $B(T)$ is the artificial pre-stress, and the stress $A(T)$ is 80% of the $B(T)$ value. The purpose of setting appropriate holding time is to eliminate possible initial stress and enhance deformation memory [9,14,20]. The application of multi-level stress is to eliminate possible initial in situ stress.

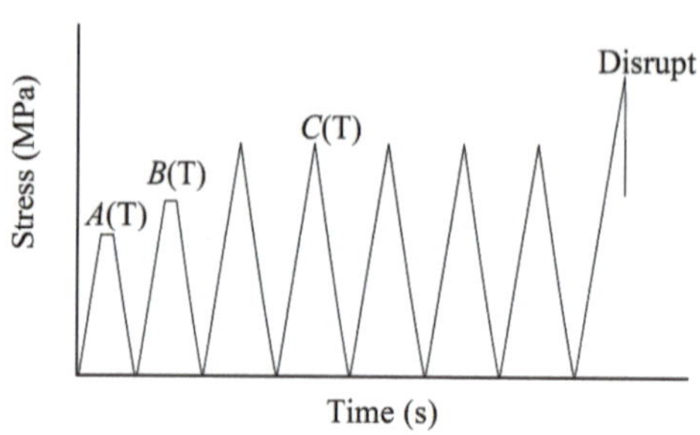

Figure 2. Cyclic loading and unloading schemes.

3. Test Results

In addition, the comparison between horizontal and longitudinal strains has not been paid enough attention. Some scholars have proposed that the accuracy of the two is consistent [24]. More effective experimental support is needed for this view.

Calculating the horizontal strain difference and longitudinal strain difference with Equation (1) to draw the DRA curve, and the deflection point is recognized as calculated pre-stress. The tensile strain is positive, and the compression strain is negative. By comparing the DRA results, it is found that the strain difference gradually increases as loading cycles increase. For example, it is found that $\Delta\varepsilon_{51}$ is larger than $\Delta\varepsilon_{21}$. Consequently, the strain difference between the fifth and first cycles is selected as an index to draw the DRA curves.

3.1. Horizontal Strain DRA

The horizontal strain difference–stress relationship curves of the specimen are shown in Figure 3. During the initial loading stage, the original cracks in the rock sample are compressed closed, the elastic modulus increases, and the DRA curve has a downward trend; as the stress increases, the DRA curve tends to be gentle, and a relatively smooth bending occurs at the DRA inflection point. There are two inflection points in the horizontal DRA curve of the G1-1 specimen, which are 6.77 MPa and 16.94 MPa, respectively. The latter inflection point is selected as the inflection point of the DRA curve [9], and the corresponding stress value is 16.94 MPa. The horizontal DRA curve of the G2-1 specimen (Figure 4b) is relatively gentle. The hyperbolic regression analysis method [25] is used to determine the inflection point, which is 26.36 MPa. The horizontal DRA curves of G3-1 and G4-1 specimens (Figure 4c,d) are similar. Obvious inflection points, 47.03 MPa and 60.79 MPa, can be recognized, respectively.

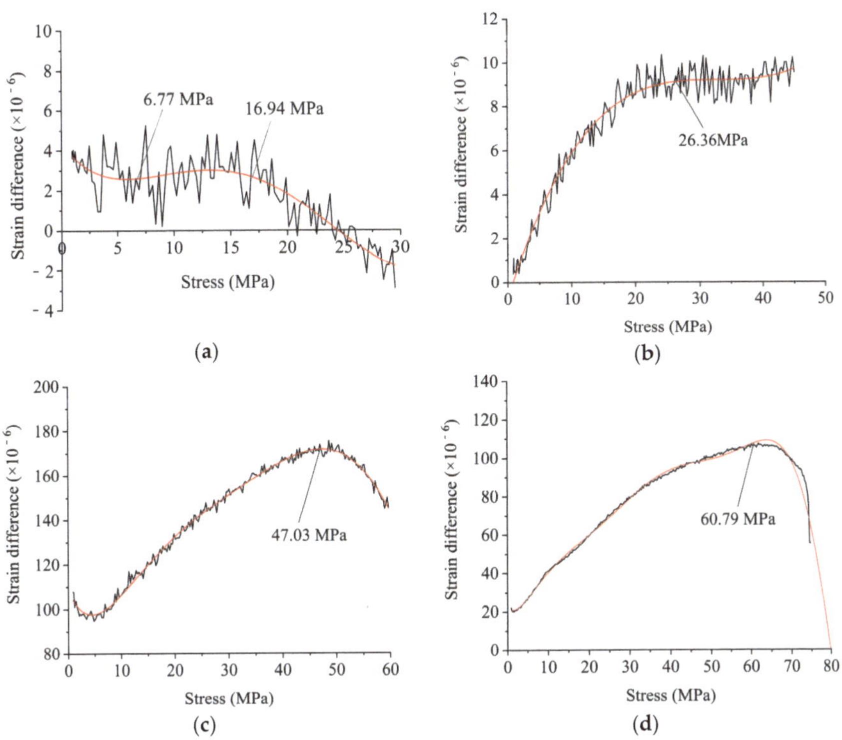

Figure 3. DRA curves calculated by horizontal strain when the pre-stress levels are (**a**) 15%UCS, (**b**) 30%UCS, (**c**) 45%UCS, and (**d**) 60%UCS, respectively.

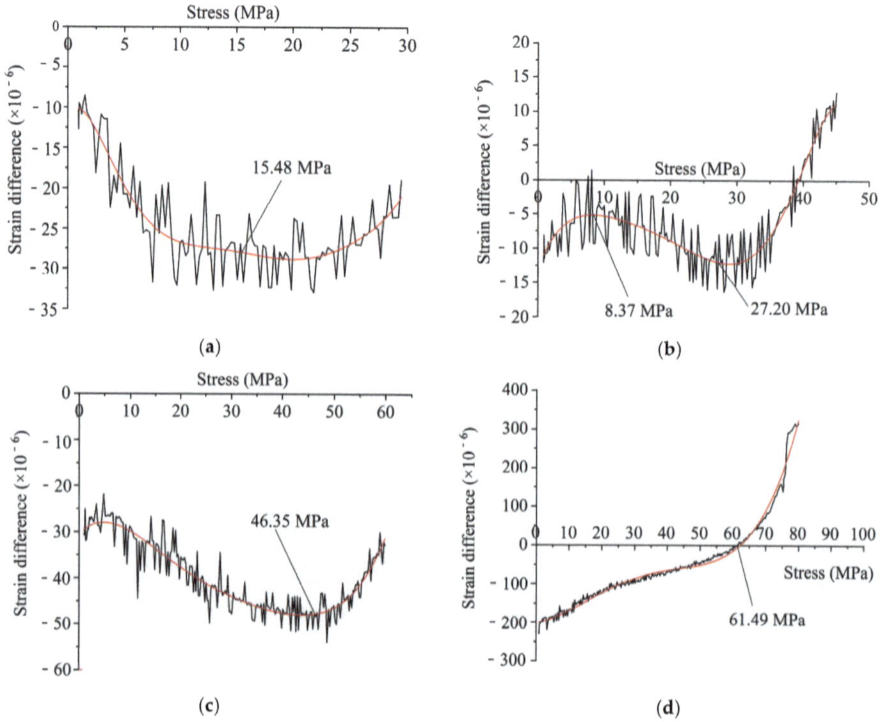

Figure 4. DRA curves drawn by longitudinal strain when the pre-stress levels are (**a**) 15%UCS, (**b**) 30%UCS, (**c**) 45%UCS, and (**d**) 60%UCS, respectively.

The *FR* values of each specimen are calculated by comparing the applied pre-stress σ_{max} with the DRA-identified pre-stress σ_{DRA}, as shown in Table 2. The *FR* values of all specimens in the four groups are within the effective range of [0.85, 1.15], indicating that the DRA method with horizontal strain is successful. When the pre-stress is relatively small, i.e., 15% and 30% UCS, only one specimen's *FR* value falls in the range of [0.90, 1.10]. For the other loading conditions, there is a certain gap between the recognition result and the true value when the pre-stress is relatively small. The maximum deviation is 13.74% (G2-1 specimen). When the initial stress was 45% and 60% UCS, the *FR* values of all specimens fell within the interval of [0.90, 1.10]. The results show that the horizontal DRA method is suitable for the in situ stress test in the range of 15–60% UCS, and with the increase in the initial stress level, the test results of the horizontal DRA method are closer to the accurate value.

Table 2. *FR* values and evaluation of the DRA method by considering horizontal strain.

No.	Pre-Stress Level	Pre-Stress /MPa	DRA Inflection Point Stress /MPa	*FR* ($\sigma_{DRA}/\sigma_{max}$)	Effectiveness
G1-1	15%UCS	15.28	16.94	1.11	good
G1-2			16.38	1.07	accurate
G2-1	30%UCS	30.56	26.36	0.86	good
G2-2			33.94	1.11	good
G3-1	45%UCS	45.84	47.03	1.03	accurate
G3-2			46.76	1.02	accurate
G4-1	60%UCS	61.12	60.79	1.09	accurate
G4-2			64.71	1.06	accurate

3.2. Longitudinal Strain DRA

Figure 5 is the longitudinal strain difference–stress relationship curve of the specimen under the fifth loading and the first loading under different pre-stress conditions.

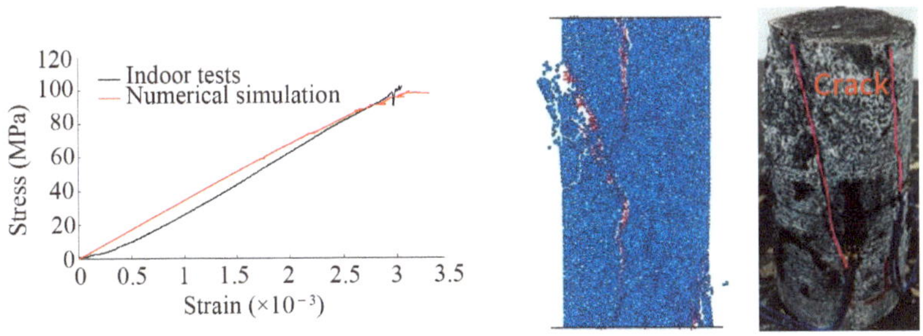

(a) (b) (c)

Figure 5. Calibration of the numerical model with uniaxial compression test specimen. (**a**) Stress–strain curves, (**b**) calculation results, (**c**) test results.

The longitudinal DRA curve of the G1-1 specimen (Figure 4a) is relatively gentle with a clear inflection point, and the stress value of the inflection point is 15.48 MPa. The longitudinal DRA curve of the G2-1 specimen (Figure 4b) has two inflection points, and 27.20 MPa is taken as the stress value of the inflection point [16]. The longitudinal DRA curves of G3-1 and G4-1 specimens (Figure 4c,d) have obvious inflection points, and the corresponding inflection point stress values are 46.35 MPa and 61.49 MPa, respectively.

The *FR* values of each specimen calculated with the longitudinal strain are summarized in Table 3. When the stress level is low, the longitudinal DRA method has poor test results. For example, the reorganization results of G1-2 and G2-2 are inaccurate, and those of G2-1 are good. As the stress level increases to 45%UCS and 60%UCS, the success rate of the longitudinal DRA method is obviously improved, and all pre-stress values are accurately tested. Consequently, the pre-stress can also be detected by the DRA method with longitudinal strain.

Table 3. *FR* values and evaluation of the DRA method by considering the longitudinal strain.

Specimen	Pre-Stress Level	Pre-Stress /MPa	DRA Inflection Point Stress/MPa	*FR* ($\sigma_{DRA}/\sigma_{max}$)	Effectiveness
G1-1	15%UCS	15.28	15.48	1.01	accurate
G1-2			19.17	1.25	inaccurate
G2-1	30%UCS	30.56	26.36	0.89	good
G2-2			24.12	0.79	inaccurate
G3-1	45%UCS	45.84	46.35	1.01	accurate
G3-2			46.75	1.02	accurate
G4-1	60%UCS	61.12	61.49	1.01	accurate
G4-2			61.42	1.00	accurate

4. Mechanism of the DRA Method

In order to further reveal the influence mechanism of microcrack activities on rock deformation memory, the PFC numerical model was established using the discrete element method to simulate the uniaxial compression failure process of granite under different pre-stress levels.

4.1. Establishment of Numerical Model

PFC is developed based on the meshless discrete element particle flow method, and it is widely used in the cracking behaviors studied in rock mass by simulating the motion and interactions of granular media [25]. The motion of particles follows Newton's law of motion, and the connection between particles follows the force–displacement law. An explicit time integration algorithm that directly solves equations at each time step is adopted, and the convergence problem essentially is the accumulation of time. The numerical calculation model is established based on the built-in parallel bond model.

The numerical calculation model is established based on the built-in parallel bond model, as shown in Figure 5a. The model size is Φ50 mm × 100 mm (as shown in Figure 5a), the particles are randomly generated, the particle radius is 0.5~0.75 mm, the density is 2500 kg/m^3, and the porosity is 0.1.

The parameters are calibrated by the single uniaxial compression test of the M1 specimen, and the specific parameters are shown in Table 4. The uniaxial compression simulation and test results are shown in Figure 5b,c. The failure of granite is controlled by tensile and shear cracks. Figure 6 shows the stress–strain curves of the test and simulation results. The average UCS of the numerical model is 100 MPa, and the crack initiation stress is 60 MPa.

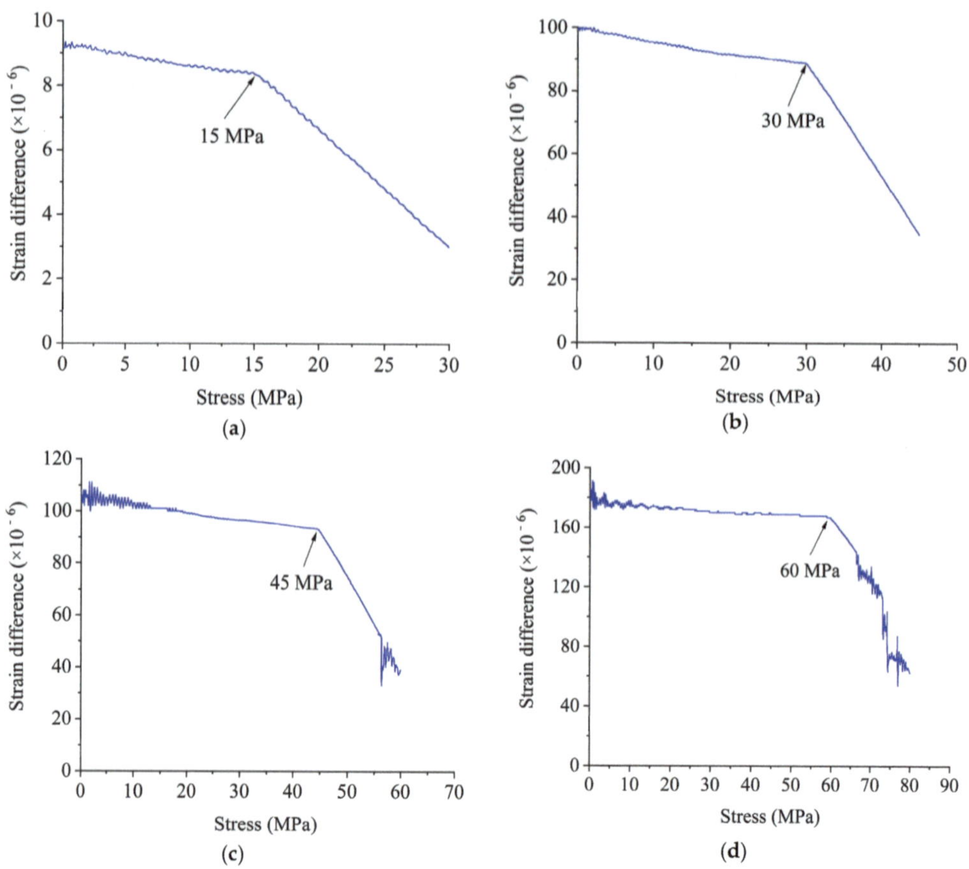

Figure 6. DRA curves of the numerical specimens when the pre-stress levels are (**a**) 15%UCS, (**b**) 30%UCS, (**c**) 45%UCS, and (**d**) 60%UCS, respectively.

Table 4. Parameters of the model.

Parameter	Value	Unit
Specimen diameter	50	mm
Specimen height	100	mm
Particle density	2500	kg/m^3
Particle elastic modulus	1.5	GPa
Stiffness ratio of particle	1.0	/
Friction coefficient of particles	0.577	/
Bond stiffness ratio	1.0	/
Bond elastic modulus	1.6	GPa
Contact tensile strength	30	MPa
Contact cohesion	60	MPa
Contact internal friction angle	30	°

4.2. Numerical Simulation Results Analysis

To illustrate the influences of initial stress levels, the pre-stress is set to 15%, 30%, 45%, and 60% of the numerical UCS, that is, 15 MPa, 30 MPa, 45 MPa, and 60 MPa, respectively. Similar loading and unloading cycles to the experiment in Figure 3 are imposed on the numerical specimens. In the process of numerical simulation, the axial stress and horizontal strain are monitored in real time, and the horizontal strain difference–stress curves are plotted, as shown in Figure 6.

Under different initial stress levels, the inflection points on DRA curves are apparent. The numerical-identified pre-stress values are 15 MPa, 30 MPa, 45 MPa, and 60 MPa, respectively, which are completely consistent with the initial stress. The reason may be that the crack initiation position, propagation path, and penetration mode of the microcrack are not affected by random distribution pores and natural cracks in numerical tests. Similar cracking processes and failure modes under different pre-stress levels (Figure 7) also provide evidence for this inference.

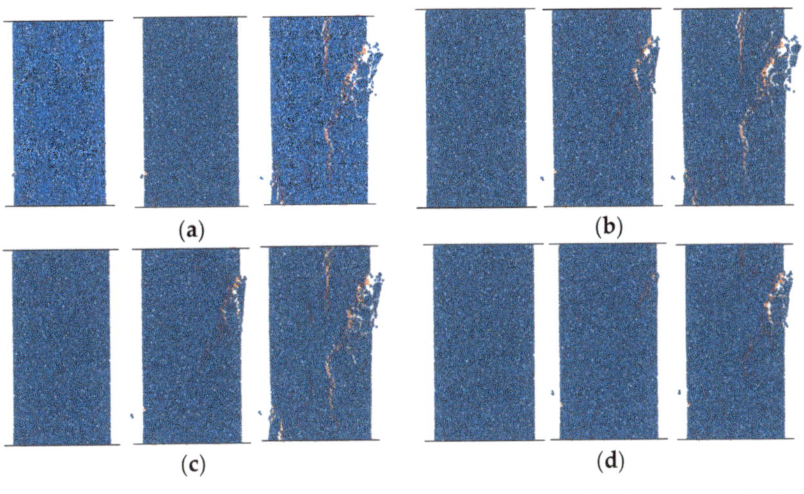

Figure 7. Cracking process and failure modes of each specimen when the pre-stress levels are (a)15%UCS, (b) 30%UCS, (c) 45%UCS, and (d) 60%UCS, respectively.

The relationship between stress cycles and crack number increases is shown in Figure 8. In Figure 8a, when the pre-stress is 15% UCS, nearly no new crack occurs before the uniaxial compressive stress reaches 56 MPa. The number of cracks increases sharply after the uniaxial compressive stress reaches 56 MPa, indicating that new cracks are generated and expanded. This result shows that when the pre-stress is low (15% UCS), the pre-stress measured by the DRA method does not depend on the generation and expansion of new cracks but depends on the closure of natural cracks. A similar situation is also seen when the pre-stress is 30% UCS, as shown in Figure 8b,c.

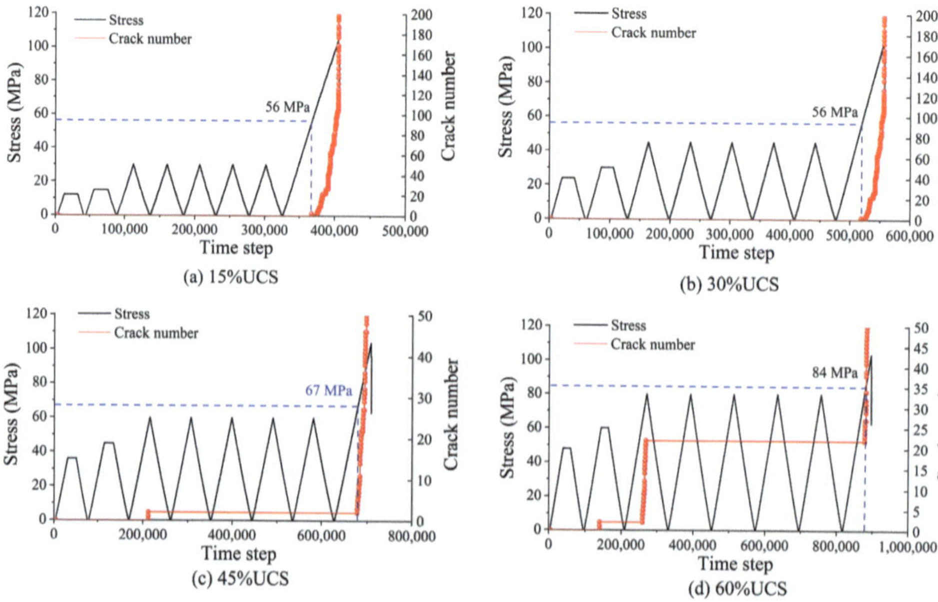

Figure 8. Relationship between stress–time step–crack quantity. (**a**)15%UCS, (**b**) 30%UCS, (**c**) 45%UCS, and (**d**) 60%UCS, respectively.

In Figure 8d, when the initial stress is 60% UCS, the number of cracks increases sharply when the axial compressive stress reaches 60 MPa, and a large number of new cracks initiate and propagate. Therefore, when the initial stress is 60% UCS, the formation mechanism of the DRA method is due to the generation of new cracks.

5. Discussion

The DRA detection is consistent with the studies of Tang et al. [12] and Wang [26], indicating that the rock deformation memory effect is available in low-stress regions. The *FR* values evolution trends with different pre-stress levels are shown in Figure 9. All *FR* values are distributed within the "accurate" region in the high-stress region (σ = 45%UCS or σ = 60%UCS), while the identification error is much higher in the low-stress region (σ = 15% UCS or σ = 30% UCS). It can be concluded that the DRA-identification effect is better in high-stress regions than in low-stress regions. The in-depth mechanism analysis shows that this is highly related to microcrack closure and development behaviors under stress.

What is more, the DRA calculation accuracy with horizontal strain and longitudinal strain are compared. As shown in Figure 9b, the overall identification error curves of DRA with longitudinal strain severe fluctuate in low-stress regions, indicating relatively significant discreteness. While in the high-stress region, the pre-stress identification accuracies are similar for both horizontal and longitudinal strains.

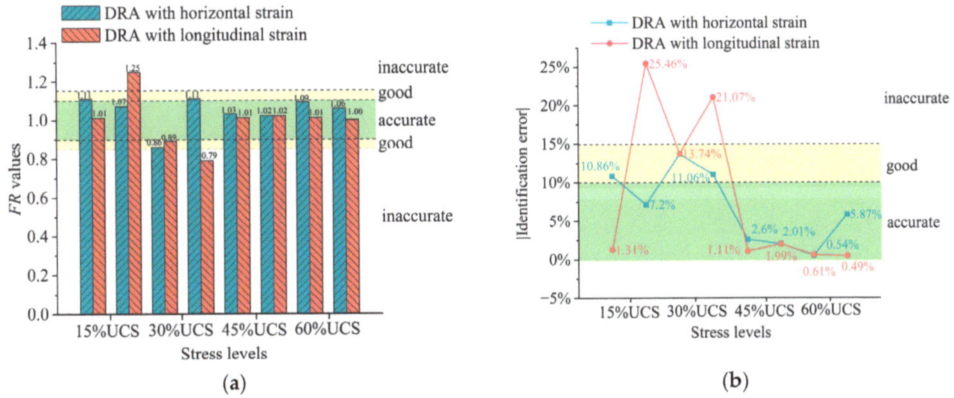

Figure 9. Distribution of *FR* values with different pre-stress levels (**a**) and the absolute values of identification error detected with horizontal and longitudinal strains (**b**).

In summary, the DRA method is suitable for testing rock in situ stress during 15–60%UCS. Although the pre-stress can be identified in low-stress regions with horizontal strain, the accuracy is not as good as that in high-stress regions. As revealed by numerical calculation, it is proposed that the deformation memory in low-stress regions is related to microcrack closure and that in high-stress regions, it is induced by microcrack initiation and propagation. It can be concluded that new microcrack initiation and propagation can strengthen the deformation memory effect in rock samples. Thus, in shallow buried areas, it is recommended to test horizontal strain when using the DRA method, and the results should be verified by other in situ stress methods; in high-stress areas, horizontal or longitudinal strains can be selected for DRA testing according to the difficulty of testing.

There is always uncertainty in any kind of geotechnical measurement parameters. The measuring methods of in situ stress, including in situ testing methods and laboratory estimate methods, are affected by many factors. To further increase the identification accuracy of DRA methods, studies have shown that the interval time between rock coring and laboratory tests should be shortened. What is more, the influences of changes in core humidity on the DRA-identification accuracy are still unclear, and thus, further research is required to determine whether moisturizing measures are needed. The influence of cyclic loading times on the deformation memory effect is another challenge for the DRA method. The cyclic loading times should be controlled within a reasonable region to avoid specimen fatigue.

6. Conclusions

Uniaxial compression tests are conducted on granite specimens to study the influences of pre-stress levels on the DRA method identification accuracy. The mechanism is revealed from the aspect of microcrack behaviors. What is more, the calculation accuracies of the DRA method with horizontal and longitudinal strains are compared. The main conclusions are as follows:

1. The DRA method is valid for detecting pre-stress during 15–60%UCS. The pre-stress level has a significant effect on the accuracy of the DRA test. The detecting accuracy is higher in the high-stress region (45–60%UCS) than in the low-stress region (15–30%UCS);
2. Both horizontal and longitudinal strains can be used as parameters of DRA curves. At low stress levels, the success rate of calculating DRA with horizontal strain is higher than that with longitudinal strain. However, as the stress level increases, both methods are effective;

3. The DRA method is related to the generation and development of internal microcracks in rock specimens. Under lower pre-stress levels, the crack initiation and propagation of the new crack lag behind the stress value measured by the DRA method, and the DRA method is affected by the natural crack closure. At higher pre-stress levels, the DRA method depends on crack initiation and propagation.

Author Contributions: Conceptualization, Q.X.; data curation, J.D. and X.X.; investigation, J.T.; methodology, Y.B.; software, J.T.; supervision, X.F.; validation, X.F.; writing—review & editing, Y.B. All authors have read and agreed to the published version of the manuscript.

Funding: This research was funded by the Natural Science Foundation of Chongqing (CSTB2023NSCQ-MSX0864), and State Grid Chongqing Electric Power Company Project (SGTYHT/24-JS-001).

Data Availability Statement: The data presented in this study are available on request from the corresponding author.

Conflicts of Interest: The authors declare no conflict of interest.

References

1. Xie, H.; Li, C.; He, Z.Q.; Li, C.; Lu, Y.; Zhang, R.; Gao, M.; Gao, F. Experimental study on rock mechanical behavior retaining the in situ geological conditions at different depths. *Int. J. Rock Mech. Min. Sci.* **2021**, *138*, 104548. [CrossRef]
2. Xiao, L.; Peng, Y.; Chen, M. Optimization analysis of excavation procedure design of underground powerhouses under high in situ stress in China. *Appl. Sci.* **2021**, *11*, 10252. [CrossRef]
3. Pei, Q.; Wu, C.; Ding, X.; Huang, S. A weight factor based backward method for estimating ground stress distribution from the point measurements. *Bull. Eng. Geol. Environ.* **2023**, *82*, 365. [CrossRef]
4. Wang, J.C.; Wang, C.Y.; Huang, J.F.; Han, Z.; Zeng, W.; Wang, Y. In Situ stress measurement method of deep borehole based on multi-array ultrasonic scanning technology. *Front. Earth Sci.* **2022**, *10*, 933286. [CrossRef]
5. Piao, S.H.; Huang, S.; Wang, Q.; Ma, B. Experimental and numerical study of measuring In-Situ stress in horizontal borehole by hydraulic fracturing method. *Tunn. Undergr. Space Technol.* **2023**, *141*, 105363. [CrossRef]
6. Sjöberg, J.; Christiansson, R.; Hudson, J.A. ISRM suggested methods for rock stress estimation part 2: Overcoring methods. *Int. J Rock Mech. Min. Sci.* **2003**, *40*, 999–1010. [CrossRef]
7. Fu, X.; Ban, Y.X.; Xie, Q.; Abdullah, R.A.; Duan, J. Time delay mechanism of the Kaiser effect in sandstone under uniaxial compressive stress conditions. *Rock Mech. Rock Eng.* **2021**, *54*, 1091–1108. [CrossRef]
8. Ban, Y.X.; Fu, X.; Xie, Q.; Duan, J. Time-sensitivity mechanism of rock stress memory properties under tensile stress. *J. Rock Mech. Geotech. Eng.* **2020**, *12*, 528–540. [CrossRef]
9. Yamshchikov, V.S.; Shkuratnik, V.L.; Lavrov, A.V. Memory effects in rocks (review). *J. Min. Sci.* **1994**, *30*, 463–473. [CrossRef]
10. Zhong, L.W.; Wang, H.J.; Ren, X.H.; Tang, L. Influence of the cyclic loading path on rock deformation memory effect. *Geofluids* **2020**, *2020*, 8896344. [CrossRef]
11. Fraser, D.; Gholami, R.; Sarmadivaleh, M. Deformation Rate Analysis: How to determine In-Situ stresses in unconventional gas reservoirs. *Int. J. Rock Mech. Min. Sci.* **2021**, *146*, 104892. [CrossRef]
12. Tang, Q.Q.; Zhou, Y.; Zhao, D. Numerical simulation of deformation memory effect of rock materials in low-stress condition using discrete element method. *Energy Sci. Eng.* **2020**, *8*, 3027–3046. [CrossRef]
13. Eberhardt, E.; Stead, D.; Stimpson, B.; Read, R.S. Identifying crack initiation and propagation thresholds in brittle rock. *Can. Geotech. J.* **1998**, *35*, 222–233. [CrossRef]
14. Alkan, H.; Cinar, Y.; Pusch, G. Rock salt dilatancy boundary from combined acoustic emission and triaxial compression tests. *Int. J. Rock Mech. Min. Sci.* **2007**, *44*, 108–119. [CrossRef]
15. Yamamoto, I.; Kuwahara, Y.; Kato, N.; Hirasawa, T. Deformation rate analysis: A new method for In Situ stress estimation from inelastic, deformation of rock samples under uniaxial compression. *Sci. Rep. Tohoku Univ. Fifth* **1990**, *33*, 34–37.
16. Hunt, S.; Meyers, A.; Louchnikov, V.; Oliver, K.J. Use of the DRA technique, porosimetry and numerical modelling for estimating the maximum In-Situ stress in rock from core. *S. Afr. Inst. Min. Metall.* **2003**, *37*, 17–22.
17. Seto, M.; Villaescusa, E.; Utagawa, M.; Katsuyama, K. In Situ stress evaluation from rock cores using AE method and DRA. *J. Min. Mater. Process. Inst. Jpn.* **1998**, *114*, 845–855.
18. Park, P.; Park, N.; Hong, C.; Jeon, S.; Kim, Y. The influence of delay time and confining pressure on In-Situ stress measurement using AE and DRA. In *DC Rocks 2001, the 38th U.S. Symposium on Rock Mechanics (USRMS), Proceedings of the Human Factors and Ergonomics Society Annual Meeting, Washington, DC, USA, 7 July 2001*; SAGE Publications: Washington, DC, USA, 2001; pp. 1259–1260.
19. Aghill, A.; Nikkhah, M. An experimental study on the stress memory retrieval in rocks using deformation rate analysis method. *Iran. J. Earth Sci.* **2023**, *15*, 170–178.
20. Wang, H.J.; Ren, X.H.; Zhang, J.X. Deformation memory effect identification using fractal dimension in the stress region above crack initiation threshold. *Appl. Mech. Mater.* **2011**, *90–93*, 2332–2338. [CrossRef]

21. Wang, H.Y.; Dyskin, A.; Pasternak, E. Triaxial deformation rate analysis (DRA). *Rock Mech. Rock Eng.* **2024**, *57*, 1939–1962. [CrossRef]
22. Zhao, K.; Zhang, L.; Yang, D.X.; Jin, J.; Zeng, P.; Wang, X.; Ran, S.; Deng, D. Cyclic impact damage and water saturation effects on mechanical properties and kaiser effect of red sandstone under uniaxial cyclic loading and unloading compression. *Rock Mech. Rock Eng.* **2024**, *57*, 181–195. [CrossRef]
23. Chen, Y.L.; Meng, Q.B.; Li, Y.C.; Pu, H.; Zhang, K. Assessment of appropriate experimental parameters for studying the Kaiser effect of rock. *Appl. Sci.* **2020**, *10*, 7324. [CrossRef]
24. Cundall, P. A computer model for simulating progressive large-scale movement in block rock systems. In Proceedings of the International Symposium Fracture, ISRM, Nancy, France, 4–6 October 1971; pp. 8–11.
25. Xie, Q.; Cao, Z.L.; Tian, R.J.; Sun, W.C.; Alessio Fumagalli; Peng, H.Y.; Fu, X.; Luo, H.Y. Complex sliding characteristics of landslides and evaluation of the reinforcement with arched anti-slide piles based on 3D discrete element method: A case study. *Nat. Hazards* **2024**, *120*, 8983–9007. [CrossRef]
26. Wang, H.; Lei, T.; Ren, X.; Zhong, L.; Si, F.; Hsieh, A. Rock deformation memory effect: Applications, experiments and theories. *Chin. J. Geotech. Eng.* **2018**, *40*, 1571–1583. (In Chinese)

buildings

MDPI

Article

Fluid Flow Modeling and Experimental Investigation on a Shear Thickening Fluid Damper

Shiwei Chen [1], Xiaojiao Fu [1], Peiling Meng [1], Lei Cheng [1,2], Lifang Wang [1,2] and Jing Yuan [1,2,*]

[1] School of Civil Engineering and Architecture, Chongqing University of Science and Technology, Chongqing 401331, China; 2013001@cqust.edu.cn (S.C.); 2022206096@cqust.edu.cn (X.F.); mpl638@163.com (P.M.); zrhd244@163.com (L.C.); 308207283@icould.com (L.W.)

[2] Chongqing Institute of Metrology and Quality Testing, Chongqing 402260, China

* Correspondence: yuanjing38@163.com

Abstract: Shear Thickening Fluid (STF) is a specialized high-concentration particle suspension capable of rapidly and reversibly altering its viscosity when exposed to sudden impacts. Consequently, STF-based dampers deliver a self-adaptive damping force and demonstrate significant potential for applications in structural vibration control. This study presents both a modeling and experimental investigation of a novel double-rod structured STF damper. Initially, a compound STF is formulated using silica particles as the dispersed phase and polyethylene glycol solution as the dispersing medium. The rheological properties of the STF are then experimentally evaluated. The STF's constitutive rheological behavior is described using the G-R model. Following this, the flow behavior of the STF within the damper's annular gap is explored, leading to the development of a two-dimensional axisymmetric fluid simulation model for the damper. Based on this model, the dynamic mechanism of the proposed STF damper is analyzed. Subsequently, the STF damper is optimally designed and subjected to experimental investigation using a dynamic testing platform under different working conditions. The experimental results reveal that the proposed STF damper, whose equivalent stiffness can achieve a nearly threefold change with excitation frequency and amplitude, exhibits good self-adaptive capabilities. By dividing the damper force into two parts: the frictional damping pressure drop, and the osmotic pressure drop generated by the "Jamming effect". A fitting model is proposed, and it aligns closely with the nonlinear performance of the STF damper.

Keywords: shear thickening fluid; smart damper; shear thickening fluid-based damper

1. Introduction

Dampers, as crucial devices in engineering for vibration reduction and buffering, possess the capabilities to dissipate and store vibrational energy in engineering structures, effectively mitigating issues such as structural fatigue, damage, collapse, and buckling caused by external dynamic excitations [1–4]. In recent years, researchers have developed various types of smart dampers with controllable damping and stiffness based on novel intelligent materials whose mechanical properties can change with external physical field excitations, including magnetorheological fluids [5], electrorheological fluids [6], magnetorheological elastomers [7], and shape memory alloys [8], etc. These novel dampers can alter their own dynamic characteristics according to the structural vibration state, significantly enhancing the safety and stability of engineering structures under various vibration conditions. Nevertheless, due to the requirement of external energy for these novel dampers to realize their controllable functions, they have the disadvantages of being bulky, having complex structures, and requiring external power sources, making them difficult to apply in field environments that require long-term power supply. Therefore, there is an urgent need for innovative smart dampers that possess energy-free features.

In recent years, the evolution of preparation theories for Shear Thickening Fluid (STF) has paved the way for the potential realization of energy-free smart dampers. STF materials, categorized as a quintessential non-Newtonian fluid, comprise a suspension medium with microscopic particles dispersed within [9,10]. Upon exposure to shear forces, the interplay of relative friction, collisions, chemical bond restructuring, and alterations in particle configurations among these microscopic constituents manifests as a pronounced increase in the material's apparent viscosity in response to varying shear rates [11]. Scholarly investigations have elucidated that factors such as the choice of dispersed particles [12–15], their concentration [16,17], size [18,19], along with the nature of the dispersion medium and additives [14,20], exert profound influences on the shear thickening behavior exhibited by the material. Through meticulous modulation of these parameters during the STF preparation process, the material's apparent viscosity can be tuned to exhibit variations up to 1000-fold in response to external shear rate changes [15]. This remarkable capacity to adjust its viscosity across significant magnitudes without necessitating energy input positions STF as an emergent focal point within the realm of engineering vibration mitigation.

Research on applying STF (shear-thickening fluid) to liquid dampers can be traced back to the work of Hleber's team [21], who developed a vibration damping device based on STF materials, where the feedback force increases with the frequency of external excitation. Following this, Laun's group [22] proposed an intelligent damper design based on STF materials. In recent years, scholars have begun improving material compositions, resulting in reports of STF dampers with varying dynamic properties [23–25]. For example, Zhang XZ et al. [26] designed and fabricated a novel STF damper using STF material composed of nanoscale SiO_2 mixed with an ethylene glycol solution. Zhou et al. [27] reported an STF damper based on modified polystyrene discrete phases, whose equivalent stiffness can rapidly increase with the external load frequency. Overall, compared to other field-induced smart dampers, STF dampers require no energy input or sensing systems to achieve rapid dynamic changes. This feature demonstrates significant potential in vibration reduction applications for large outdoor structures, such as overhead high-voltage transmission lines.

At present, research on STF dampers is still in its early stages, with existing studies primarily focusing on changes in viscous damping. Apart from the ref. [27], There are few reports on STF dampers capable of achieving rapid and significant changes in equivalent stiffness. Furthermore, the optimization of damper structural parameters based on the rheological properties of STF materials, aiming to ultimately achieve the utmost variation in its equivalent stiffness, still necessitates comprehensive and thorough investigation. Therefore, this paper presents the development of a composite STF material based on a mixture of SiO_2 and a compound polyethylene glycol solution. By testing the rheological behavior of the material, the strain-rate-sensitive region was identified. Using a combination of FEM simulation and mechanical design, the internal flow field distribution within the annular gap was investigated, leading to the design and prototyping of an STF damper. Experimental tests were conducted using a dynamic testing platform, and the results demonstrate that the proposed STF damper can achieve a nearly threefold change in equivalent stiffness. This novel STF-based damper is expected to exhibit high stiffness when subjected to impact loads, maintaining structural stability, while offering significant flexibility under low-frequency loads to reduce internal fatigue stress, showing great potential for applications in fields such as seismic mitigation for structures and wind vibration suppression for power transmission lines.

2. The Preparation and Rheological Testing of STF Samples

2.1. Preparation of STF Samples

Considering the chemical and physical stability requirements as well as the cost demands for the damper's STF material, the STF in this study employs silicon dioxide (SiO_2) with a diameter of 500 nm (procured from Sinopharm Chemical Reagent Co., Ltd., Shanghai, China) as the dispersed phase, and a compounded polyethylene glycol (PEG) solution (PEG200:PEG400 = 1:2, purchased from Xilong Scientific Co., Ltd., Shenzhen,

China) as the dispersing medium. The detailed preparation process of the STF material is shown in Figure 1a. As illustrated in Figure 1a, the SiO_2 particles are first placed in a blast drying oven at 70 °C for 30 min to dry and dehumidify, ensuring that the dispersed phase particles remain dry. Then, the SiO_2 particles are mixed with the compounded polyethylene glycol solution at a mass ratio of 1:3, and anhydrous ethanol (purchased from Sinopharm Chemical Reagent Co., Ltd., Shanghai, China) is added to form a high-concentration particle suspension. Subsequently, the suspension is stirred for 60 min using an electric stirrer to ensure that the SiO_2 particles are uniformly dispersed in the dispersing medium. The suspension is then transferred to a beaker and placed in an ultrasonic cleaner for 20 min, followed by placement in an electric vacuum drying oven for 120 min, completing the STF material preparation.

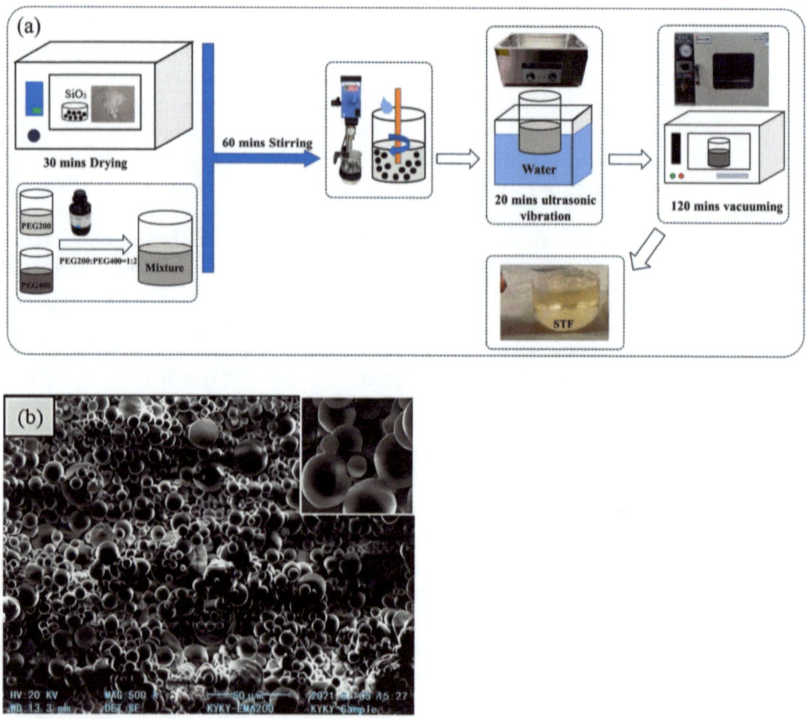

Figure 1. The fabrication of the STF material (**a**) The preparation process of STF material (**b**) The SEM picture of SiO_2.

Additionally, Figure 1b shows the SEM picture of the STF material. From the picture, it can be observed that the SiO_2 particles have smooth surfaces and regular shapes, which aids their uniform dispersion in the polyethylene glycol medium. This allows the material to exhibit low viscosity at low shear rates while demonstrating Newtonian fluid characteristics [28]. Moreover, the particle clusters show a significant degree of agglomeration, which helps to increase inter-particle friction within the clusters. This enhances the likelihood of jamming at high shear rates, thus enabling the liquid-solid phase transition region.

2.2. Rheological Testing of STF Samples

At a stable ambient temperature of 20 °C, the rheological properties of the prepared STF material were meticulously evaluated using the compact modular rheometer (MCR102, Anton Paar Company, Duisburg, Germany). The experiment adhered to the manufacturer's recommended rheological testing protocol, and a pre-shear process was conducted before initiating the test. During the oscillatory shear rheological test, the shear rate was varied within a range of 1 to 100 1/s. The corresponding test results are presented in Figure 2.

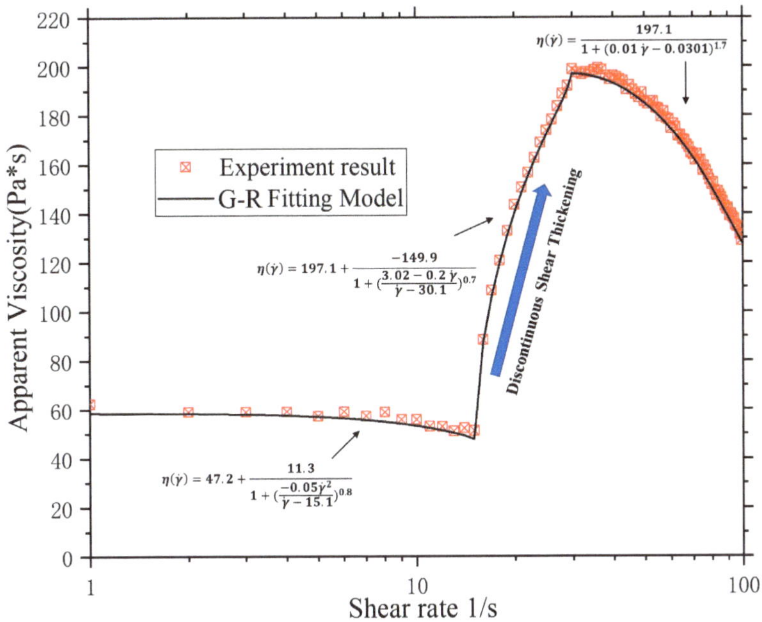

Figure 2. The rheological properties of the STF.

As shown in Figure 2, the critical shear rate for the compounded STF material's shear thickening behavior occurs at 15.1 1/s. Before the shear rate reaches this value, the fluid's viscosity changes insignificantly, and it can be approximately considered as a Newtonian fluid, with an apparent viscosity of 62.5 Pa·s. Upon reaching the critical shear rate, a marked discontinuous shear thickening phenomenon was observed, with the apparent viscosity at 47.8 Pa·s. The viscosity rapidly increased from 47.8 Pa·s to 197.2 Pa·s, with a change rate of 412%, clearly demonstrating discontinuous shear thickening.

As the shear rate continued to rise and reached 30.1 1/s, the material yielded, and the apparent viscosity began to decrease with increasing shear rate. By the time the shear rate reached 100 1/s, the viscosity dropped from 197.2 Pa·s to 132.1 Pa·s, exhibiting shear thinning behavior. Therefore, to design a self-adaptive smart damper with a large range of variable damping force, it is essential to ensure that the STF material inside the damper operates within the shear thickening state, specifically maintaining a shear rate between 15.1 1/s to 30.1 1/s during damper operation.

Based on the previous analysis, the viscosity variation in STF materials can be roughly categorized into three regions: the low-shear rate region, the shear thickening region, and the shear thinning region. Galindo-Rosales [29] provided a phenomenological model for these three regions, and its expression is as follows:

$$\begin{cases} \eta(\dot\gamma) = \eta_c + \dfrac{(\eta_0-\eta_c)}{1+\left[K_1\left(\frac{\dot\gamma^2}{\dot\gamma-\dot\gamma_c}\right)\right]^{n_1}} & for\ \dot\gamma \le \dot\gamma_c \\ \eta(\dot\gamma) = \eta_{max} + \dfrac{\eta_c-\eta_{max}}{1+\left[K_2\left(\frac{\dot\gamma-\dot\gamma_c}{\dot\gamma-\dot\gamma_{max}}\right)\right]^{n_2}} & for\ \dot\gamma_c \le \dot\gamma \le \dot\gamma_{max} \\ \eta(\dot\gamma) = \dfrac{\eta_{max}}{1+\left[K_3(\dot\gamma-\dot\gamma_{max})\right]^{n_3}} & for\ \dot\gamma_{max} \le \dot\gamma \end{cases} \quad (1)$$

In the equation, η_0 represents the initial apparent viscosity of the shear thickening fluid, with a fitted value of 58.5 Pa·s; η_c is the minimum apparent viscosity of the shear thickening fluid, with a fitted value of 47.2 Pa·s; and η_{max} is the maximum viscosity of the shear thickening fluid, with a fitted value of 197.1 Pa·s. $\dot{\gamma}_c$ and $\dot{\gamma}_{max}$ represent the critical shear rates for the onset of shear thickening and shear thinning phenomena, respectively, with values of 15.1 1/s and 30.1 1/s. The parameters n_1, n_2, n_3, K_1, K_2, and K_3 are fitting constants, with values of 0.8, 0.7, 1.7, −0.05, −0.2, and 0.01, respectively. As shown in Figure 2, the G-R fitting model effectively captures the steady-state rheological properties of the compounded STF material reported in this study.

3. The Design and Simulation of STF Dampers

3.1. Designing of the STF Dampers

Considering the advantages of a double-rod structure, such as fewer seals and the loss of requirement for an air compensation chamber, a double-rod intermittent STF damper based on the compounded STF material is proposed. The schematic diagram, 3D models, and prototype are shown in Figure 3a, Figure 3b, and Figure 3c, respectively.

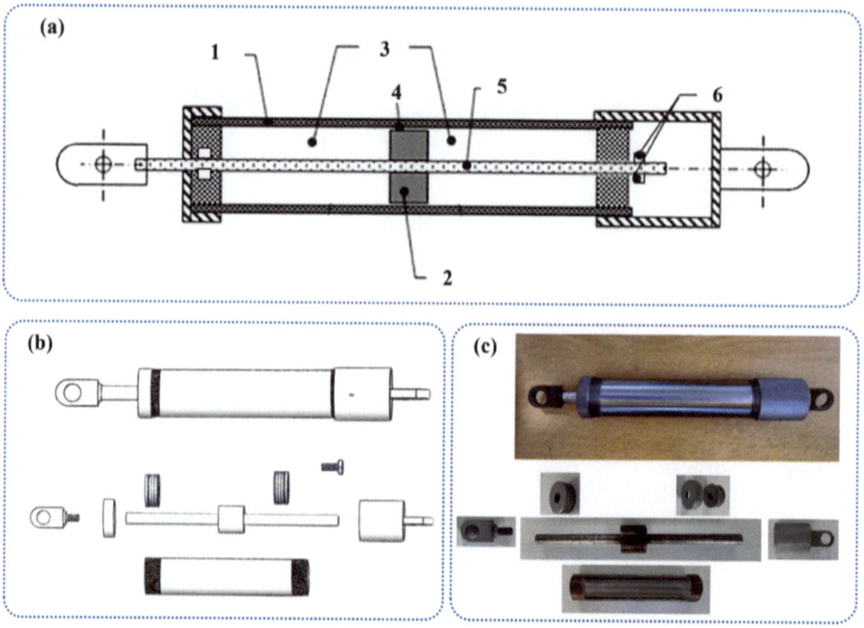

Figure 3. Designing of the STF damper (**a**) schematic, 1. Damper tube, 2. Piston head, 3. Cylinder chamber, 4. Annular gap, 5. Piston rod, 6. Seal rings (**b**) 3D model (**c**) Prototype.

As illustrated, the proposed STF damper primarily consists of front and rear end caps, a damper tube, and a piston rod. The damper tube and push rod (including the piston) are made of Q235 stainless steel. Inside the STF cylinder chamber, a piston rod (with piston) is installed. To ensure the smooth movement of the piston within the sealed cylinder chamber, sealing rings are placed above and below the piston rod, while the damper tube is sealed by the front and rear end caps. By adopting the double-rod structure, an annular gap is formed between the cylinder chamber and the piston.

The compounded STF material is filled into the cylinder chamber of the damper. When subjected to external loads, the push rod moves, causing the piston to compress the STF material within the annular gap. This forces the STF from the cylinder chamber on one side of the piston through the annular gap into the cylinder chamber on the other side of the piston. During this process, as the piston movement speed changes, the apparent viscosity of the STF material also changes accordingly, resulting in variations in the damping force exerted on the piston rod. This ultimately leads to the damper filled with STF exhibiting a nonlinear dynamic property that changes with external loads. Considering the structural sealing and the overall machining complexity of the damper, the initial design dimensions of each component are listed in Table 1.

Table 1. Parameters of the proposed STF damper.

Parameters	Value (mm)
Diameter of push rod	15
Length of push rod	295
Length of piston	15
Diameter of piston	30–40
Outer diameter of damper	58
Wall thickness	4
Length of cylinder chamber	255
Diameter of cylinder chamber	54

3.2. Simulation and Optimization of the STF Damper

As shown in Figure 2, the STF material exhibits shear thickening behavior only within a suitable range of shear rates. Therefore, this section establishes a FEM model of the STF damper to analyze the shear rate of the STF in the annular gap. The goal is to ensure that the STF material operates at its optimal shear rate within the intermittent region of the impact load, ultimately completing the optimization of the piston diameter in the damper. The FEM model of the STF damper used in this study is shown in Figure 4. Based on the symmetry of the flow field inside the damper, a two-dimensional axisymmetric model is adopted, with the geometric central axis of the damper as the axis of symmetry. The model dimensions are provided in Table 1.

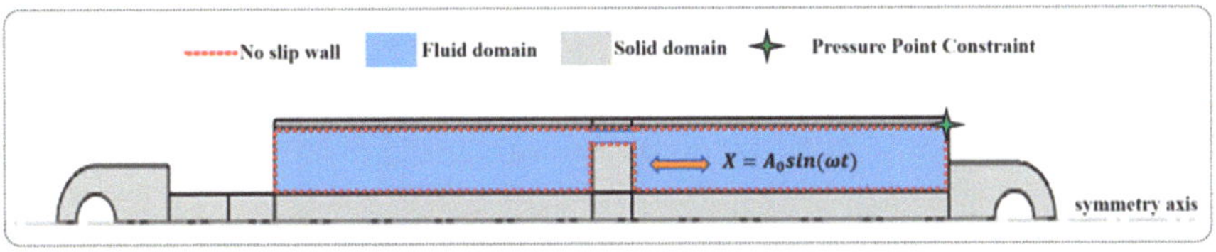

Figure 4. Flow Field Computation Model STF damper.

As shown in Figure 4, the flow field of the damper was simulated using the Laminar Flow Module of the commercial finite element software COMSOL Multiphysics 6.0. In the computation model, the cylinder chamber of the damper was defined as the fluid domain, while the other regions of the damper were set as solids. The interface between the fluid and solid domains was configured as a no-slip wall, and the upper-right corner of the fluid domain was set as a zero-pressure boundary. On this basis, the Lagrangian–Eulerian method was employed to simulate the cyclic motion of the STF damper under sinusoidal excitation. By incorporating a moving mesh module in the software, the fluid domain was set as a freely moving mesh domain, with the piston movement defined by the equation $X = A_0 sin(\omega t)$. And the density of the STF (Shear Thickening Fluid) in the fluid domain is set at 1.870 g/mm^3, while its dynamic viscosity, which varies based on the shear rate, is determined using fitting data from Equation (1).

Leveraging the developed model, Figure 5 meticulously depicts the average shear rate and its distribution within the STF-filled annular gap, under conditions where the piston diameter is 36 mm, the displacement amplitude is 3 mm, and the frequency is 5 Hz. The figure vividly illustrates that the piston's cyclic motion induces considerable shear deformation within the STF in the annular gap, with the highest shear rate concentrated near the edges of the piston in the annular gap region. As the piston reaches its central position, the velocity attains its peak, resulting in a maximum shear rate at the periphery of the annular gap, signifying that the nonlinear shear thickening behavior of the STF will predominantly emerge in this region. Furthermore, Figure 5 reveals a periodic pattern in the average shear rate variation within the annular gap, closely corresponding to the oscillatory movement of the piston. The shear rate achieves its maximum value of 62.1 1/s when the piston displacement is at zero and diminishes to 0.1 1/s at the maximum displacement.

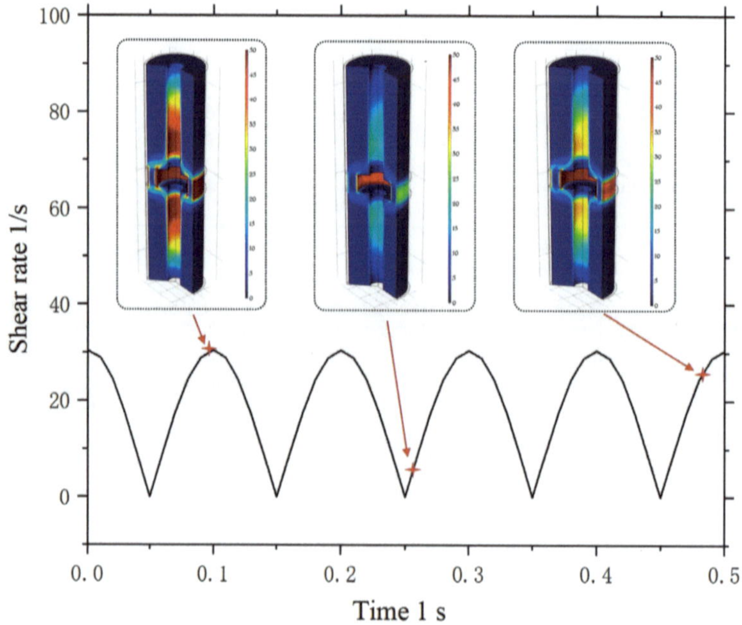

Figure 5. The average shear rate of STF in the annular gap when the damper subjected a A_0 = 3 mm, f = 5 Hz sine displacement excitation.

Building on the FEM model, Figure 6 demonstrates the variation in the amplitude of the average shear rate of the STF within the annular gap, as the piston diameter ranges from 30 to 40 mm, with a piston displacement amplitude of 3 mm and varying frequencies of 1, 3, and 5 Hz. The figure reveals a clear trend: as the piston diameter increases, the average shear rate of the STF escalates accordingly. Under the excitation of a 1 Hz sinusoidal displacement, the shear rate increases from 3.4 1/s to 11.9 1/s; when subjected to a 2 Hz excitation, it rises from 10.3 1/s to 35.9 1/s; and at 3 Hz excitation, it surges from 13 1/s to 55.9 1/s.

As depicted in Figure 2, the shear thickening behavior of the STF manifests within the shear rate range of 15.1 to 30.1 1/s. When the shear rate surpasses 30.1 1/s, the material experiences shear yielding and transitions into a shear thinning state, thereby diminishing the variable damping performance of the STF damper. Accordingly, as shown in Figure 6, with a piston diameter of 36 mm, the maximum average shear rate of the STF within the annular gap fluctuates between 5.3 and 30.1 1/s. This ensures that the average shear rate of the fluid within the annular gap resides in the optimal shear rate range, allowing the STF to exhibit low viscosity under low vibration energy excitation, which places the damper in a low damping state. This significantly mitigates alternating fatigue stress imposed on the structure. Conversely, under high vibration energy excitation, the fluid transitions into a shear thickening high-viscosity state, thereby amplifying the damper's damping force and enhancing the structural stability. Accordingly, a piston with a diameter of 36 mm was selected for the fabrication of the STF damper prototype.

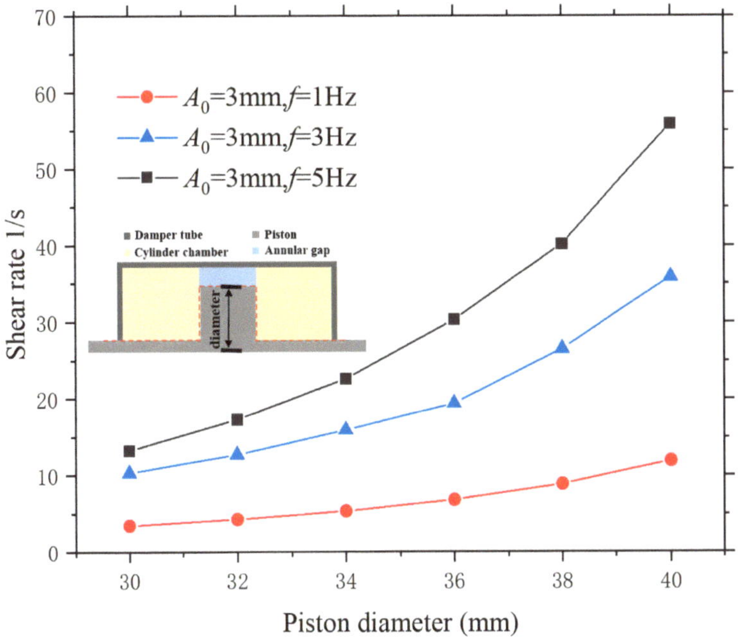

Figure 6. Amplitude variation of STF shear rate with piston diameter (30–40 mm) when the damper subjected a A_0 = 3 mm, f = 5 Hz sine displacement excitation.

4. The Dynamic Property of STF Dampers

4.1. Dynamic Testing of STF Damper

To investigate the dynamic properties of the STF damper, a dynamic testing platform was constructed. Figure 7 illustrates the experimental picture and its schematic diagram. As depicted in Figure 7a, the entire system comprises an electric loading device, a spatial adjustment frame, and fixture. The electric loading device integrates a high-precision planetary ball screw and a Shandong Bangce AC servo motor (PWD-10), capable of handling a maximum dynamic load of ±10 kN, with a dynamic frequency range of 0 to 10 Hz and an amplitude property of ±10 mm. The spatial adjustment frame includes a lifting beam, guide rails, and a lift cylinder to precisely adjust the height required for the STF damper experiments. The fixture, specifically designed in a "horn" shape to accommodate the structural configuration of the STF damper, provide efficient, accurate, and user-friendly testing capabilities. The damper is secured to the fixture through washers, which enhance the overall stability between the line clip and the fixture.

As illustrated in Figure 7b, the STF damper is firmly mounted onto the testing platform using a fixture. A force sensor is installed at the upper part of the platform to capture the output force, with the data collected and uniformly stored in a computer via a controller for seamless post-processing. The dynamic characteristic experiments on the STF damper were conducted using a sinusoidal excitation signal as the displacement input, with excitation frequencies set at 1 Hz, 3 Hz, and 5 Hz, and displacement amplitudes of 1 mm, 2 mm, and 3 mm. To provide a meaningful comparison, an ordinary particle suspension lacking shear thickening properties was introduced into the damper to form a control group, simulating a conventional viscous damper. The dynamic properties of this control damper were tested at a displacement amplitude of 3 mm, with frequencies of 1 Hz, 3 Hz, and 5 Hz, to compare against the STF damper under identical conditions.

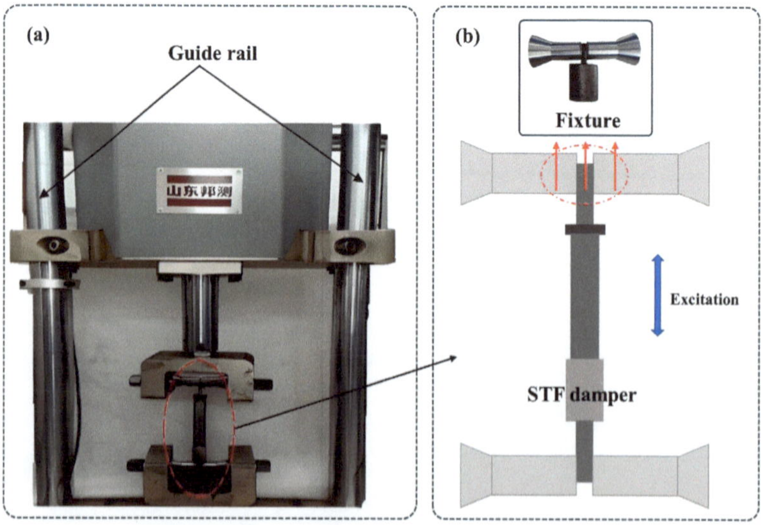

Figure 7. Dynamic testing platform of the STF damper (**a**) experimental picture (**b**) enlarged schematic diagram of the STF damper.

Figure 8 presents the force-displacement hysteresis curves for both the STF damper and the control damper under various frequencies and amplitudes. As illustrated in Figure 8a, the feedback force of the STF damper demonstrates heightened sensitivity to changes in excitation frequency and displacement amplitude, underscoring the pronounced rate-dependent nature of the dynamic properties exhibited by the shear thickening STF damper. At lower excitation frequencies, the piston velocity of the STF damper is relatively slow, leading to a significant overlap in the 1 Hz curve with a gentle slope, indicative of the STF damper's limited energy dissipation capability and correspondingly low output force at these frequencies.

However, as both the excitation frequency and amplitude increase, the rapid motion of the piston elevates the shear rate, ultimately surpassing the critical shear rate, which triggers a sharp escalation in the viscosity of the STF damper. As a result, the output force surges from 13 N to 58 N, representing an increase in more than four times compared to the output force at lower frequencies. Furthermore, as depicted in Figure 8b,c, the reduction in amplitude from 3 mm to 1 mm is accompanied by a marked contraction in the area of the hysteresis loops, indicating a substantial decrease in the damper's energy dissipation capacity per cycle. The observed narrowing of the curve width at lower amplitudes further highlights a reduced range of response forces, signifying a diminished control capacity of the damper in these conditions.

Figure 8d portrays the force-displacement hysteresis curves of the control group at a displacement amplitude of 3 mm. Much like the STF damper, the control group displays a pronounced sensitivity to frequency variations, with the amplitude of the damping force fluctuations expanding in tandem with increasing frequency. This effect is particularly notable at 5 Hz, where the force response becomes significantly more prominent. However, a closer examination of the curve's slope reveals that there is virtually no observable change in slope under displacement loading, indicating that the adjustable stiffness capability of this damper is exceedingly limited.

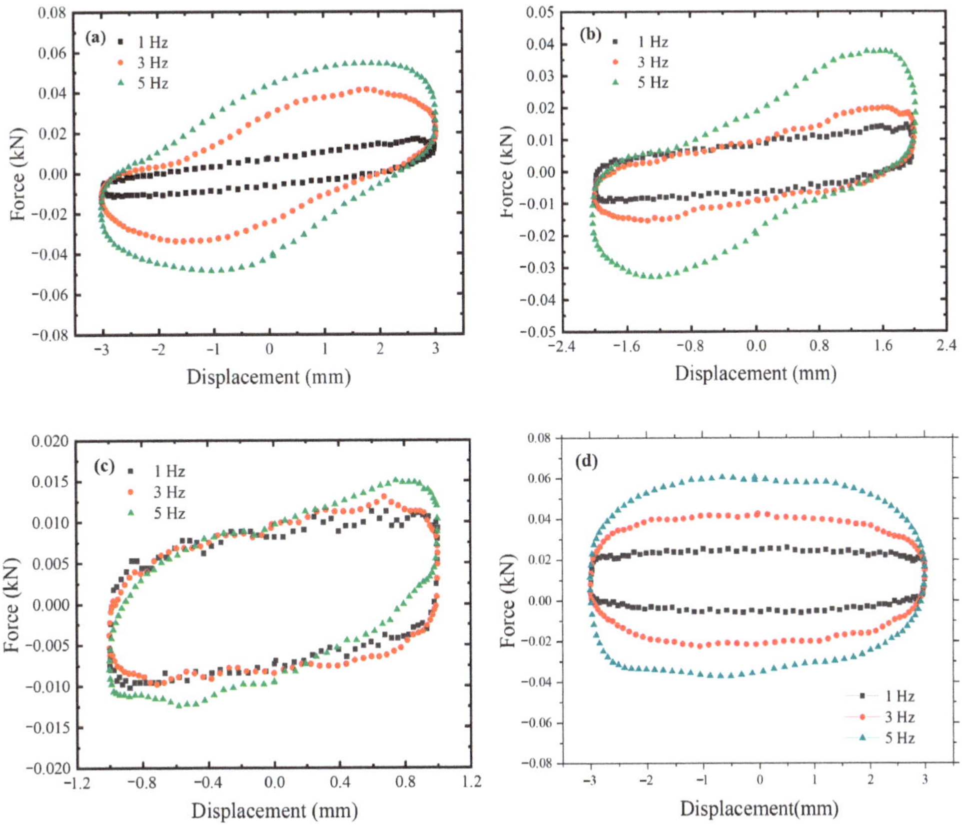

Figure 8. Force-displacement hysteresis curves at different frequencies and amplitudes (**a**) STF damper, A = 3 mm (**b**) STF damper, A = 2 mm (**c**) STF damper, A = 1 mm (**d**) Control group damper, A = 3 mm.

By employing Equations (2) and (3) to articulate the definitions of equivalent stiffness and viscous coefficient [27], one can compute the dynamic properties of both the STF damper and the control group damper.

$$k_e = (F_+ - F_-)/2A_{max} \tag{2}$$

$$c_e = w_d/\pi\omega A_0^2 \tag{3}$$

In the equation: F_+, F_- represent the output forces at the maximum piston displacement A_0, w_d denotes the energy dissipated during each cycle, and ω is the excitation angular frequency.

As depicted in Figure 9a, the control group damper exhibits an equivalent stiffness of zero under reciprocating loading conditions. In stark contrast, the equivalent stiffness of the STF damper undergoes a substantial increase with rising excitation frequency, ascending from 3.1 N/mm to 9.2 N/mm under dynamic excitation with a displacement amplitude of 3 mm. This phenomenon, which is also observed in reference [27], can be attributed to the "jamming effect" caused by the internal particles of the STF liquid as it flows through the annular gap [30]. During high-speed motion, friction and collisions occur among the dispersed particles within the STF material, ultimately leading to the formation of particle clusters. This process triggers a phase transition in the material at the macroscopic level, causing it to shift from a liquid to a solid phase. As the piston moves, the deformation of the solid-phase STF material within the model causes the damper to exhibit elastic properties. With an increase in piston speed, the average shear rate of the STF material within the chamber rises, resulting in more STF material undergoing phase transition. Consequently, the STF damper exhibits variable stiffness characteristics that are intrinsically linked to the excitation frequency.

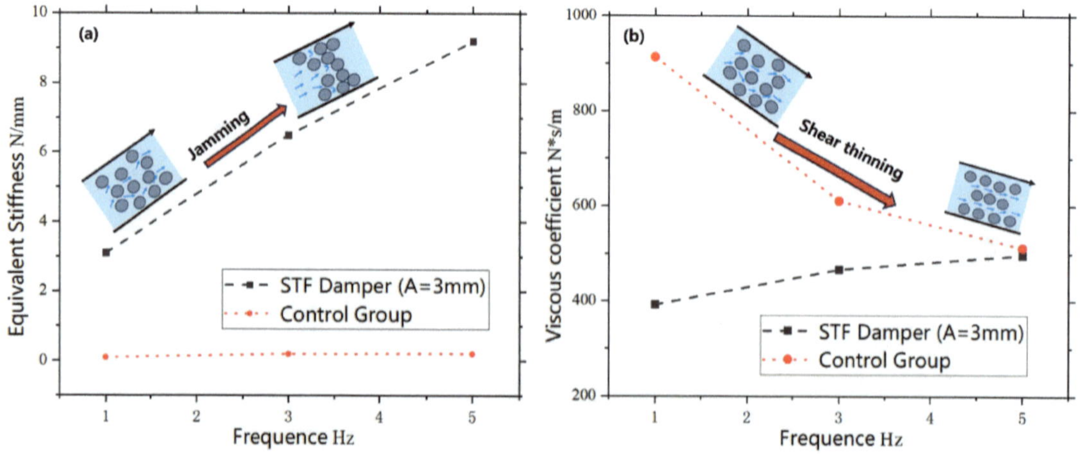

Figure 9. The dynamic properties of the proposed dampers, the blue arrows represent the trajectory of an infinitesimal fluid element, the black arrow represent the direction of fluid flow (**a**) Equivalent stiffness (**b**) Viscous coefficient.

Furthermore, as illustrated in Figure 9b, the viscous coefficient of the control group damper experiences a pronounced decline as the excitation frequency diminishes, plummeting from 913 N·s/m to 512 N·s/m, which signifies a clear shear thinning phenomenon. Conversely, the viscous coefficient of the STF damper exhibits a gradual increase in response to escalating excitation frequencies.

4.2. Nonlinear Fitting Model

As demonstrated in Figure 8, the dynamic properties of the STF damper are influenced by fluctuations in the frequency and amplitude of external excitation. However, employing both frequency and amplitude as fitting parameters for the equivalent stiffness and viscosity coefficient of the damper introduces the challenge of an excessive number of fitting variables. Consequently, this chapter will initially utilize the theory of non-Newtonian fluid dynamics to analyze the nonlinear behavior of the fluid within the STF damper. Building upon this foundational analysis, a quantitative assessment of the correlation between the dynamic properties of the STF damper and the parameters of the excitation load will be conducted, culminating in the establishment of a nonlinear fitting model aimed at accurately capturing the complex nonlinear dynamics of the STF damper.

This study posits that the STF is incompressible, continuous, homogeneous, and maintained in a laminar flow state, while disregarding both inertial and gravitational forces. The flow of the STF material within the annular gap is conceptualized as a one-dimensional gap flow distribution, as depicted in Figure 10a.

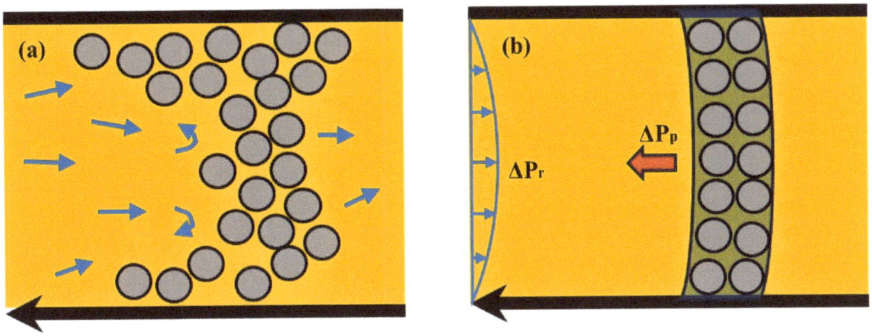

Figure 10. Fluid state between the gap of the damping spacer bar plate, the blue arrows represent the trajectory of an infinitesimal fluid element (**a**) Micro-scale particle blockage phenomenon (**b**) Flow model.

As depicted in Figure 10b, the feedback force F_f of the STF damper is intrinsically defined by two principal components: the frictional damping pressure drop ΔP_r that the STF material encounters while navigating the annular gap, and the osmotic pressure drop ΔP_p resulting from the "jamming effect". This relationship can be articulated mathematically as follows:

$$F_f = \Delta P_r + \Delta P_p \tag{4}$$

When the STF damper performs sinusoidal motion characterized by $A_0 sin(2\pi f t)$, the frictional damping pressure drop within the STF damper can be rigorously analyzed through the flow model established by parallel plate theory [31]. By employing the momentum equation for the fluid, the governing equation for the flow of the STF can be expressed as follows:

$$\frac{\partial \sigma}{\partial x} + \frac{\partial \tau}{\partial z} = \rho\left(\frac{\partial v_x}{\partial t} + v_x\frac{\partial v_x}{\partial x} + v_z\frac{\partial v_x}{\partial z}\right) \tag{5}$$

In the context of the equation, σ and τ represent the normal stress and shear stress of the STF liquid in the x-direction, respectively. Concurrently, v_x and v_z denote the flow velocities of the STF along the x and z axes, respectively, while ρ signifies the liquid density of the STF material. Considering that the fluid exhibits one-dimensional flow in the x-direction and neglecting the effects of inertial forces, Equation (4) can be succinctly simplified to:

$$\frac{\partial \sigma}{\partial x} + \frac{\partial \tau}{\partial z} = 0 \tag{6}$$

According to the definition of the pressure gradient, $dp/dx = -\partial\sigma/\partial x$. Thus, Equation (5) can be further simplified to:

$$\frac{\partial \tau}{\partial z} = \frac{dP_r}{dx} \tag{7}$$

In this equation, P_r represents the frictional damping pressure of the liquid. Upon integration, the result is:

$$\tau(z) = \frac{dP_r}{dx}z + C \tag{8}$$

According to the definition of viscosity, the relationship between shear stress and shear rate is as follows:

$$\tau(z) = \eta\frac{dv_x}{dz} \tag{9}$$

By jointly solving Equation (8) and Equation (7), integrating with respect to z, and subsequently applying the boundary conditions, the flow velocity of the STF material in the x-direction can be expressed as:

$$v_x = \begin{cases} -\frac{1}{2\eta}\frac{dP_r}{dx}\left[\frac{h}{4}^2 - \left(\frac{h}{2} - z\right)^2 - \dot{X}(t)\right] & (0 \le z < \frac{h}{2}) \\ -\frac{1}{2\eta}\frac{dP_r}{dx}\left[\frac{h}{4}^2 - \left(z - \frac{h}{2}\right)^2\right] & (\frac{h}{2} \le z < h) \end{cases} \tag{10}$$

According to the definition of volumetric flow rate:

$$\pi r \int_0^h v_x dz = S_p\, \dot{X}(t) \tag{11}$$

In this context, r denotes the radius of the parallel plate, equivalent to half the circumference of the annular gap, while S_p represents effective cross-sectional area of the STF. $X(t)$ corresponds to the shear displacement of the parallel plate (i.e., the piston), with $\dot{X}(t)$ indicating the shear rate of the parallel plate. By substituting Equations (8) and (9) into Equation (10), and applying $dP_r/dx = -\Delta P_r/l$ (under conditions of uniform laminar flow), the frictional damping pressure drop can be derived through integration [31].

$$\Delta P_r = \frac{12{S_p}^2\dot{X}(t)L\eta}{\pi r h^3} + \frac{6S_p\dot{X}(t)L\eta}{h^2} \tag{12}$$

In this equation, L denotes the length of the shear domain, and the aforementioned represents the frictional force arising from the flow of the STF liquid. The parameter η refers to the viscosity of the STF liquid. Moreover, at high shear rates, the particles within the STF become jammed in the shear zone, inducing a "jamming effect" that leads to solidification around the piston, which, on a macroscopic scale, triggers a phase transition in the STF liquid. To model this phenomenon, this study simplifies the post-phase-transition material as a simply supported beam undergoing bending deformation in the flow field. Consequently, the osmotic pressure drop in the liquid region, induced by the permeation of the elastic solidified phase, can be expressed as:

$$\Delta P_p = \frac{48EI}{h^3}X(t) + f_N sign(\dot{X}(t)) \tag{13}$$

In this equation, EI represents the equivalent shear stiffness, and f_N denotes the frictional force exerted by the piston on the "jammed" phase. By substituting Equations (12) and (13) into Equation (4), the damping force relationship for the STF damper can be derived as follows:

$$F = (\frac{12{S_p}^2L}{\pi r h^3} + \frac{6S_pL}{h^2})\eta\dot{X}(t) + \frac{48EI}{h^3}X(t) + f_N sign(\dot{X}(t)) \tag{14}$$

In this equation, h, L, S_p and r represent the structural constants of the annular gap, while the STF viscosity η, equivalent shear stiffness EI and the frictional force f_N are related to the degree of "jamming" within the STF's internal micro particles. The "jamming" phenomenon of the particles is directly correlated with the peak piston velocity v_{max}. Therefore, the damping force model of the STF damper, as expressed in Equation (14), can be further simplified into the following phenomenological model:

$$F = c_e(v_{max})\dot{X}(t) + k_e(v_{max})X(t) + \mu_e(v_{max})sign(\dot{X})$$

$$c_e = a + bv_{max} + cv_{max}^2$$

$$k_e = e + fv_{max} + gv_{max}^2$$

$$\mu_e = h + iv_{max} + jv_{max}^2 \tag{15}$$

In this equation, c_e and k_e denote the dynamic viscous and equivalent stiffness of the damper, respectively. Drawing from the experimental data, Figure 11 presents the relationship between these two parameters and the peak velocity of the piston motion.

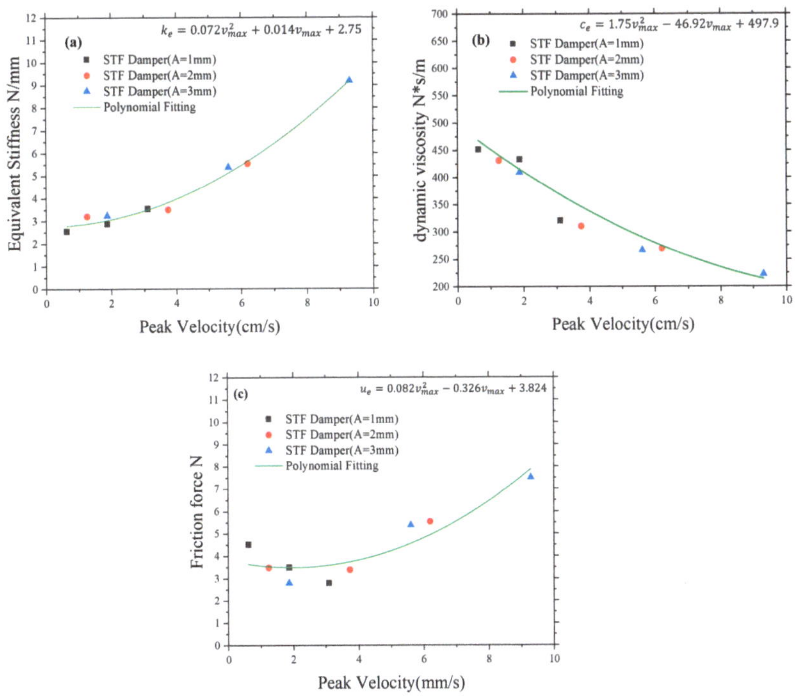

Figure 11. The dynamic properties of the proposed dampers (**a**) Equivalent stiffness, e = 2.75, f = 0.014, g = 0.072 (**b**) Dynamic viscous a = 497.9, b = −46.92, c = 1.75 (**c**) Friction force, h = 3.824, i = −0.326, j = 0.082.

As illustrated in Figure 11, fitting the damper's dynamic viscous and equivalent stiffness to the piston's peak velocity yields a highly effective representation of the material's inherent nonlinearity. In Figure 11a, the equivalent stiffness of the damper demonstrates an initial slow increase, followed by a rapid escalation as piston velocity intensifies, signifying a nonlinear correlation between equivalent stiffness response and piston velocity. The gradual increase in equivalent stiffness at lower speeds may be attributed to the fact that the material's rheological properties are not fully activated at lower shear rates. Conversely, at higher velocities, the growth rate of equivalent stiffness markedly accelerates, revealing that the shear thickening effect becomes considerably more pronounced under conditions of high shear.

Additionally, a comparative analysis of equivalent stiffness trends at different amplitudes reveals that the stiffness increases most rapidly at an amplitude of 3 mm. This suggests that, at larger amplitudes, the STF damper can more effectively harness its shear thickening properties to respond to greater external forces. As amplitude increases, the material inside the damper undergoes more substantial deformation, leading to significant microstructural changes and molecular rearrangements, thereby enabling the damper to exhibit greater stiffness under dynamic loading at higher amplitudes.

Simultaneously, in Figure 11b, the dynamic viscosity reveals a nonlinear relationship with piston peak velocity; however, its trend contrasts with that of the equivalent stiffness, exhibiting a decrease as the piston peak velocity increases. This observation suggests that during the initial phase of velocity escalation, the dynamic viscosity of the STF material responds promptly, indicating that its internal structure can adapt swiftly to velocity variations at lower velocity thresholds, which leads to a pronounced reduction in dynamic viscosity. Nonetheless, as the piston peak velocity reaches a specific threshold, the rate of increase in the dynamic viscosity gradually diminishes, entering a more stable regime. Additionally, a comparative analysis of the dynamic viscosity trends at varying amplitudes indicates that, at lower amplitudes, the decline is most significant within the low-speed range, although the maximum dynamic viscosity is still observed at 1 mm.

Figure 11c illustrates the nonlinear relationship between friction force and piston peak velocity. At an amplitude of 1 mm, the friction force decreases as piston peak velocity increases. At an amplitude of 2 mm, the friction force initially decreases before increasing, while at an amplitude of 3 mm, the friction force continuously rises. This demonstrates that at higher amplitudes, the shear thickening characteristics of the STF material are fully activated. This sustained increase in friction force highlights the exceptional performance of the STF damper at larger amplitudes, allowing it to deliver effective energy dissipation and control across a broader range of velocities.

Figure 12 illustrates the force-displacement curves derived from both experimental measurements and theoretical calculations under varying frequencies and amplitudes. The figure clearly reveals that the force-displacement response undergoes significant changes as both amplitude and frequency increase. As the amplitude rises, the hysteresis loops expand, indicating greater energy dissipation. Similarly, an increase in frequency results in a broader force response within the hysteresis loops, signifying that the STF damper exhibits stronger damping characteristics under higher dynamic loads.

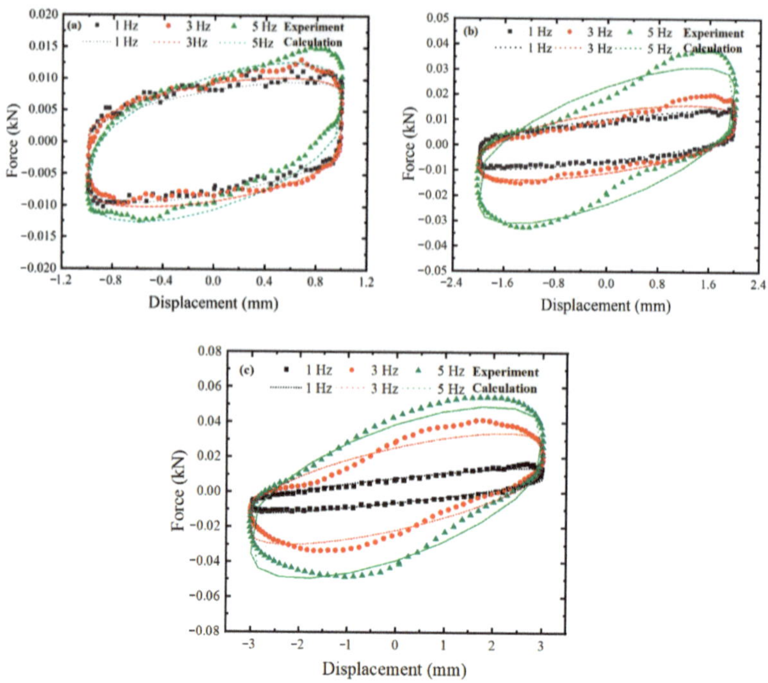

Figure 12. Force-displacement curves from both experiment and calculation under different frequencies and amplitudes. (**a**) STF damper, A = 1 mm (**b**) STF damper, A = 2 mm (**c**) STF damper, A = 3 mm.

As for the discrepancies between the experimental measurements and theoretical calculations, these deviations may partially stem from the compression and expansion of air within the damper. During dynamic loading, especially at higher frequencies, the behavior of the air may not have been fully captured by the theoretical model, resulting in the experimental data showing greater force variations than anticipated.

5. Conclusions

This study developed a self-adaptive STF material that rapidly and reversibly modifies its viscosity in response to shear rates, utilizing a 25% mass fraction of SiO_2 and a compounded polyethylene glycol solution (PEG200:PEG400 = 2:1). The study optimized and validated the design of the STF damper, demonstrating its significant potential in variable stiffness applications.

(1) The proposed STF material exhibits a pronounced shear thickening behavior, with a 412% increase in viscosity. This thickening initiates at a shear rate of 15.1 1/s and peaks at 30.1 1/s, beyond which the material transitions into a shear thinning state.
(2) Simulation results indicate that as the excitation frequency, excitation amplitude, and piston radius increase, the average shear rate of the STF in the annular gap region also gradually increases. When the piston diameter is 36 mm, the average shear rate of the STF material in the annular gap region does not exceed 30.1 1/s, ensuring that the STF material does not extensively enter the shear thinning stage when the damper is subjected to sinusoidal excitation within the range of 1–5 Hz and 1–3 mm.
(3) A STF damper was fabricated based on the optimized design. Experimental results showed that the stiffness of the STF damper varies with changes in excitation frequency. When the external frequency increases from 1 Hz to 5 Hz at an amplitude of 3 mm, the equivalent stiffness of the STF spacer increases from 3.2 N/mm to 9.1 N/mm, nearly tripling.

Author Contributions: S.C.: Conceptualization, Investigation, Writing—original draft; X.F.: Software, Methodology; P.M.: Formal analysis, Experiment; L.C. and L.W.: Validation; J.Y.: Funding. All authors have read and agreed to the published version of the manuscript.

Funding: This study is supported by the Science and Technology Research Program of Chongqing Municipal Education Commission (Grant No. KJZD-K202401505), and the Natural Science Foundation of Chongqing Science and Technology Commission (Grant No. QCSTB2023NSCQ-MSX0029).

Data Availability Statement: The data presented in this study are available from the corresponding author upon reasonable request.

Conflicts of Interest: The authors declare no conflicts of interest.

References

1. Zhang, Y.; Guo, J.; Yang, J.; Li, X. Recent Structural Developments and Applications of Magnetorheological Dampers (MRD): A Review. *Magnetochemistry* **2023**, *9*, 90. [CrossRef]
2. Li, H.; Li, L.; Hu, R.; Ye, M. Simplified design of nonlinear damper parameters and seismic responses for long-span cable-stayed bridges with nonlinear viscous dampers. *Front. Struct. Civ. Eng.* **2024**, *18*, 1103–1116. [CrossRef]
3. Mohtasim, S.M.; Ahammed, R.; Rahman, M.M.; Rashid, M.M.; Roy, R.; Aziz, M.A. Recent developments of regenerative magnetorheological (RMR) damper: A review. *Korea-Aust. Rheol. J.* **2021**, *33*, 201–224. [CrossRef]
4. Aziz, M.A.; Aminossadati, S.M. State-of-the-art developments of bypass Magnetorheological (MR) dampers: A review. *Korea-Aust. Rheol. J.* **2021**, *33*, 225–249. [CrossRef]
5. Deng, H.; Lian, X.; Gong, X. A brief review of variable stiffness and damping magnetorheological fluid dampers. *Front. Mater.* **2022**, *9*, 1019426. [CrossRef]
6. Chen, C.; Chan, Y.S.; Zou, L.; Liao, W.-H. Self-powered magnetorheological dampers for motorcycle suspensions. *Proc. Inst. Mech. Eng. Part D J. Automob. Eng.* **2017**, *232*, 921–935. [CrossRef]
7. Bastola, A.K.; Li, L.; Paudel, M. A hybrid magnetorheological elastomer developed by encapsulation of magnetorheological fluid. *J. Mater. Sci.* **2018**, *53*, 7004–7016. [CrossRef]
8. Cao, S.; Hu, F.; Zhang, G. Superelastic Shape Memory Alloy Honeycomb Damper. *Appl. Sci.* **2023**, *13*, 13154. [CrossRef]

9. Liu, H.; Fu, K.; Cui, X.; Zhu, H.; Yang, B. Shear Thickening Fluid and Its Application in Impact Protection: A Review. *Polymer* **2023**, *15*, 2238. [CrossRef]
10. Fu, K.; Wang, H.; Zhang, Y.X.; Ye, L.; Escobedo, J.P.; Hazell, P.J.; Friedrich, K.; Dai, S. Rheological and energy absorptio characteristics of a concentrated shear thickening fluid at various temperatures. *Int. J. Impact Eng.* **2020**, *139*, 103525. [CrossRef
11. Hasanzadeh, M.; Mottaghitalab, V. The Role of Shear-Thickening Fluids (STFs) in Ballistic and Stab-Resistance Improvement o Flexible Armor. *J. Mater. Eng. Perform.* **2014**, *23*, 1182–1196. [CrossRef]
12. Aşkan, A.; Çapkurt, M.; Acar, E.; Aydın, M. Influence of dispersion liquid and silica concentration on rheological properties o shear thickening fluids (STFs). *Rheol. Acta* **2023**, *62*, 447–460. [CrossRef]
13. Sun, L.; Liang, T.; Zhang, C.; Chen, J. The rheological performance of shear-thickening fluids based on carbon fiber and silic nanocomposite. *Phys. Fluids* **2023**, *35*, 032002. [CrossRef]
14. Madraki, Y.; Ovarlez, G.; Hormozi, S. Transition from Continuous to Discontinuous Shear Thickening: An Excluded-Volum Effect. *Phys. Rev. Lett.* **2018**, *121*, 108001. [CrossRef] [PubMed]
15. Fu, K.; Wang, H.; Chang, L.; Foley, M.; Friedrich, K.; Ye, L. Low-velocity impact behaviour of a shear thickening fluid (STF) an STF-filled sandwich composite panels. *Compos. Sci. Technol.* **2018**, *165*, 74–83. [CrossRef]
16. James, N.M.; Han, E.; de la Cruz, R.A.L.; Jureller, J.; Jaeger, H.M. Interparticle hydrogen bonding can elicit shear jamming i dense suspensions. *Nat. Mater.* **2018**, *17*, 965–970. [CrossRef]
17. Xu, Y.-l.; Gong, X.-l.; Peng, C.; Sun, Y.-q.; Jiang, W.-q.; Zhang, Z. Shear Thickening Fluids Based on Additives with Differen Concentrations and Molecular Chain Lengths. *Chin. J. Chem. Phys.* **2010**, *23*, 342–346. [CrossRef]
18. Moriana, A.; Pérez-López, D.; Gómez-Rico, A.; Salvador, M.d.l.D.; Olmedilla, N.; Ribas, F.; Fregapane, G. Irrigation schedulin for traditional, low-density olive orchards: Water relations and influence on oil characteristics. *Agric. Water Manag.* **2007**, *87* 171–179. [CrossRef]
19. Antosik, A.; Głuszek, M.; Żurowski, R.; Szafran, M. Effect of SiO_2 Particle Size and Length of Poly(Propylene Glycol) Chain o Rheological Properties of Shear Thickening Fluids. *Arch. Metall. Mater.* **2016**, *61*, 1511–1514. [CrossRef]
20. Gürgen, S.; Sofuoğlu, M.A.; Kuşhan, M.C. Rheological compatibility of multi-phase shear thickening fluid with a phenomenologi cal model. *Smart Mater. Struct.* **2019**, *28*, 035027. [CrossRef]
21. Helber, R.; Doncker, F.; Bung, R. Vibration attenuation by passive stiffness switching mounts. *J. Sound Vib.* **1990**, *138*, 47–57 [CrossRef]
22. Laun, H.M.; Bung, R.; Schmidt, F. Rheology of extremely shear thickening polymer dispersionsa) (passively viscosity switchin fluids). *J. Rheol.* **1991**, *35*, 999–1034. [CrossRef]
23. Gong, X.; Zhang, J.; Xuan, S. Multi-Functional Systems Based on Shear Thickening Fluid. In *Shear Thickening Fluid*; Springer Cham, Switzerland, 2023; pp. 53–75. [CrossRef]
24. Sheikhi, M.R.; Gürgen, S.; Altuntas, O.; Sofuoğlu, M.A. Anti-impact and vibration-damping design of cork-based sandwic structures for low-speed aerial vehicles. *Arch. Civ. Mech. Eng.* **2023**, *23*, 71. [CrossRef]
25. Yay, Ö.; Diltemiz, S.F.; Kuşhan, M.C.; Gürgen, S. Shear Thickening Fluid-Based Vibration Damping Applications. In *Smart System with Shear Thickening Fluid*; Springer: Cham, Switzerland, 2024; pp. 59–69. [CrossRef]
26. Zhang, X.Z.; Li, W.H.; Gong, X.L. The rheology of shear thickening fluid (STF) and the dynamic performance of an STF-fille damper. *Smart Mater. Struct.* **2008**, *17*, 035027. [CrossRef]
27. Zhou, H.; Yan, L.; Jiang, W.; Xuan, S.; Gong, X. Shear thickening fluid–based energy-free damper: Design and dynami characteristics. *J. Intell. Mater. Syst. Struct.* **2014**, *27*, 208–220. [CrossRef]
28. Srivastava, A.; Majumdar, A.; Butola, B.S. Improving the Impact Resistance of Textile Structures by using Shear Thickening Fluids A Review. *Crit. Rev. Solid State Mater. Sci.* **2012**, *37*, 115–129. [CrossRef]
29. Galindo-Rosales, F.J.; Rubio-Hernández, F.J.; Sevilla, A. An apparent viscosity function for shear thickening fluids. *J. Non-Newton Fluid Mech.* **2011**, *166*, 321–325. [CrossRef]
30. Brown, E.; Jaeger, H.M. Shear thickening in concentrated suspensions: Phenomenology, mechanisms and relations to jamming *Rep. Prog. Phys.* **2014**, *77*, 046602. [CrossRef]
31. Wei, M.; Lin, K.; Guo, Q.; Sun, H. Characterization and performance analysis of a shear thickening fluid damper. *Meas. Control* **2019**, *52*, 72–80. [CrossRef]

buildings

MDPI

Article

Examination of Damage Evolution in Slurry Masonry Schist Subjected to Biaxial Compressive Stresses

Jie Dong [1], Siwu Cheng [1], Hongyun Chen [1,*], Hongfeng Zhang [1], Yadong Zhao [1], Guoxiang Zhang [2] and Fengwu Gong [3]

1 College of Civil Engineering, Hebei University of Architecture, Zhangjiakou 075000, China; dongjie1003@hotmail.com (J.D.); cheng1651049@outlook.com (S.C.); zhf17861127976@outlook.com (H.Z.); zyd18336466820@outlook.com (Y.Z.)
2 China Railway Design Group Limited, Tianjin 300380, China; 13512499160@163.com
3 No. 3 Geological Brigade, Hebei Provincial Geological Prospec Ting Bureau, Zhangjiakou 075000, China; gfw15830301682@126.com
* Correspondence: chy2180@hebiace.edu.cn; Tel.: +86-198-9899-9275

Abstract: This study used a static bidirectional multifunctional loading system. The system conducted bidirectional compression tests on scaled specimens of slurry masonry schist under freeze–thaw cycling conditions. This study aimed to investigate the influence of bidirectional stress coupling with freeze–thaw cycles on the mechanical properties of slurry masonry schist. The results indicate that lateral pressure can increase the peak stress of slurry masonry schist, while freeze–thaw cycles have an adverse effect on the material's internal pore structure, counteracting the gain effect of lateral pressure. This study also employed acoustic emission (AE) technology to analyze the evolution of slurry masonry schist failure characteristics. The findings reveal that freeze–thaw cycles accelerate the failure of slurry masonry schist during loading, and lateral pressure to some extent mitigates the damage development of slurry masonry schist. The synergistic effect of lateral pressure and freeze–thaw cycles alters the fracture mode of slurry masonry schist. Acoustic emission signal localization demonstrates numerous AE localization points in the interface transition zone, forming a coherent signal band where cracks propagate toward complete interface penetration. The crack extension process of the slurry masonry schist was investigated using the digital image correlation (DIC) method. The results indicated that macroscopic cracks formed in the strain localization zone, resulting in fracture damage to the specimens, with interfacial debonding identified as the primary failure mode for slurry masonry schist structures.

Keywords: slurry masonry schist; biaxial stress; freeze–thaw cycle; damage evolution; interface debonding

1. Introduction

In the context of high-speed railroad and highway construction in China, slurry masonry schist lattice berms are extensively employed because of their convenient and cost-effective attributes. In service, slurry masonry schist often experiences biaxial forces; damage to slurry masonry schist under two-way coupling mechanics can occur prematurely, leading to potential safety hazards and, possibly, serious economic losses. In addition, frost damage has become a major problem for slope protection in cold regions. Understanding the effects of coupled biaxial stresses and freeze–thaw cycles on the damage evolution of slurry masonry schist can help to evaluate and predict the performance of slurry masonry structures under a variety of stresses and environmental conditions and improve the safety and durability of these structures.

Current research on slurry masonry schist slope protection has focused on the overall stability of slopes [1,2], and several scholars have established slope models through numerical simulation software to study the effects of freeze–thaw cycles, herbaceous root systems, and other factors on slope stability [3–6]. Several scholars have achieved good results by designing lattice beams and anchor cables to improve the overall stability of slopes [7–10]. Slurry masonry schist, a combination of schist and mortar, is different from ordinary concrete. The aggregate size of slurry masonry schist is usually greater than 150 mm [11], and the proportion of aggregate is greater than 70%. Consequently, the fracture properties of slurry masonry schist diverge from those of ordinary concrete. Caliskan et al. [12] reported that the bonding behavior between schist and mortar is mechanical rather than chemical in nature. Li et al. [13] concluded that the damage in slurry masonry schist structures occurs mainly in the form of interfacial debonding damage and that the interfacial transition zone (ITZ) between schist and mortar is the weakest place. Furthermore, several scholars have studied the debonding behavior of the interface transition zone between schist and mortar [14,15].

Numerous studies have analyzed the bidirectional force properties of cementitious materials to investigate the effect of coupled bidirectional forces on the mechanical properties of cementitious materials. Several scholars have investigated the effect of particle size on cementitious materials under biaxial loading [16–18]. Murad et al. [19] developed a model for predicting the shear strength of concrete columns under biaxial stresses by employing pertinent technical methods. Van Mier [20] examined the impact of structural effects on concrete softening during multiaxial compression. Shang et al. [21] explored the influence of freeze–thaw cycles on the strength and deformation of concrete under biaxial compression, proposing an equation to determine biaxial compressive strength in principal stress space. Zhu et al. [22] innovatively simulated the process of damage development in concrete subjected to biaxial loading by employing a correlation model grounded in elastic mechanics. Tschegg et al. [23] investigated type I fracture behavior of concrete subjected to biaxial loading. The dynamic mechanical properties of cementitious materials under biaxial loading have also been investigated by several scholars [24–26]. Li Z et al. [27] investigated the impact resistance of mortar–rock berm structures based on fracture mechanics theory.

Structural damage and crack formation under loading typically occur gradually. Therefore, continuous monitoring of damage and deformation throughout the loading process is imperative. Nondestructive testing techniques have been used by many scholars to study the damage mechanisms of materials. The current nondestructive testing techniques for cementitious materials mainly include digital image correlation (DIC) and acoustic emission analysis. DIC acquires the displacement field of the surface by analyzing the changes in the three-dimensional coordinates of each point in the measurement area at various deformation states and subsequently computes the strain field of the surface. Crack expansion has also been monitored using DIC during the deformation of cementitious materials to provide information on the evolution of the interface in the cracked zone [28–32]. Boulekbache et al. [33] employed DIC to analyze the fracture mechanism of steel fiber-reinforced concrete during tensile splitting. Huang et al. [34] investigated the deformation field of concrete under axial pressure based on DIC. The acoustic emission technique is extensively employed in the nondestructive evaluation of engineering structures. Li D et al. [35] investigated the evolution of the fracture process zone (FPZ) under static and fatigue loading using DIC analysis. Lakavath C et al. [36] employed DIC technology to identify the loading of interface shear cracks and potential failure modes, proposing a simplified model for load-bearing interface shear. Additionally, several scholars have conducted nondestructive testing to assess damage in concrete structures by analyzing acoustic emission data. Ohno et al. [37] classified concrete cracks based on acoustic emission data. Grosse [38] et al. quantitatively evaluated the fracture process of concrete based on acoustic emission data. Sagar [39] et al. investigated the fracture mechanism of concrete and mortar specimens based on the acoustic emission b-value. Zhang H et al. [40] investigated the dynamic tensile fracture characteristics of steel fiber-reinforced concrete using AE technology, enabling early warning of the critical stage under freeze–thaw cycling conditions. Barbosh M et al. [41] proposed a damage identification method combining multi-dimensional empirical mode decomposition (MEMD) and a Gaussian mixture model (GMM) using AE as the basis.

Several scholars have studied the fracture behavior of concrete by combining DIC and acoustic emission techniques [42–44].

In this study, porous basalt, which is common in North China, was selected as the coarse aggregate for slurry masonry schist, and biaxial compression tests with different lateral pressures (LP) coupled with freeze–thaw cycles (FT) were carried out. Acoustic emission and DIC techniques were used to monitor the whole test process. The damage mode and damage mechanism of the slurry masonry schist were evaluated by closely monitoring the changes in the strain field on the surface of the specimen and collecting the acoustic signals generated by internal damage occurring in the material; these data were statistically processed and analyzed.

2. Test Profile

2.1. Test Preparation

The experiment employed porous basalt as the coarse aggregate for the slurry schist, utilizing Chinese standard (GB 175-99) PO.42.5 ordinary Portland cement as the cement [13] and medium sand as the fine aggregate. Since the grain sizes of the schist specimens were larger than 150 mm, the slurry schist structure was scaled down by a ratio of 1:2 according to similarity theory; hence, the grain size of the schist was set to 60~90 mm. The chemical composition of the porous basalt was measured via X-ray fluorescence analysis. The relevant technical indexes are shown in Tables 1–3. Under actual engineering conditions, the coarse aggregate in slurry masonry schist accounts for a large proportion of the total coarse aggregate; hence, the coarse aggregate content was set to 75%. The slurry masonry schist mix ratios are shown in Table 4. The specimen preparation was divided into five steps as follows:

a. Water, sand, and cement were mixed in the appropriate proportions with a mixer until the water was thoroughly incorporated and the consistency was suitable; at this point, the mortar mixing was complete.
b. To ensure that the sample was released from the mold smoothly, the bottom of the 300 mm × 300 mm × 80 mm mold was brushed with oil in advance.
c. The mold was filled with the mixed cement mortar in layers. First, a layer of mortar was applied to the bottom of the mold, and the mortar surface was smoothed. Subsequently, the moistened schist was placed in the mold. A concrete vibration table was used to vibrate the mold, and the samples were allowed to harden for 24 h after demolding.
d. The demolded specimens were transferred to a standard curing room set at a temperature of 20 °C and a relative humidity of 95% for the duration of 28 days.
e. The surface of each slurry masonry specimen was polished, and Figure 1 shows a drawing of a polished slurry masonry specimen.
f. The slurry masonry schist was subjected to freeze–thaw cycles using a high- and low-temperature alternating test chamber. The cyclic temperature range was set to −20 °C to 20 °C, with each freeze–thaw cycle lasting 24 h. The number of freeze–thaw cycles was set to 0, 40, and 80, aligning with the climatic patterns in the North China region.

Table 1. Basic performance parameters of P.O42.5 cement.

Standard Consistency/%	Fineness/mm	Initial Fluidity/mm	20 min Flow Rate/mm	Solidification Time (min) 3 d	Solidification Time (min) 7 d	24 h Compressive Strength/MPa	24 h Flexural Strength/MPa
29.8	0.03	>195	>180	180	370	25.5	6.2

Table 2. Sand fineness and density.

Aperture Size	7.5 mm	4.5 mm	2.5 mm	1.25 mm	600 μm	300 μm	≤150 μm
Subtotal sieve residue %	0.1	0.3	10.2	18.5	26.4	44.2	0.3
Cumulative sieve rate %	0.1	0.4	10.6	29.1	55.5	99.7	100
Fineness modulus	2.5 belongs to the medium sand aggregate						
Packing density (g/cm^3)	1.45						

Table 3. Chemical composition of the porous basalt.

Rock Type	Chemical Composition (%)				
Porous basalt	SiO_2	CaO	Al_2O_3	MgO	Fe_2O_3
	47.45	6.75	15.18	5.42	16.03
	Na_2O	K_2O	TiO_2	SO_3	-
	4.30	1.74	3.08	0.05	-

Table 4. Slurry masonry schist mixing ratio.

Mortar:Aggregate	Water (kg/m^3)	Cement (kg/m^3)	Sand (kg/m^3)	Aggregate (kg/m^3)
1:3	280	250.2	1208.3	5215.4

Figure 1. Slurry masonry schist specimen.

2.2. Experimental Design

The stress states studied in this experiment included uniaxial and biaxial forces, as shown in Figure 2. In this study, constant lateral pressure loading was used, and the vertical pressure direction was specified as the main axial direction of loading. The compressive stress in the main axis direction was y, the horizontal direction was the direction of the loaded subaxis, and the compressive stress in the subaxis direction was x. To clarify the effect of lateral pressure on the mechanical properties of the slurry masonry schist, the horizontal pressure was set to 0, 0.2 fc, 0.4 fc, and 0.6 fc, where fc is the uniaxial compressive strength of the slurry masonry schist.

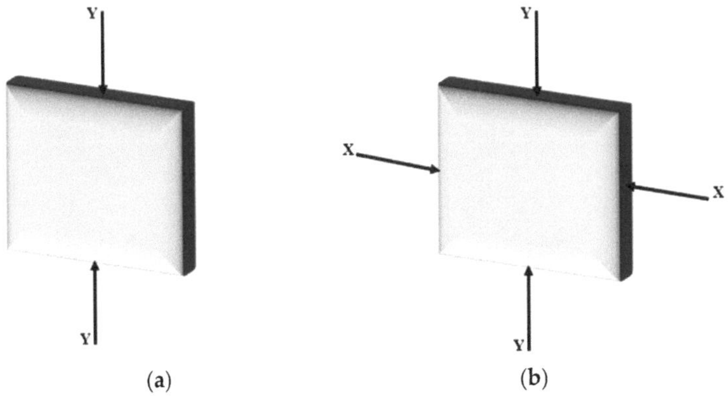

Figure 2. Schematic representation of stress states: (**a**) uniaxial compression and (**b**) biaxial compression.

In this study, the fixed lateral pressure loading method was adopted. First, load control is adopted for the horizontal direction x. After reaching the set stress, the fixed lateral pressure was unchanged, and y was loaded according to the static loading rate (0.5 mm/min) until damage occurred in the specimen. The loading schematic is shown in Figure 3. In the process of slurry masonry schist pressurization, because of the different stiffnesses between the specimen and the loading plate, shear stress, resulting in uneven force on the specimen, affected the test results. To determine the real compressive strength, according to the research by Shiming et al. [45], two layers of butter were brushed between three layers of plastic film to reduce friction on the four sides.

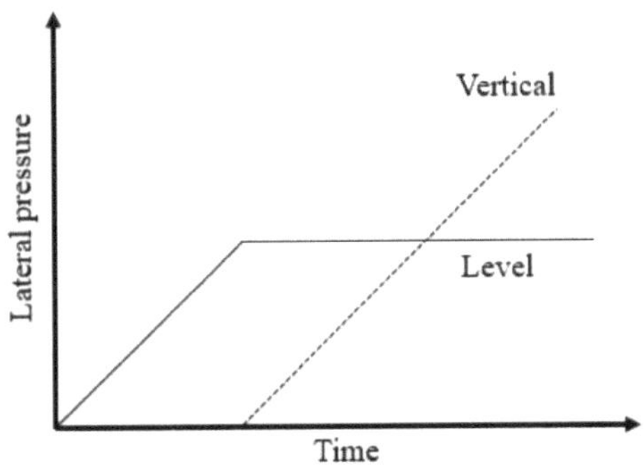

Figure 3. Bidirectional loading path schematic.

2.3. Test Methods

To enhance the reliability of the acquired data, biaxial loading was conducted using a YDW-300/2 static two-way multifunctional loading test system manufactured by Jilin Guanteng Automation Technology Co. (Changchun, China). An XTDIC scattering deformation measurement system from Shenzhen Xintuo 3D Co. (Shenzhen, China). was employed in each test. Additionally, the DS-8 digital acoustic emission system manufactured by Beijing Soft Island Times Technology Co. (Beijing, China) was utilized. Figure 4 shows a schematic diagram depicting the test layout.

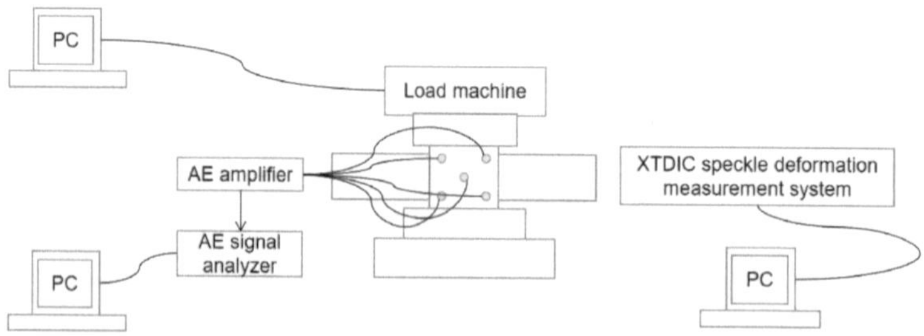

Figure 4. Test layout schematic.

2.3.1. DIC Test Method

The DIC speckle deformation measurement system employs two high-precision cameras to capture real-time speckle images of the object at various deformation stages. It utilizes digital image correlation to match deformation points on the object's surface and reconstructs the three-dimensional coordinates of surface calculation points based on the disparity data of each point and the pre-calibrated camera parameters. By comparing the changes in the three-dimensional coordinates of each point within the measurement area for each deformation state, the displacement field of the object's surface is obtained, further calculating the strain field of the surface. The displacement and strain of the specimen were determined using Equations (1)–(3) [46]. The parameters of the DIC speckle deformation measurement system were configured as shown in Table 5. The testing standards were referenced from the study on the failure mechanism of slurry masonry schist conducted by Li et al. [13].

$$\varepsilon_{xx} = \frac{1}{2}\left[2\frac{\partial u}{\partial x} + \left(\frac{\partial u}{\partial x}\right)^2 + \left(\frac{\partial v}{\partial x}\right)^2\right] \tag{1}$$

$$\varepsilon_{yy} = \frac{1}{2}\left[2\frac{\partial v}{\partial y} + \left(\frac{\partial u}{\partial y}\right)^2 + \left(\frac{\partial v}{\partial y}\right)^2\right] \tag{2}$$

$$\varepsilon_{xy} = \frac{1}{2}\left(\frac{\partial u}{\partial y} + \frac{\partial v}{\partial x}\right) + \frac{1}{2}\left(\frac{\partial u}{\partial x}\frac{\partial u}{\partial y} + \frac{\partial v}{\partial x}\frac{\partial v}{\partial y}\right) \tag{3}$$

where ε_{xx}, ε_{yy}, and ε_{xy} are the transverse strain, longitudinal strain, and shear strain, respectively, and u and v denote the displacement fields in the x- and y-axes, respectively.

Table 5. The main parameters of the DIC scattering deformation measurement system.

Pixel Size	Capture	Matching Method	Search Range Radius	Background Check Threshold	Continuity Threshold
1960 × 1280	2 Hz	Horizontal matching	75	1	1

2.3.2. Acoustic Emission Test Method

The acoustic emission system consisted of a digital acoustic emission system, transducer, and preamplifier. During each test, the acoustic emission equipment collected, processed, and displayed the acoustic emission signal parameters in real time. The main parameters of the acoustic emission system are shown in Table 6. The testing standards were referenced from the study on the failure mechanism of slurry masonry schist conducted by Li et al. [13].

Table 6. Main parameters of the acoustic emission system.

Sampling Frequency	PDT	HDT	HLF	High Pass On-Board Filter	Low Pass On-Board Filter
2.5 MHz	100 μs	300 μs	200 μs	100 kHz	20 kHz

To enhance the conduction between the sensor and the slurry schist, the sensor was coupled to the slurry schist with petroleum jelly, and then the sensor was secured with electrical tape. A minimum of four sensors were needed to determine the spatial location of an acoustic emission source. To enhance the localization effect, five sensors were arranged on the surface of the slurry schist. The locations of the sensors are shown in Figure 5.

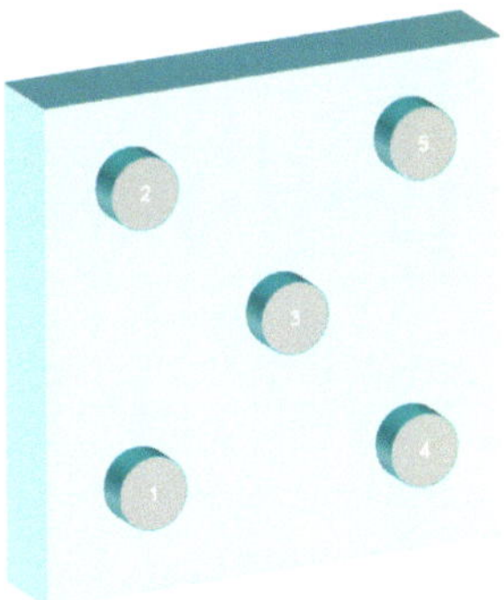

Figure 5. Acoustic emission sensor layout.

3. Test Results and Analysis

3.1. Peak Stress and Peak Strain

Figure 6a illustrates the relationship between the compressive strength of the slurry masonry schist and the number of freeze–thaw cycles under different lateral pressure conditions. This study indicates that with an increase in the number of freeze–thaw cycles, the compressive strength of the slurry masonry schist exhibits a decreasing trend. This phenomenon is attributed to the expansion of internal moisture in the slurry masonry schist during freezing, leading to pore enlargement. Subsequently, upon thawing, new pores and microcracks form and continue to absorb water, causing the internal structure of the slurry masonry schist to become loose and the bonds among aggregates to weaken, thus reducing the compressive strength. Additionally, the experiment reveals that increasing lateral pressure enhances the peak stress of the slurry masonry schist specimens, but the internal structural damage caused by freeze–thaw cycles weakens their compressive strength. Therefore, within the range of 0 to 0.6 fc lateral pressure, despite the enhancement in peak stress due to lateral pressure, the overall decrease in compressive strength is still caused by the damaging effects of freeze–thaw cycles. This suggests a competitive relationship between lateral pressure and freeze–thaw effects, where the gain effect of lateral pressure is counteracted by the damage and weakening induced by freeze–thaw cycles.

Figure 6b illustrates the relationship between the peak strain of the slurry masonry schist and the number of freeze–thaw cycles under different lateral pressure conditions. As the number of freeze–thaw cycles increases, the peak strain of the slurry masonry schist specimens continues to rise. This is because with the increase in freeze–thaw cycles, the porosity inside the slurry masonry schist increases, weakening the bonds among aggregates and leading to the accumulation of microscopic damage and cracks within the slurry masonry schist. This diminishes the inherent brittleness of the slurry masonry schist, which no longer exhibits a trend of sudden failure but gradually deteriorates instead. Furthermore, for the same degree of freeze–thawing, the peak strain of the slurry masonry schist exhibits a trend of increase followed by a decrease with increasing lateral pressure. This is because at lower lateral pressures, the lateral expansion caused by vertical compression of the slurry masonry schist is less constrained, and, at this point, the peak strain of the slurry masonry schist is mainly influenced by the vertical load pressure. However, under higher lateral pressure, when the slurry masonry schist specimens are compressed vertically to the point of peak stress, the lateral pressure constrains them, preventing significant deformation.

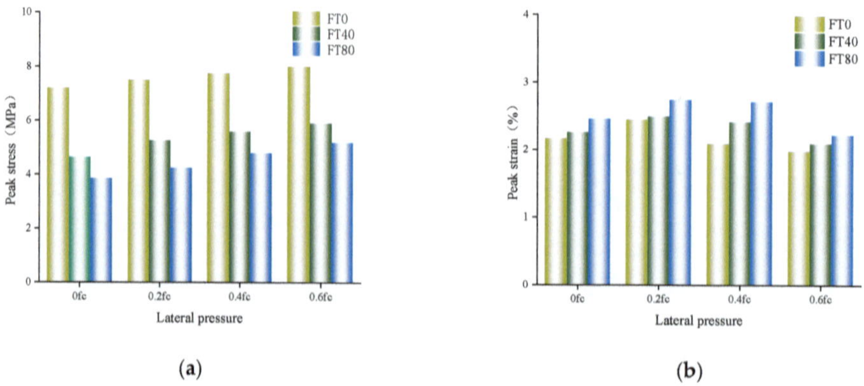

Figure 6. Peak stress and peak strain of the slurry masonry schist samples under different lateral pressures: (**a**) peak stress and (**b**) peak strain.

3.2. Acoustic Emission Signal Analysis

3.2.1. Damage Path Analysis

The damage variable D is defined as the ratio of the area of microdefects A_d on a material section to the area of undamaged section A [47], and this variable can be expressed in terms of the amount of acoustic emission energy:

$$D = N/N_m \tag{4}$$

where N is the cumulative number of acoustic emission counts at a certain stage of destruction of a cross-sectional area and N_m is the cumulative number of acoustic emission counts at complete destruction.

Figure 7a depicts the relationship between the stress level and damage degree under different freeze–thaw cycle numbers with a lateral pressure of 0.4 fc, while Figure 7b illustrates the relationship between the stress level and damage degree under different lateral pressures with 80 freeze–thaw cycles. In Figure 7a, it can be observed that at low stress levels, the damage degree of the slurry masonry schist after freeze–thaw cycles is greater than those without freeze–thaw cycles. This may be attributed to the gradual compaction of the internal pores in the slurry masonry schist at low stress levels, leading to increased ductility of the slurry masonry schist after freeze–thaw cycles [48,49] and resulting in relatively reduced damage. As the stress level increases and exceeds the compaction point of the slurry masonry schist, under the same stress level, the damage degree of the slurry masonry schist increases with the number of freeze–thaw cycles. This is because, at this stage, the factor influencing the damage degree shifts to the bond strength between stones and mortar, while freeze–thaw cycles weaken the bonds among aggregates. At peak stress, the damage degree values of the slurry masonry schist specimens with 0, 40, and 80 freeze–thaw cycles are 0.71, 0.9, and 0.938, respectively. The analysis reveals that the slurry masonry schist without freeze–thaw cycles has not reached complete failure at the peak stress, while those subjected to freeze–thaw cycles tend to be completely damaged at the peak stress, indicating that freeze–thaw cycles accelerate the damage development in slurry masonry schist. In Figure 7b, it can be observed that at lateral pressures of 0, 0.2 fc,

0.4 fc, and 0.6 fc, the damage degree (*D*) of the slurry masonry schist at peak stress is 0.969, 0.958, 0.938, and 0.91, respectively, indicating that lateral pressure to some extent inhibits damage development in slurry masonry schist.

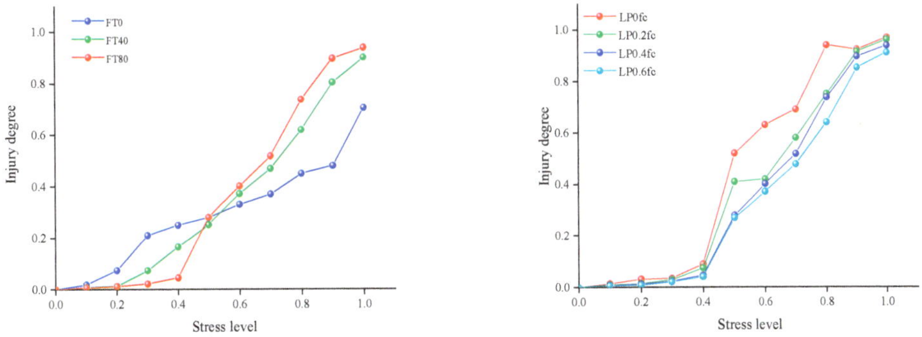

(**a**) (**b**)

Figure 7. Relationship between stress level and damage: (**a**) LP0.4 fc and (**b**) FT80.

According to the research by Ohtsu, M. et al. [50,51], there is a relationship between the level of applied stress $V(\%)$ and the total number of acoustic emission events N.

$$N = cV^a \exp(bV) \tag{5}$$

where a, b, and c are material fitting parameters.

With the combination of Equations (4) and (5), the expression for the damage can be found:

$$D = \frac{cV^a \exp(bV)}{N_m} \tag{6}$$

Hence, there exists a corresponding relationship between the stress level $V(\%)$ and the damage value (*D*) during the loading process of slurry masonry schist. The fitting results obtained using Equation (6) are shown in Figure 8, and the fitting parameters are listed in Table 7. Figure 8a illustrates the fitting graph of the stress level and damage degree under different freeze–thaw cycle numbers with a lateral pressure of 0.4 fc, while Figure 8b depicts the fitting graph of the stress level and damage degree under the effect of different lateral pressures with 80 freeze–thaw cycles. The analysis results indicate that the fitting effect of the damage model established based on the relationship between the damage value (*D*) and stress level is significant, effectively reflecting the relationship between the stress level and damage during the loading process of slurry masonry schist.

Table 7. Parameter fitting table.

Test Conditions	a	b	c/N_m	R^2
FT0	0.689	0.825	0.282	0.938
FT40	2.758	−1.222	3.099	0.991
FT80	24.241	−22.318	4.631	0.926
LP0fc	5.661	−6.086	415.227	0.962
LP0.2fc	4.659	−4.007	53.253	0.975
LP0.4fc	6.018	−5.432	217.341	0.989
LP0.6fc	5.092	−4.076	54.396	0.986

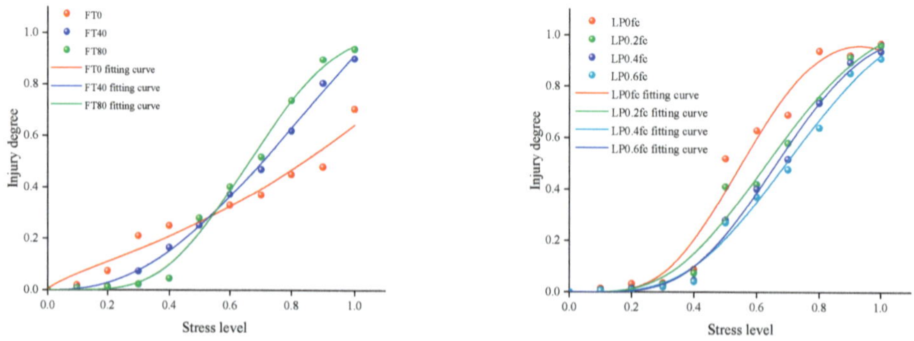

(**a**) (**b**)

Figure 8. Damage prediction models: (**a**) LP0.4 fc and (**b**) FT80.

3.2.2. Stress, Acoustic Emission Ringing Count, and Cumulative Impact Curve Characteristics

Figure 9 displays the acoustic emission characteristic curve of the slurry schist under uniaxial compressive load. From the acoustic emission characteristic curve, it can be observed that the initial value of ringing times is relatively small. After surpassing the initial load, the amplitude of the acoustic emission signal increases, consistent with the findings of Nosov et al. [52]. As the load increases, the ringing times sharply increase when the specimens are about to fail. The cumulative impact count of AE shows a continuously increasing trend, which can be divided into three stages as follows: firstly, the initial compaction stage, during which, because of the absence of cracks, AE signals are few, and the cumulative impact count increases slowly. Secondly, the crack generation stage, where AE signals increase because of crack formation, and the cumulative impact count rises more rapidly. Finally, as the load continues to increase, the slurry masonry schist enters the crack expansion and failure stage, during which, because of crack expansion, AE signals significantly increase, and the cumulative impact count growth rate significantly accelerates. When the specimen is about to fail, the cumulative impact count sharply rises and reaches a peak, and then the growth rate slows down. During this period, AE signals mainly come from shear slip in the interface transition zone. These three stages reveal the variation law of acoustic emission activity in slurry masonry schist specimens under different loading stages.

The cumulative number of impacts and ringing counts of the slurry masonry schist specimens decrease with an increasing number of freeze–thaw cycles. This phenomenon is attributed to the freeze–thaw cycles causing a loosening of the internal structure of the slurry masonry schist, leading to gradual expansion and penetration of internal cracks. The ingress of moisture from the air reduces the friction among particles within the slurry masonry schist, thereby diminishing acoustic emission activity.

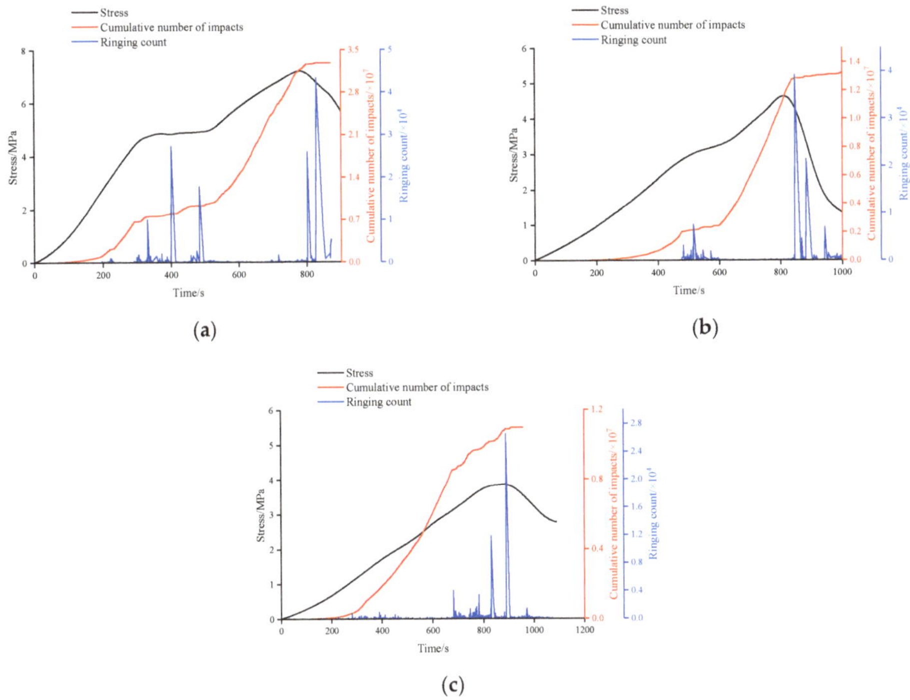

Figure 9. Stress, acoustic emission ringing count, and cumulative impact curves: (**a**) FT0; (**b**) FT40; and (**c**) FT80.

3.2.3. Crack Classification Analysis Based on Acoustic Emission Parameters

In acoustic emission monitoring, RA and AF values are often used to characterize the rupture mechanism, and, in general, tensile damage corresponds to acoustic emission events with lower RA values and higher AF values. In contrast, the acoustic emission events corresponding to shear damage usually have higher RA values and lower AF values. These acoustic emission characteristic parameters reflect the different features of acoustic emission events for different damage types. The damage in slurry schist is caused mainly by the expansion and penetration of internal microcracks. This type of damage mainly includes tensile damage and shear damage, as shown in Figure 10. Studying the ratio of tensile to shear cracks is important for understanding the damage mechanism of slurry schist because different types of cracks reflect different stress states and damage mechanisms when the material is subjected to stresses. RA denotes the ratio of the rise time of an acoustic emission event to its amplitude, and AF denotes the ratio of the ringing counts of an acoustic emission event to its duration. Researchers usually denote the ratio of RA to AF as the k value, which, according to LI et al. [53], is taken to be 50 to help characterize acoustic emission events.

Figure 11 presents scatter plots depicting the behavior of the slurry masonry schist specimens labeled RA-AF under uniaxial and biaxial compression conditions. These specimens exhibit two distinct failure modes, namely, tensile and shear failure, under compressive loading. Statistical analysis of RA-AF acoustic emission events reveals the relationship between the proportion of tensile cracks and lateral pressure, as shown in Figure 12. Tensile fracture predominates during specimen failure, with relatively fewer instances of shear failure. With increasing lateral pressure, there is a notable decreasing trend in the proportion of tensile cracks during specimen fracture. This trend is attributed to the lateral pressure's imposition of transverse expansion constraints on the vertical compression of the slurry masonry schist, thereby intensifying the constraint on tensile cracks with higher lateral pressure. Additionally, freeze–thaw cycles disrupt the internal structure of the slurry masonry schist, resulting in fluctuations in the proportion of tensile cracks. These findings indicate that both lateral pressure and freeze–thaw cycles induce changes in the fracture pattern of the slurry masonry schist specimens, consistent with the findings of Li et al. [53].

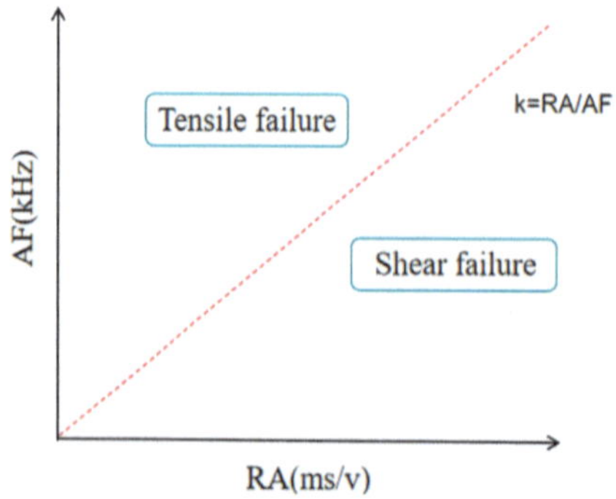

Figure 10. Crack classification.

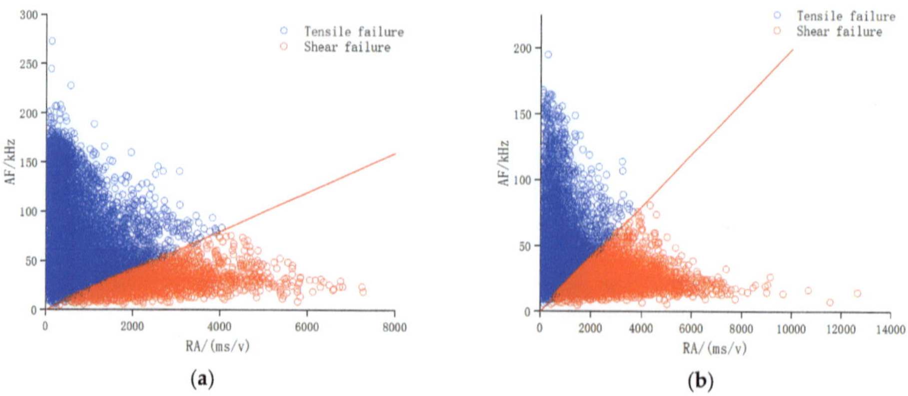

Figure 11. Characteristic diagrams of RA–AF: (**a**) uniaxial compression and (**b**) biaxial compression.

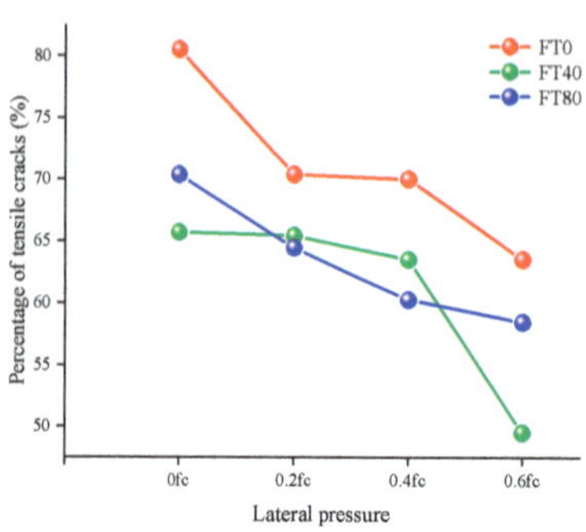

Figure 12. Schematic diagram of the percentage of tensile cracks.

3.2.4. (B) Value Analysis

The (b) value in acoustic emission experiments is a crucial parameter, as it reflects the material's fracture process. Specifically, it represents the ratio of small to large events in acoustic emission events, which is vital for understanding the material's fracture behavior. The variation in (b) values during specimen fracture is considered a significant precursor to rupture. In cement-based materials, (b) values can also reflect the average stress level and the degree of proximity to the ultimate strength. The change in (b) values can provide information about the stress state and the scale of internal microcracks, thus possessing spatiotemporal evolution characteristics. By monitoring the variation in (b) values, a better understanding of the material's behavior under different stress conditions, including fracture mechanisms and precursors, can be achieved.

Based on the Gutenberg–Richter relationship and the raw acoustic emission data, the acoustic emission (b) value can be calculated [39] with the following equation:

$$\begin{aligned} \lg N &= a - bM \\ M &= A_{dB}/20 \end{aligned} \tag{7}$$

where M is the earthquake magnitude, N is the number of earthquakes, a and b are constants, b is the value of b in seismology, and A_{dB} is the amplitude of the acoustic emission. In seismology, it is common to use the amplitude of the acoustic emission divided by 20 instead of the earthquake magnitude.

Figure 13 depicts the fluctuation in the (b) value as the load increases. Throughout various loading phases on the slurry masonry schist specimen, changes in acoustic emission (b) values offer crucial insights into the internal propagation of cracks and damage. Initially, minor oscillations in the (b) values indicate uniform crack propagation within the specimen. As the load rises, slight declines in (b) values suggest significant crack penetration within the specimen, albeit with minor damage. As loading progresses, the (b) values of the slurry masonry schist demonstrate an upward trajectory, indicating continuous generation and spread of internal microcracks. A notable difference between biaxial and uniaxial compression lies in a sudden surge in (b) values following a period of axial loading, likely because of lateral pressure impeding crack development during axial loading, resulting in the accumulation of numerous microcracks within the slurry masonry schist. Subsequently, there is stable expansion of microcracks, accompanied by fluctuating (b) values. Upon reaching peak stress load, a significant decrease in (b) values is observed, indicating substantial internal damage within the slurry masonry schist at this stage. The fluctuation in (b) values elucidates the progression of damage within the specimen across different loading stages, consistent with the findings of Xu et al. [54] on cement-based materials.

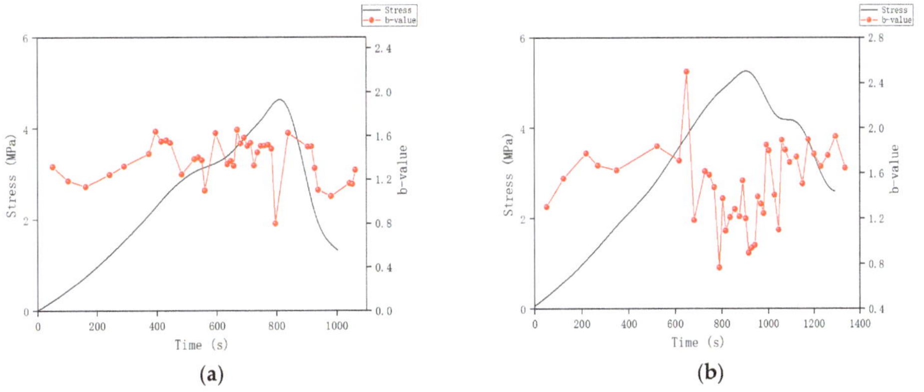

Figure 13. Plots of (b) value changes: (**a**) uniaxial compression; (**b**) biaxial compression.

3.2.5. Analysis of Damage Localization Points

Figure 14 illustrates the evolution of damage localization points in the slurry masonry schist during uniaxial compression at different stress stages. Initially, damage localization points tend to occur at weak locations. With increasing vertical load, the number of acoustic emission localization points significantly rises. At 50% of P_{max}, a large number of acoustic emission localization points appear in the interface transition zone, forming coherent signal bands. As the load reaches P_{max}, the quantity of acoustic emission signals sharply increases, forming a distinct damage band. Overall, AE signals first manifest at both ends of the specimen, with slightly more AE signals at the loading end, which is attributed to the "Poisson effect" of the cement-based material at the loading end. With increasing vertical load, AE signals gradually propagate towards the middle of the specimen; upon reaching the peak load P_{max}, cracks develop along the interface transition zone, penetrating the entire interface.

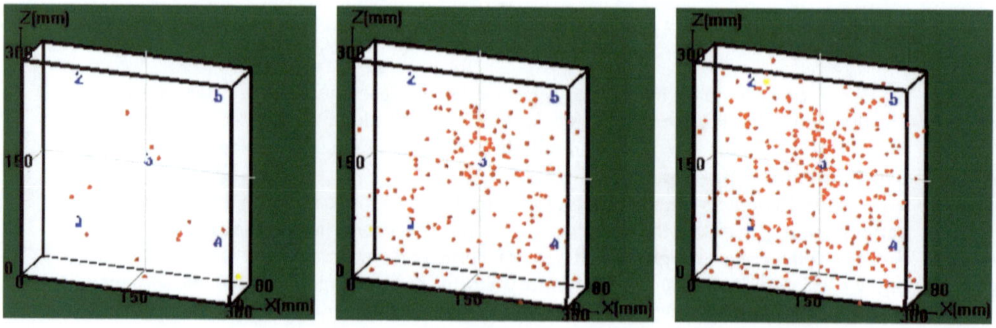

(**a**) 0.01 P (**b**) 0.5 P (**c**) 1.0 P

Figure 14. Analysis of localized damage points: The blue symbols indicate sensor positions, while the red symbols represent acoustic emission signals.

3.2.6. Maximum Principal Strain Analysis

The strain results of the slurry masonry schist under a lateral pressure of 0.6 fc are shown in Figure 15, corresponding to the maximum principal strain cloud maps and failure modes at peak strengths of 0%, 50%, and 100%. When the load reaches 50% of P_{max}, the specimen exhibits visible longitudinal crack extension with prominent color gradients, accompanied by gradual development and enlargement of interface cracks. As the load reaches P_{max}, the cracks further propagate, leading to the phenomenon of crack propagation along the interface between aggregate and mortar. Sudden strain changes indicate interface debonding between aggregate and mortar, resulting in the formation of macroscopic cracks along localized strain bands and ultimately leading to specimen failure. These results indicate that interface debonding is the primary failure mode for slurry masonry schist structures, consistent with the acoustic emission findings. The typical failure states of the specimens are illustrated in Figure 16.

To clarify the evolution of strain at the interfaces in slurry masonry schist, a truncation line was set along the horizontal direction in the strain cloud map to ensure that the truncation line passed through the tip of the notch with a mature crack. The location of the truncation line is shown in Figure 17. Figure 18 shows the distribution pattern of the maximum principal strains along the truncation line. The maximum principal strains of the slurry schist along the truncation line tend to decrease with an increasing number of freezing and thawing cycles. This may be because water expands as it freezes during the freezing and thawing cycle, which leads to greater stresses in the pore space, thus causing microcracks in the slurry schist. When the specimen thaws, these microcracks may expand, leading to changes in the pore properties within the concrete. These changes in pore properties result in a decrease in the compressive properties of the slurry masonry schist. As the freeze–thaw cycles are repeated, microcracks accumulate within the slurry schist. These microcracks weaken the overall strength and deformability of slurry masonry schist, resulting in a reduction in the maximum principal strain.

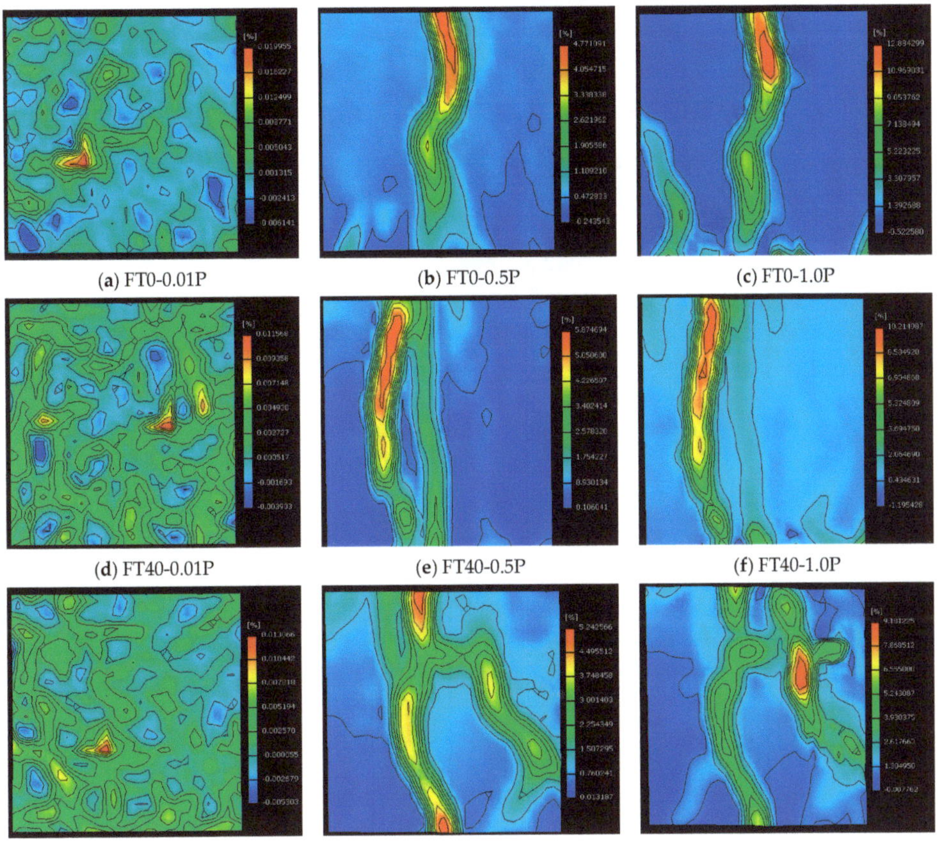

(**a**) FT0-0.01P (**b**) FT0-0.5P (**c**) FT0-1.0P

(**d**) FT40-0.01P (**e**) FT40-0.5P (**f**) FT40-1.0P

(**g**) FT80-0.01P (**h**) FT80-0.5P (**i**) FT80-1.0P

Figure 15. Main strain cloud diagrams of the slurry masonry schist samples (this figure shows the main strain clouds of the slurry masonry schist samples in different stress stages with the number of freezing and thawing cycles at a lateral pressure of 0.6 fc).

Figure 16. Damaged state of a slurry masonry schist sample.

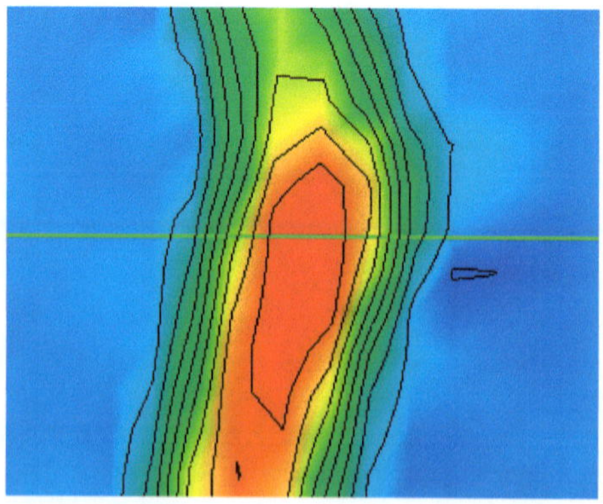

Figure 17. Intersecting line position.

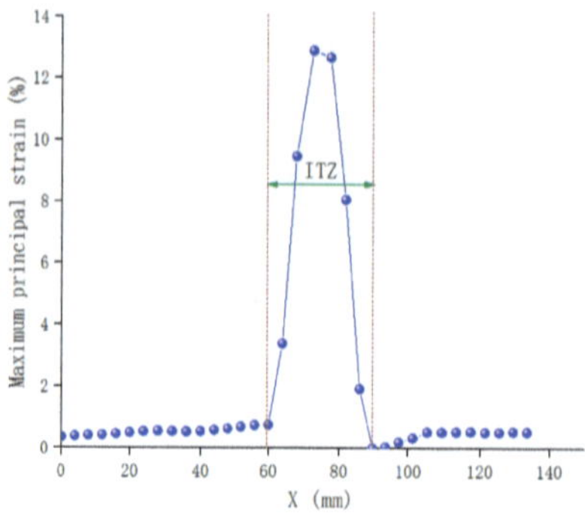

(**a**) FT0

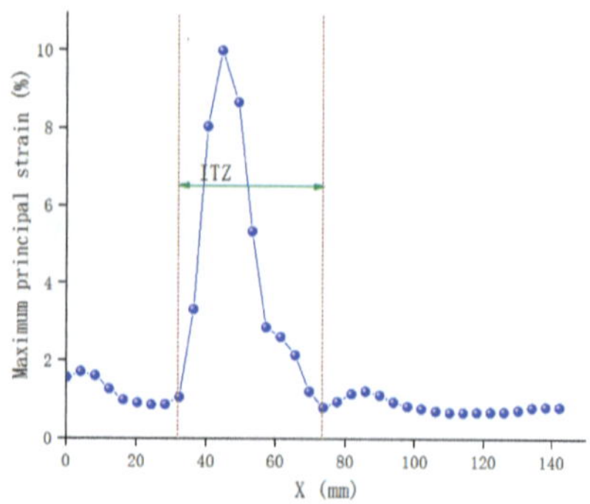

(**b**) FT40

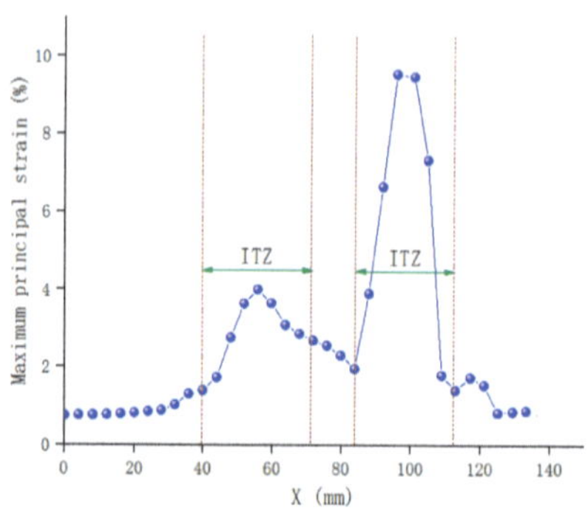

(**c**) FT80

Figure 18. Plot of the maximum principal strain variation along the truncation direction (this figure shows the strain variation along the truncation direction at a lateral pressure of 0.6 fc).

4. Conclusions

In this study, the effect of coupled lateral pressure and freeze–thaw cycles on the performance of slurry masonry schist specimens was investigated, and the whole test process was monitored using acoustic emission and DIC techniques. By organizing and analyzing the test data, we obtained the following conclusions:

1. Freezing and thawing cycles expedite the damage process in slurry masonry schist during loading. As the number of freeze–thaw cycles increases, the damage evolution

path shifts forward. Despite the overlap of the starting and ending points of the damage curve, the curve's trajectory significantly shortens, resulting in earlier damage to the slurry masonry schist during the loading process. The presence of lateral pressure partially inhibits the development of damage in slurry masonry schist.

2. The fracturing of the slurry schist is dominated by tensile damage and, to a lesser extent, by shear damage. The presence of lateral pressure produces lateral expansion constraints during the vertical compression of the slurry schist, and the greater the lateral pressure is, the stronger the constraints on tensile cracks. Freeze–thaw cycles damage the internal structure of the slurry schist, and the percentage of tensile cracks fluctuates. Lateral pressure and freeze–thaw cycles induce alterations in the fracture pattern of the slurry schists.
3. The fluctuation in the acoustic emission (b) value offers crucial insights into crack ex-tension and damage within slurry masonry schist. An increase in the (b) value signifies a prevalence of microcracks within the material, whereas a decrease indicates a dominance of larger crack extension and penetration. Fluctuations in the (b) value reveal the damage evolution of the slurry masonry schist during various loading stages.
4. Under loading, crack propagation occurs in the interface zone between the aggregate and mortar in slurry masonry schist, and the sudden change in strain indicates that the interface between the aggregate and mortar experiences debonding. During the loading process, the cracks produce macroscopic cracks along the localized zone of strain, leading to damage to the specimens by fracturing. The findings from digital image correlation (DIC) strain monitoring and acoustic emission (AE) signal localization suggest that the primary mode of damage in slurry schist structures is interfacial debonding.

Author Contributions: Conceptualization, J.D.; data curation, S.C.; formal analysis, Y.Z.; funding acquisition, J.D.; investigation, J.D. and Y.Z.; methodology, J.D., H.C. and H.Z.; resources, H.C., supervision, G.Z.; validation, H.C.; visualization, G.Z. and F.G.; writing—original draft, S.C.; writing—review and editing, H.Z. and F.G. All authors have read and agreed to the published version of the manuscript.

Funding: The research described in this paper was financially supported by the Natural Science Foundation of China (NO. 51878242), the Natural Science Foundation of Hebei Province of China (NO. E2020404007), the Research Project of Young Top Talent in Hebei Province (BJK2024116), and the Project of Research Start-up Fund (B-202307).

Data Availability Statement: The data presented in this study are available on request from the corresponding author.

Conflicts of Interest: Author Guoxiang Zhang was employed by the company China Railway Design Group Limited. The remaining authors declare that the research was conducted in the absence of any commercial or financial relationships that could be construed as a potential conflict of interest.

References

1. Kokutse, N.K.; Temgoua, A.G.T.; Kavazović, Z. Slope stability and vegetation: Conceptual and numerical investigation of mechanical effects. *Ecol. Eng.* **2016**, *86*, 146–153. [CrossRef]
2. Li, J.; Zhou, K.; Liu, W.; Zhang, Y. Analysis of the effect of freeze–thaw cycles on the degradation of mechanical parameters and slope stability. *Bull. Eng. Geol. Environ.* **2018**, *77*, 573–580. [CrossRef]
3. Su, H.; Wu, D.; Lu, Y.; Peng, X.; Wang, X.; Chen, W.; Wang, S. Experimental and numerical study on stability performance of new ecological slope protection using bolt-hinge anchored block. *Ecol. Eng.* **2021**, *172*, 106409. [CrossRef]
4. Gentilini, C.; Govoni, L.; de Miranda, S.; Gottardi, G.; Ubertini, F. Three-dimensional numerical modelling of falling rock protection barriers. *Comput. Geotech.* **2012**, *44*, 58–72. [CrossRef]
5. Yang, Y.; Wang, J.; Duan, Q.; Su, C.; Yan, M.; Dong, Y. The Investigation and 3D Numerical Simulation of Herb Roots in Reinforcing Soil and Stabilizing Slope. *KSCE J. Civ. Eng.* **2018**, *22*, 4909–4921. [CrossRef]
6. Yu, S.Y.; Ren, X.H.; Zhang, J.X.; Wang, H.J.; Sun, Z.H.; Zhou, Y. Numerical simulation on the stability of rock slope based on an improved SPH Method. *J. Mt. Sci.* **2021**, *18*, 1937–1950. [CrossRef]

7. Liu, Y.; Han, D.; Liu, N.; Wang, W. Reinforcement mechanism analysis of lattice beam and prestressed anchor rod system for loess slope. *Front. Earth Sci.* **2023**, *11*, 1121172. [CrossRef]
8. Zhu, X.; Chen, Z.; Ren, Y. The Application of Ecological Treatment Technology of Bolt and Lattice Beam in Highway Mountain Slope Support. *Geotech. Geol. Eng.* **2019**, *37*, 1891–1896. [CrossRef]
9. Zhang, J.J.; Niu, J.Y.; Fu, X.; Cao, L.C.; Xie, Q. Shaking table test of seismic responses of anchor cable and lattice beam reinforced slope. *J. Mt. Sci.* **2020**, *17*, 1251–1268. [CrossRef]
10. Li, J.; Hu, D.; Li, L.; Li, C. Research on Application of Lattice Anchor System Design and Construction Monitoring in Slope Protection. *IOP Conf. Series Earth Environ. Sci.* **2021**, *781*, 22049. [CrossRef]
11. Dong, J.; Li, Z.Q.; Yan, X.; Liu, Y.; Zhao, S.R.; Qian, R.; Jiang, T. Effects of coarse aggregates on the mechanical properties, durability and microscopic behaviour of mortar rubble. *Constr. Build. Mater.* **2022**, *354*, 129187. [CrossRef]
12. Caliskan, S.; Karihaloo, B.L.; Barr, B.I.G. Study of rock—Mortar interfaces. Part I: Surface roughness of rock aggregates and microstructural characteristics of interface. *Mag. Concr. Res.* **2002**, *54*, 449–461. [CrossRef]
13. Li, Z.Q.; Chen, H.; Dong, J.; Yan, X.; Zhao, S.R.; Zheng, Y.H.; Liu, Y. Study of the Failure Mechanism of Mortar Rubble Using Digital Image Correlation, Acoustic Emission and Scanning Electron Microscopy. *Buildings* **2022**, *12*, 1313. [CrossRef]
14. Scrivener, K.L.; Crumbie, A.K.; Laugesen, P. The interfacial transition zone (ITZ) between cement paste and aggregate in concrete. *Interface Sci.* **2004**, *12*, 411–421. [CrossRef]
15. Pan, J.; Shen, Y.; Yang, G.; Zhang, H.; Yang, H.; Zhou, Z. Debonding behaviors and micro-mechanism of the interface transition zone in sandstone-concrete interface in response to freeze-thaw conditions. *Cold Reg. Sci. Technol.* **2021**, *191*, 103359. [CrossRef]
16. Jin, L.; Fan, L.; Li, P.; Du, X. Size effect of axial-loaded concrete-filled steel tubular columns with different confinement coefficients. *Eng. Struct.* **2019**, *198*, 109503. [CrossRef]
17. Jin, L.; Li, J.; Yu, W.; Du, X. Size effect modelling for dynamic biaxial compressive strength of concrete: Influence of lateral stress ratio and strain rate. *Int. J. Impact Eng.* **2021**, *156*, 103942. [CrossRef]
18. Jin, L.; Li, J.; Yu, W.; Du, X. Mesoscopic simulations on the strength and size effect of concrete under biaxial loading. *Eng. Fract. Mech.* **2021**, *253*, 107870. [CrossRef]
19. Murad, Y.Z. Predictive model for bidirectional shear strength of reinforced concrete columns subjected to biaxial cyclic loading. *Eng. Struct.* **2021**, *244*, 112781. [CrossRef]
20. Van Mier, J.G.M.; Vonk, R.A. Fracture of concrete under multiaxial stress-recent developments. *Mater. Struct.* **1991**, *24*, 61–65. [CrossRef]
21. Shang, H.S.; Song, Y.P. Experimental study of strength and deformation of plain concrete under biaxial compression after freezing and thawing cycles. *Cem. Concr. Res.* **2006**, *36*, 1857–1864. [CrossRef]
22. Zhu, W.C.; Teng, J.G.; Tang, C.A. Numerical simulation of strength envelope and fracture patterns of concrete under biaxial loading. *Mag. Concr. Res.* **2002**, *54*, 395–409. [CrossRef]
23. Tschegg, E.K.; Elser, M.; Kreuzer, H. Model fracture behaviour of concrete under biaxial loading. *J. Mater. Sci.* **1995**, *30*, 235–242. [CrossRef]
24. Wang, H.T.; Zhou, Y.; Shen, J.Y. Experimental study of dynamic biaxial compressive properties of full grade aggregate concrete after freeze thaw cycles. *Cold Reg. Sci. Technol.* **2023**, *205*, 103710. [CrossRef]
25. He, Z.J.; Ding, M.J.; Zhang, X.J.; Zhang, X.S. The Biaxial Compressive Mechanical Properties and Strength Criterion of Recycled Aggregate Concrete Under Different Dynamic Strain Rates. *Iran. J. Sci. Technol. Trans. Civ. Eng.* **2021**, *45*, 125–146. [CrossRef]
26. Gao, S.; Hu, G. Experimental Study on Biaxial Dynamic Compressive Properties of ECC. *Materials* **2021**, *14*, 1257. [CrossRef] [PubMed]
27. Li, Z.; Dong, J.; Jiang, T.; Feng, K.; Cheng, S.; Liu, Y.; Tian, X. Experimental Study on the Dynamic Fracture Characteristics of Mortar–Rock Interface Zones with Different Interface Inclinations and Shapes. *Materials* **2023**, *16*, 5475. [CrossRef] [PubMed]
28. Wu, Z.; Rong, H.; Zheng, J.; Xu, F.; Dong, W. An experimental investigation on the FPZ properties in concrete using digital image correlation technique. *Eng. Fract. Mech.* **2011**, *78*, 2978–2990. [CrossRef]
29. Fayyad, T.M.; Lees, J.M. Application of Digital Image Correlation to Reinforced Concrete Fracture. *Procedia Mater. Sci.* **2014**, *3*, 1585–1590. [CrossRef]
30. Alam, S.Y.; Loukili, A.; Grondin, F. Monitoring size effect on crack opening in concrete by digital image correlation. *Eur. J. Environ. Civ. Eng.* **2012**, *16*, 818–836. [CrossRef]
31. Jing, G.; Yunchang, D.; You, R.; Siahkouhi, M. Comparison study of crack propagation in rubberized and conventional prestressed concrete sleepers using digital image correlation. *Proc. Inst. Mech. Eng. Part F J. Rail Rapid Transit* **2022**, *236*, 350–361. [CrossRef]
32. Thirumalaiselvi, A.; Sindu, B.S.; Sasmal, S. Crack propagation studies in strain hardened concrete using acoustic emission and digital image correlation investigations. *Eur. J. Environ. Civ. Eng.* **2022**, *26*, 4346–4373. [CrossRef]
33. Boulekbache, B.; Hamrat, M.; Chemrouk, M.; Amziane, S. Failure mechanism of fibre reinforced concrete under splitting test using digital image correlation. *Mater. Struct.* **2015**, *48*, 2713–2726. [CrossRef]
34. Huang, Y.; He, X.; Wang, Q.; Xiao, J. Deformation field and crack analyses of concrete using digital image correlation method. *Front. Struct. Civ. Eng.* **2019**, *13*, 1183–1199. [CrossRef]
35. Li, D.; Huang, P.; Chen, Z.; Yao, G.; Guo, X.; Zheng, X.; Yang, Y. Experimental study on fracture and fatigue crack propagation processes in concrete based on DIC technology. *Eng. Fract. Mech.* **2020**, *235*, 107166. [CrossRef]

36. Lakavath, C.; Prakash, S.S. Interface Shear Behavior of Ultrahigh-Performance Fiber-Reinforced Concrete Using Digital Image Correlation Technique. *J. Mater. Civ. Eng.* **2024**, *36*, 04023589. [CrossRef]
37. Ohno, K.; Ohtsu, M. Crack classification in concrete based on acoustic emission. *Constr. Build. Mater.* **2010**, *24*, 2339–2346. [CrossRef]
38. Grosse, C.U.; Finck, F. Quantitative evaluation of fracture processes in concrete using signal-based acoustic emission techniques. *Cem. Concr. Compos.* **2006**, *28*, 330–336. [CrossRef]
39. Sagar, R.V.; Prasad, B.R.; Kumar, S.S. An experimental study on cracking evolution in concrete and cement mortar by the b-value analysis of acoustic emission technique. *Cem. Concr. Res.* **2012**, *42*, 1094–1104. [CrossRef]
40. Zhang, H.; Ji, S.S.; Liu, H.; Liu, X.Y.; Li, X.C.; Zheng, S.Z.; Cao, Z.X. Research on dynamic splitting tensile damage characteristics of steel fiber reinforced concrete under freeze-thaw cycle environment based on AE. *J. Build. Eng.* **2024**, *84*, 108490. [CrossRef]
41. Barbosh, M.; Sadhu, A. Damage identification in concrete structures using a hybrid time–frequency decomposition of acoustic emission responses. *J. Civ. Struct. Health Monit.* **2024**, *14*, 237–253. [CrossRef]
42. Alam, S.Y.; Saliba, J.; Loukili, A. Fracture examination in concrete through combined digital image correlation and acoustic emission techniques. *Constr. Build. Mater.* **2014**, *69*, 232–242. [CrossRef]
43. Fan, X.; Li, S.; Chen, X.; Liu, S.; Guo, Y. Fracture behaviour analysis of the full-graded concrete based on digital image correlation and acoustic emission technique. *Fatigue Fract. Eng. Mater. Struct.* **2020**, *43*, 1274–1289. [CrossRef]
44. Feiteira, J.; Tsangouri, E.; Gruyaert, E.; Lors, C.; Louis, G.; De Belie, N. Monitoring crack movement in polymer-based self-healing concrete through digital image correlation, acoustic emission analysis and SEM in-situ loading. *Mater. Des.* **2017**, *115*, 238–246. [CrossRef]
45. Shang, S.; Song, Y. Dynamic biaxial tensile–compressive strength and failure criterion of plain concrete. *Constr. Build. Mater.* **2013**, *40*, 322–329. [CrossRef]
46. Sun, Y.; Pang, J.H.; Wong, C.K.; Su, F. Finite element formulation for a digital image correlation method. *Appl. Opt.* **2005**, *44*, 7357–7363. [CrossRef] [PubMed]
47. Kachanov, L.M. Rupture time under creep conditions. *Int. J. Fract.* **1999**, *97*, 11–18. [CrossRef]
48. Lian, S.; Zheng, K.; Zhao, Y.; Bi, J.; Wang, C.; Huang, Y.S. Investigation the effect of freeze–thaw cycle on fracture mode classification in concrete based on acoustic emission parameter analysis. *Constr. Build. Mater.* **2023**, *362*, 129789. [CrossRef]
49. Zhao, Y.-R.; Wang, L.; Lei, Z.-K.; Han, X.-F.; Shi, J.-N. Study on bending damage and failure of basalt fiber reinforced concrete under freeze-thaw cycles. *Constr. Build. Mater.* **2018**, *163*, 460–470. [CrossRef]
50. Ohtsu, M. Rate process analysis of acoustic emission activity in core test of concrete. *Doboku Gakkai Ronbunshu* **1992**, *1992*, 211–217. [CrossRef]
51. Ohtsu, M.; Watanabe, H. Quantitative damage estimation of concrete by acoustic emission. *Constr. Build. Mater.* **2001**, *15*, 217–224. [CrossRef]
52. Nosov, V.V.; Chaplin, I.E.; Gilyazetdinov, E.R.; Grigoriev, E.V.; Pavlenko, I.A. Micromechanics, nanophysics and non-destructive testing of the strength of structural materials. *Mater. Phys. Mech.* **2019**, *42*, 808–824.
53. Li, Z.; Dong, J.; Chen, H.; Wu, Z.; Feng, K.; Zhang, G.; Jiang, T. Mechanical behaviour and acoustic emission characteristics of basalt fibre mortar rubble under uniaxial cyclic compression. *Constr. Build. Mater.* **2023**, *393*, 132145. [CrossRef]
54. Xu, J.; Niu, X.; Yao, Z. Mechanical properties and acoustic emission data analyses of crumb rubber concrete under biaxial compression stress states. *Constr. Build. Mater.* **2021**, *298*, 123778. [CrossRef]

buildings

Article

Field Test and Numerical Simulation Study on Pipe Sticking of Pipe Jacking in Composite Stratum

Shilei Zhang [1], Xiaodong Xu [1], Feilun Luo [2], Tao Shi [3], Tianshuo Xu [4,*] and Peng Zhang [4]

1 Inner Mongolia Yin Chao Ji Liao Water Supply Co., Ltd., Xing'an League 137400, China; gcxy321@163.com (S.Z.)
2 Sinohydro Bureau 5 Co., Ltd., Chengdu 610066, China
3 Inner Mongolia Autonomous Region Water Resources and Hydropower Survey and Design Institute, Hohhot 010020, China
4 School of Engineering, China University of Geosciences, Wuhan 430079, China
* Correspondence: xtsxts11@cug.edu.cn

Abstract: In this study, the jacking force of a stuck pipe was explored under various contact conditions in long-distance composite strata. The Jiaoliu River jacking project in Yichao, Inner Mongolia, was selected for study, and the precise location of the pipe sticking was determined by laying strain gauges on the surface of the pipe and via integration with the measured data. Corresponding technical measures for releasing the pipe sticking were also put forward. Finally, ABAQUS 2022 software was used to establish a finite element analysis model of jacking force considering full contact conditions, and the maximum friction coefficient of the pipe that can be jacked forward in the pipe sticking state was calculated, providing a reference for related engineering cases. The results show that the position of the pipe sticking can be accurately identified through strain gauges on the surface of the pipe. The axial jacking force is transmitted more effectively in the upper section of the pipe than in the lower section, and as the jacking force increases, the length of the pipe becomes longer according to this rule. Measures were taken to adequately lubricate the pipe affected by sticking to ameliorate this condition.

Keywords: pipe jacking; pipe sticking; composite strata; field monitoring; pipe–soil contact

1. Introduction

In the laying of underground pipelines, the pipe jacking method is one of the most commonly used forms of underground excavation construction and is particularly suitable for pipeline crossings of buildings, highways, and river courses. It offers advantages such as a short construction period, high safety, and minimal disturbance to the construction environment. With the increasing popularity of pipe jacking construction, pipe jacking technology is gradually being developed with a trend toward longer distances and larger diameters. The stratum passed through during construction is also gradually developed from a single soil layer to a complex stratum and rock stratum. As its construction conditions become increasingly complex, the number of construction problems grows. Among them, the phenomenon of pipe sticking in long-distance pipe jacking construction is one of the major challenges faced by pipe jacking technology. Pipe sticking refers to the condition wherein a pipe can no longer be jacked in the rock and soil mass due to the convergence or instability of the surrounding rock. The resolution of pipe sticking is challenging, demanding significant amounts of time, manpower, and material resources, along with numerous potential safety hazards. Thus, how to safely and effectively address the issue of pipe sticking constitutes an important research topic in long-distance pipe jacking construction.

Recently, Liu et al. [1] investigated the influence of the pipe sticking phenomenon on the mechanical response during the steel pipe jacking process. Shou et al. [2] considered different over-excavation ranges and the location of the pipe sticking and used the ABAQUS finite element software to carry out a three-dimensional numerical simulation of pipe jacking construction in a gravel stratum. The results show that the location of the pipe sticking affects the stress field in the pipe. When the pipe is completely stuck, the adjacent soil and the pipeline undergo excessive deformation. Li Chao et al. [3–6] examined the mechanical effects of pipe jacking through mountain rocks under complex contact conditions. They provided crucial references for the optimal design and safe construction of pipe jacking in such strata through in-depth studies of the mechanical characteristics of pipe jacking in rocky strata. They also conducted case studies on crack situations and explored the on-site performance of reinforced concrete pipe sections during the pipe jacking process. Through numerical simulations, they evaluated the influence of the friction characteristics of the interface between the pipe section and the rock on the jacking force in ultra-long-distance pipe jacking projects. Zhong et al. [7] explored the pipe–rock friction contact characteristics based on indoor rock friction characteristic tests and identified the causes of pipe sticking in rock strata pipe jacking in combination with the pipe sticking phenomenon in the Chongqing Guanjingkou Water Conservancy Project. Deng Zhiyun et al. [8–10] studied the mechanical properties of fiber concrete pipes and their application in crack control during long-distance rock pipe jacking and analyzed and established a friction resistance calculation model for long-distance rock pipe jacking. Yang Qinghui [11,12] determined the trapped point of the pipe jamming by installing strain gauges on the concrete surface, removed the pipe section at this point, and installed a temporary intermediate jacking station, successfully proposing a technology for resolving pipe sticking in long-distance rock pipe jacking construction. In light of the problems encountered during the construction of the water conveyance tunnel of Chongqing Guanjingkou, Zhu Fayong et al. [13] addressed issues such as pipe sticking, water gushing, and cutter head jamming in faults, providing valuable experience for similar projects. Based on a pipe jacking project in a water-rich, fine sand stratum, Nayang [14] analyzed the causes and scope of pipe sticking and put forward the idea of increasing jacking force locally for large pipe diameters, which can provide a reference for accident prevention and relief in similar jacking projects. He Ziyong [15] studied the friction coefficient of the pipe–soil contact interface under seven complex contact conditions. The test results show that bentonite slurry plays an important role in reducing the friction resistance of the pipe–soil contact interface, and in the over-excavation gap, the increase in the friction resistance of the pipe–soil contact interface caused by the complex contact state of the pipe–soil is the root cause of the pipe sticking.

In summary, existing studies by scholars on pipe sticking in pipe jacking mainly focus on the causes of pipe sticking and the construction techniques for its resolution and prevention. However, there is a lack of a calculation model for the jacking force required to resolve pipe sticking in different contact states. Therefore, using the Jiaoliu River Pipe Jacking Project of Yinchuo Jiliao in Inner Mongolia as a study area, the specific pipe sticking position was determined in this research based on the measured data by arranging strain gauges on the surface of the pipe on site, and technical measures are proposed for resolving the pipe sticking. Finally, ABAQUS software was used to establish a three-dimensional finite element model under the condition of full contact, and the contact friction coefficient required for the pipe to resolve the pipe sticking at the maximum allowable jacking force was calculated. These findings provide data support for the optimization of intermediate jacking stations and the resolution of pipe sticking in pipe jacking projects.

2. Field Monitoring of Pipe Sticking

2.1. Project Profile

The Jiaoliu River non-pressure tunnel pipe jacking project of the sixth bid section of the water diversion project from Chuoji to Liaoling is located in Tuquan County, Xing'an League, Inner Mongolia Autonomous Region. It is a linear pipe jacking tunnel. From northeast to southwest, the pipe jacking non-pressure tunnel section is 940 m long, with a gradient of 1/1100. The straight pipe jacking construction method is used for one-time jacking. The ground field is shown in Figure 1.

Figure 1. Pipe jacking trajectory diagram.

The C50 precast reinforced concrete pipe is used in the construction of the pipe jacking. The inner diameter of the pipe is 3600 mm, the outer diameter is 4300 mm, the length of the single section is 2500 mm, the impermeability grade is P10, and it has a flexible "F"-type interface and a wedge double-rubber ring. The structure of the pipe is shown in Figure 2. Both the inner and outer longitudinal bars and the circumferential bars of the pipe are composed of 12 mm HRB400 steel bars. The spacing between the circumferential bars is 40 mm, and the angle interval between the longitudinal bars is 6°. There are six grouting holes. In the actual construction, only five grouting holes are utilized, and the bottom grouting hole is not used.

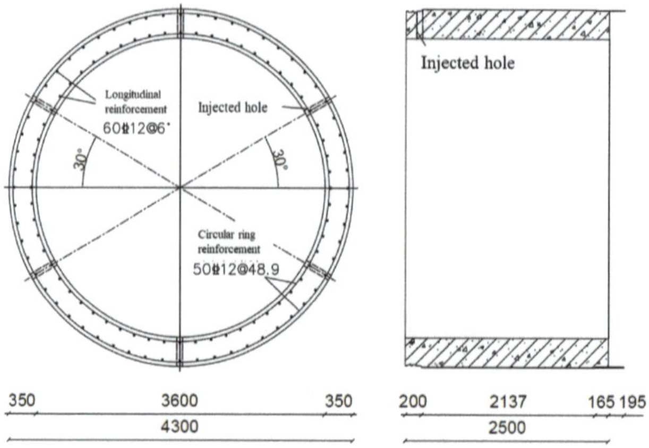

Figure 2. Pipe structure diagram.

The geological conditions of the pipe jacking crossing section are more complex, as shown in Figure 3. The main strata are tuff, muddy gravel, and granite. The saturated compressive strength of tuff is 67~98.7 MPa, and the saturated compressive strength of granite is 35~51 MPa. Generally, the pipe jacking is carried out in the full, strong weathering layer of muddy gravel and bedrock and locally in the weak weathered layer. The softness and hardness of different rock strata are different, and there are upper soft and lower hard strata composed of muddy gravel and strongly weathered tuff strata. The risk of pipe jacking construction is high under complex geological conditions [16,17].

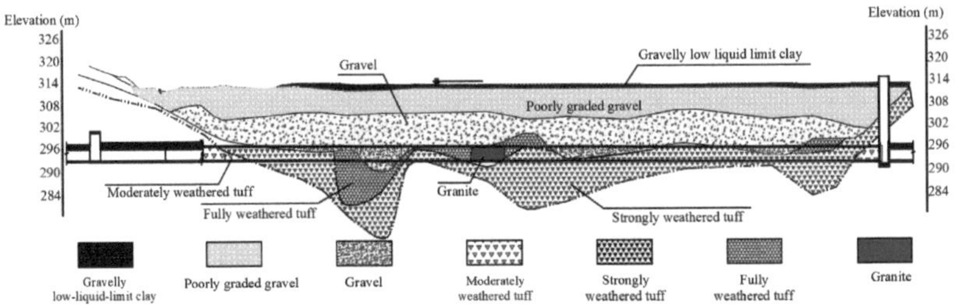

Figure 3. Geological conditions of pipe jacking cross-section of Jiaoliu River.

According to the results of the borehole acoustic wave test, the integrity of the rock mass is poor, and the buried depth of groundwater is 0.70~5.40 m. The permeability rate of tuff in this section is 2.3~65.0 Lu according to the borehole water pressure test. The tuff is weakly permeable to moderately permeable rock mass. The recommended values of the physical and mechanical parameters of the pipe jacking crossing the stratum are shown in Table 1.

Table 1. Calculation parameters of strata.

Stratum Name	Depth h (m)	Unit Weight γ (kN/m^3)	c (kPa)	φ (○)	fr (MPa)
Gravelly low-liquid-limit clay	0.80–3.30	19.0	8.5	25	-
Poorly graded gravel	4.50–11.00	21.0	0	29.5	-
Gravel	3.00–16.00	21.0	6.5	28	28.5
Fully weathered tuff	-	24.7	8	24	35.4
Granite	-	24.6	-	-	50.6
Strongly weathered tuff	-	25.5	100	28	67.5
Moderately weathered tuff	-	24.7	-	-	87.2

2.2. *Engineering Problem of Pipe Jacking*

Jacking was halted in the project from November 1 to November 5, and the warehouse was opened to pour thick mud at the head of the microtunnel boring machine. After the jacking was resumed on November 6, the jacking length was 2.96 m on that day, and the jacking length was 4.22 m on November 7. Jacking difficulty occurred at the No. 3 intermediate jacking station on November 8. During the jacking period, the resistance of the pipe side wall increased continuously, and the jacking speed decreased continuously. In order to cope with this situation, the pressure of the No. 3 intermediate jacking station cylinder was gradually increased for jacking. There are 33 jacking cylinders in the intermediate jacking station. The ultimate jacking force of a single jacking cylinder is 80 t. When the total jacking force of the No. 3 intermediate jacking station was increased to 2640 t, the jacking was still very slow, and the single-jacking length was several centimeters. After the cylinder was depressurized, the pipe retreated. Subsequently, on November 8, the number of intermediate jacking station cylinders was increased to 48, and the total jacking force reached 3500 t, which is the ultimate bearing capacity of the pipe section, but the intermediate jacking station was still non-operational.

Subsequently, a series of measures were taken on the site, such as encrypting the grouting pipeline measurements, as jacking was still difficult in this location. At this time, it was judged that the pipe between the No. 3 and No. 2 intermediate jacking stations was locked, and the pipe sticking was located in the interval between the No. 2 and No. 3 intermediate jacking stations, with a total length of 137.5 m. The location of the No. 2 intermediate jacking station is between the No. 47 and No. 48 pipes, and the location of the No. 3 intermediate jacking station is between the No. 103 and No. 104 pipes.

3. Accurate Judgment of the Position of Pipe Sticking

3.1. Calculation of Maximum Allowable Jacking Force of Pipe Jacking

Based on the field data, ABAQUS was used for modeling. The jacking force was gradually applied to the end of the pipe with an ultimate bearing capacity of 3500 t, which was calculated with Formula (1). The stress characteristics of the pipe for different friction coefficients under the conditions of one-third, one-half, and full contact between the pipe and the cutting filler were simulated, analyzed, and compared with the data measured during the pipe sticking period.

$$F_{dc} = 0.5 A_p f_c \frac{\phi_1 \phi_2 \phi_3}{\gamma_d \phi_4} \tag{1}$$

In the formula, F_{dc} is the maximum allowable jacking force of the pipe, N; A_p is the force transmission area of the pipe, and the value is 3,938,816 mm^2; f_c is the design value of the compressive strength of the concrete used in the pipe, and the value is 23.1 N/mm^2; ϕ_1, ϕ_2, and ϕ_3 are the compression strength reduction coefficient, the eccentric compression strength improvement coefficient, and the brittleness coefficient, which are 0.9, 1.05, and 0.85, respectively; γ_d is the jacking force partial coefficient, and the value is 1.3; and ϕ_4 is the adjustment coefficient, with a value of 0.79.

In order to determine the location of the pipe sticking as accurately as possible, the judgment method adopted in the field is as follows: Firstly, the pipe sticking interval is preliminarily judged using the displacement monitoring of the pipe; then, the pipe sticking interval is reduced via the strain monitoring of the inner surface of the pipe. Finally, the density of the strain monitoring of the inner surface of the pipe is increased in the reduced pipe sticking interval to accurately judge the pipe sticking interval.

3.2. Pipe Displacement Monitoring

Because there is a gluing plate directly between the pipes, when the pipe is subjected to axial pressure, it transmits jacking force, which compresses the gluing plate and causes displacement. Even if the entire pipe is not pushed, the position of pipe sticking can be judged according to the axial displacement of different positions.

After the occurrence of pipe sticking, in order to accurately identify its location, fixed measuring point sensors were installed on the No. 103, No. 92, No. 82, No. 66, and No. 60 pipes, and an electronic total station was set up in the rear visible pipe to ensure that the pipe did not move during the monitoring process. The operator only started the No. 3 intermediate jacking station and maintained the pushing state of the main cylinder of the pipe. When the jacking force of the No. 3 intermediate jacking station was set to 3500 t, the first measurement was carried out. The transmission law of jacking force and the approximate position of pipe jacking were judged by calculating the difference in pipe displacement when the No. 3 intermediate jacking station cylinder was pushed out and retracted. Then, according to the results of the first measurement, the interval where pipe jacking may occur was encrypted.

Table 2 shows the two results of the pipe displacement measurement. The displacements of the No. 103, No. 82, No. 66, and No. 62 pipes were firstly measured. It is only necessary to calculate the difference between the axial left side of the pipe when the intermediate jacking station is pushed out and retracted. Therefore, it is not necessary to consider the assumed position of the total station. The first measurement revealed that the difference at the No. 103 pipe was 0.04 m, and the difference between the front and rear of the No. 82 pipe was only 0.004 m. Therefore, it was inferred that the position of the pipe sticking was after the No. 82 pipe. The No. 103, No. 92, and No. 82 pipes were measured for the second time. The measurement difference at the No. 103 pipe remained about 0.04 m, while the measurement difference at the No. 92 pipe was 0.0118 m, and the difference at the No. 82 pipe was only 0.0036 m. It can be considered that even if the No. 3 intermediate jacking station is operational and added to the allowable jacking force of the pipe, there is essentially no displacement between the No. 82 pipe and the No. 92 pipe, so it can be inferred that the pipe sticking position occurs after the No. 92 pipe.

Table 2. Pipe displacement measurement.

Pipe Number	First Measurement			Second Measurement		
	Intermediate Jacking Station State		Differential/(m)	Intermediate Jacking Station State		Differential/(m)
	Retrieval/(m)	Propulsion/(m)		Retrieval/(m)	Propulsion/(m)	
103	166.9284	166.971	0.0426	166.9318	166.9722	0.0404
92	-	-	-	139.2991	139.3109	0.0118
82	114.6014	114.6054	0.004	114.6014	114.605	0.0036
66	73.8068	73.8072	0.0004	-	-	-
60	58.6012	58.6016	0.0004	-	-	-

3.3. Pipe Surface Strain Monitoring

It is necessary to determine the cause and location of pipe sticking as soon as possible after the occurrence of pipe sticking due to the limited amount of on-site equipment and the urgency of engineering problems. It was decided to arrange a surface strain gauge monitoring point on the left side of the No. 82 pipe to the No. 49 pipe and on the left and right sides of the No. 103 pipe to the No. 82 pipe. The installation positions were specifically arranged on the left and right sides of the No. 98, No. 93, and No. 88 pipes, and the measuring point sensors were arranged on the left side of the No. 103, No. 83, No. 72, No. 61, and No. 49 pipes. The arrangement was also conducive to the installation, reading, and protection of the measuring point sensor, as shown in Figure 4. Then, using the relevant initially measured data, the second encrypted fixed-point measurement was performed on the strain value jump interval.

Figure 4. Monitoring of pipe surface strain after pipe sticking.

A VS100 vibrating wire strain gauge and a high-precision sinusoidal frequency reader were used for monitoring. The relevant parameters are shown in Table 3 below. The monitoring equipment was operated using the manual single-point method, which fully meets the reading requirements at the engineering site.

Table 3. Surface strain gauge parameters.

Instrument Type	Vibration String Frequency Measurement Range	Temperature Measurement Range	Resolution	Temperature Accuracy	Temperature Correction Coefficient
Installation method	0~3000 Hz	−25~60 °C	≤0.03%F.S	±0.5°	11.2 με/°C

The strain can be calculated using Formula (2).

$$\Delta\varepsilon = K(F_i - F_0) + b(T_i - T_0) \tag{2}$$

In the formula, $\Delta\varepsilon$ is the strain change of the measured structure; k is the minimum reading of the strain gauge; F_i is the real-time measurement frequency modulus value of the strain gauge; F_0 is the frequency modulus value of the strain gauge reference measurement; b is the temperature compensation correction coefficient of the strain gauge; T_i is the temperature of the strain gauge at zero pressure; and T_0 is the temperature that the strain gauge is subjected to at that time.

Afterward, the jacking force of the No. 3 intermediate jacking station was raised to 3500 t, which is the ultimate bearing capacity of the pipe, and the cylinder pressure was kept stable. At this time, the handheld high-precision sinusoidal frequency reader was used for fixed-point reading.

Because there were only a few measuring point sensors arranged on the pipe, and the actual contact state between the pipes is not understood, half sums of the left and right strain values were used to estimate the average strain value of the whole pipe. The conversion results of the first measured frequency are shown in Figure 5 below. The average strain values of the No. 98, No. 93, and No. 88 pipes are 132.8ε, 81.055ε, and 65.5ε, respectively. It can be seen that the numerical value of the pipe changes greatly after strain transfer, and there is no significant law. The strain change value on the left side of the No. 103 pipe near the No. 3 intermediate jacking station is 207.36ε. When the jacking force is transmitted to the No. 98 pipe, the strain value on the left side is 90.1ε, with a difference of 117.26ε, but the strain change value on the right side of the pipe is 185.5ε. It can be judged that the pipes between the No. 98 pipe and No. 103 pipe are in a deflection stress state, and this occurs mainly on the right side. The difference between the strain value of the No. 103 pipe and the average strain value of the No. 98 pipe is 74.56ε, and the jacking force is greatly consumed in this interval. When the jacking force is transmitted to the No. 93 pipe, contrary to the interval between the No. 103 pipe and No. 98 pipe, the jacking force is mainly transmitted by the left side of the pipe, and the difference between the average strain values of the No. 93 pipe and No. 98 pipe is 51.745ε. When the jacking force is transmitted to the No. 88 pipe, the difference between the strain values of the No. 93 pipe and No. 88 pipe is 15.555ε. When the jacking force is transmitted to the No. 83 pipe, the difference between the average strain values of the adjacent two pipes is 33.56ε. When the jacking force is transmitted to the No. 61 pipe, the strain value of the No. 61 pipe is −0.98ε, and the jacking force is very small. The jacking force is essentially exhausted, and the effective transmission distance is about 107.5 m. It is judged that the main jacking force consumption interval is between the No. 83 pipe and No. 103 pipe.

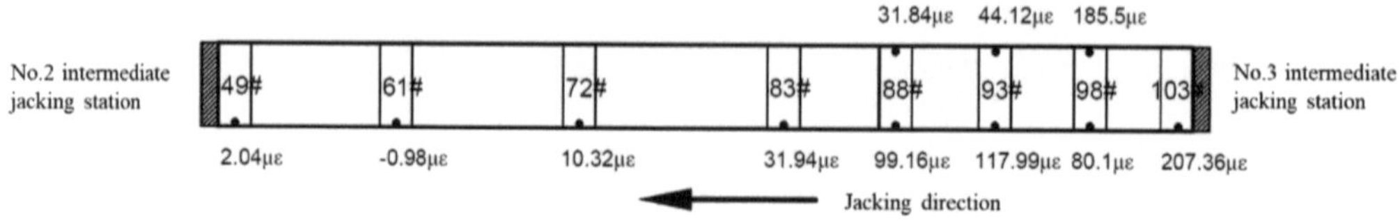

Figure 5. The strain difference measured with a surface strain gauge.

Then, the encrypted fixed-point measurement was performed in the strain numerical problem interval. The measuring point sensor and monitoring results are shown in Figure 6. The average strain values of the No. 103, No. 101, No. 98, No. 96, and No. 93 pipes were 203.725ε, 162.965ε, 206.755ε, 188.65ε, and 130.485ε, respectively. The transmission of the No. 101, No. 98, and No. 96 pipes does not conform to the law of jacking force transmission, which may be related to the change in pipe deflection in the limited jacking force state. The difference between the average strain values of the No. 103 pipe and No. 93 pipe is 73.24ε, and the top force in this interval decays rapidly. Therefore, it is speculated that pipe sticking occurs after the No. 93 pipe.

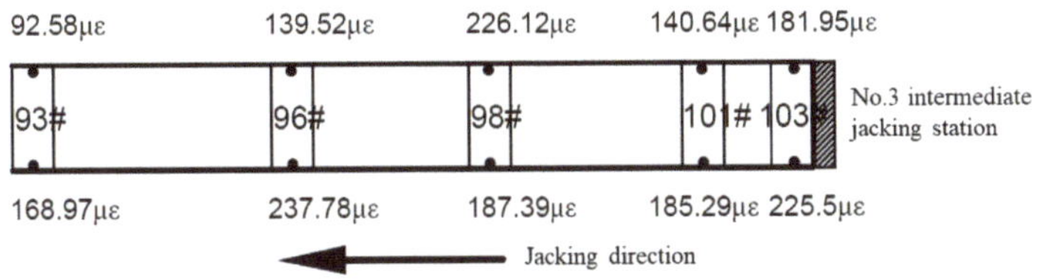

Figure 6. Encryption fixed-point measurement of strain difference.

4. Numerical Simulation of Pipe Sticking

4.1. Fundamental Assumptions

(1) It was assumed that the soil layer is a continuous, homogeneous, and isotropic elastic–plastic body, and the microtunnel boring machine and pipe are a linear elastic body. In addition to the pipe undergoing jacking, the remaining concrete pipes are continuous as a whole, regardless of the influence of the floating and deflection of the pipes.

(2) The static analysis method was used without considering the influence of seismic force and other factors.

(3) It was assumed that before jacking, the ground stress was balanced, and the deformation of the stratum had occurred. In the subsequent jacking process, the deformation of the stratum was no longer considered. Only the instantaneous settlement of the stratum caused by the jacking pipe was calculated, and the consolidation settlement of the soil caused by the time effect was not considered.

(4) The influence of steel bars in the pipe on the overall performance of the pipe was not considered separately in the simulation of reinforced concrete pipes, but it was assumed that the steel bars were evenly distributed in the pipe, and the mechanical properties of the steel bars were directly applied to the pipeline through calculation.

4.2. Establishing the Model

(1) Model size and material parameters

The pipe sticking section is a composite upper soft and lower hard stratum of boulder clay and strong weathering tuff rock, with a total length of 142.5 m. Considering the influence of the boundary effect on pipe jacking, the width of the model stratum was set to 38 m, and the height was 45 m. The length of the concrete pipe was 142.5 m, the outer diameter of the pipe was 4.3 m, the inner diameter was 3.6 m, and the wall thickness was 0.3 m. The equivalent cement and the equivalent mud layer were set to simulate the mixture of cuttings and mud around the pipe, and the length of the equivalent layer was 142.5 m; the inner diameter was 4.3 m, and the layer thickness was 0.04 m. According to the stratum, buried depth, and parameter settings of the geotechnical numerical model in the practical project, the pipe was set in the XY plane. The stratum parameters are shown in Figure 7.

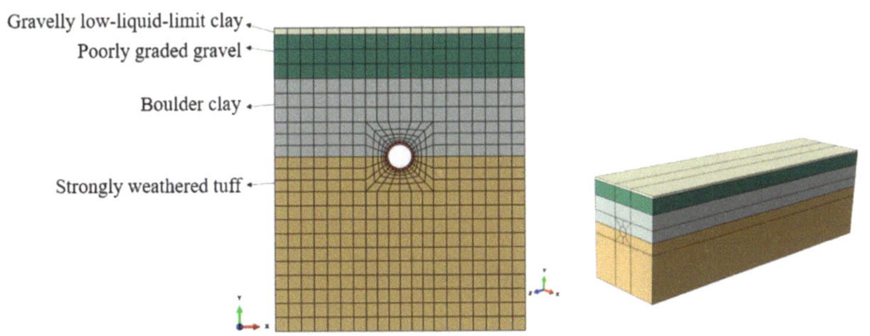

Figure 7. Stratigraphic parameter diagram.

Based on the applicability of different constitutive models, the soil is regarded as an ideal elastic–plastic material, and the linear elastic and Mohr–Coulomb models were selected for the elastic and plastic parts, respectively. The elastic constitutive model was selected due to the large elastic modulus of the microtunnel boring machine and pipe. The C3D8 R solid grid element was adopted for the solid element modeling. In the process of grid element division, the grid elements of the soil around the pipe were encrypted, as shown in Figure 8. After meshing, the soil was divided into 27,144 units, and the number of nodes was 29,786. The pipe section and pipe jacking machine were divided into 6376 units, and the number of nodes was 12,700.

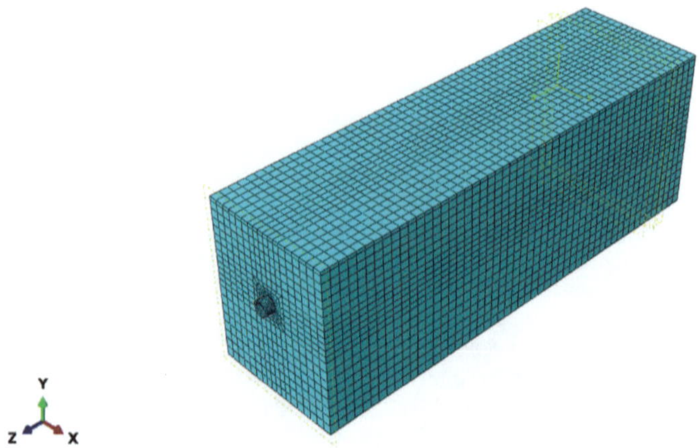

Figure 8. Model grid division.

The relevant material values are shown in Table 4, and the stratigraphic parameters were determined according to Table 1.

Table 4. Recommended value of friction coefficient of concrete pipe jacking.

Material	Unit Weight (kN/m^3)	Elastic Modulus (MPa)	Poisson Ratio
Equivalent mud layer	16.0	1.05	0.45
Equivalent cement	26.0	40	0.4
Concrete pipe	26.0	34,500	0.2

(2) Setting the analysis step and the interaction relationship.

Step 1: The stress of soil was simulated using the Abaqus automatic ground stress balance method. Figure 9 is the vertical stress cloud diagram of the stratum, which meets the calculation accuracy requirements.

Step 2: The 142.5 m concrete pipe was regarded as the same throughout, and the position of the head of the microtunnel boring machine was used as a reference. The life-and-death element method was used to invalidate the excavated soil. The pipe and the equivalent layer were placed in a predetermined position, and the equivalent layer was established. The contact between the rock and soil mass and the interaction relationship between the pipe and the equivalent layer were established at different contact positions and ranges. The tangential direction was set to the penalty function, the corresponding friction coefficient was set, and the normal direction was set to "hard" contact.

Step 3: The jacking force was gradually applied at the end of the pipe until the ultimate bearing capacity of the pipe was 3500 t so as to simulate the jacking of the pipe by the intermediate jacking station.

Considering the complex construction conditions in practical engineering, the contact between the pipe and the soil is changeable. Therefore, the contact state between the pipe and the soil was set to the full-contact state, and the stress distribution and axial displacement of the pipe section were simulated when the maximum allowable jacking force was 3500 t under the most extreme conditions.

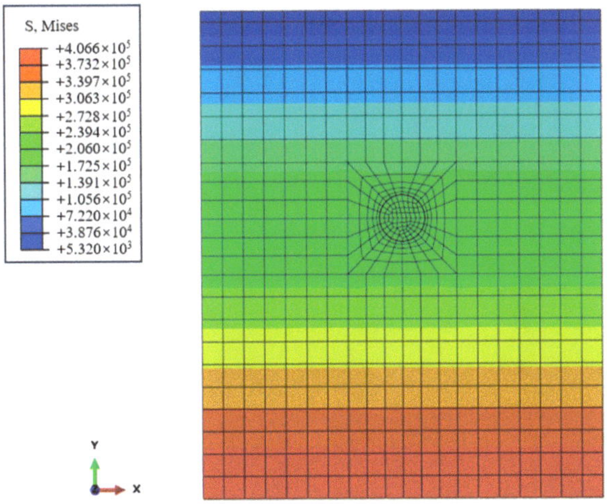

Figure 9. Soil stress diagram after in situ stress balance.

(3) Loading and boundary conditions

In the first analysis step, the gravity load was applied to the rock and soil mass. After the excavation of the soil mass, the gravity load was applied to the pipe in the second analysis step. In the third analysis step, the jacking force was applied to the pipe. In order to avoid the convergence of the model caused by the excessive single jacking load, the load was applied multiple times, with a total load of 3500 t.

In order to simulate the semi-infinite space ground stress field of the pipe jacking construction stratum more realistically, the upper surface of the soil was set to a free boundary, and the lower surface limited the displacement in three directions at the same time. The other four sides limited their respective normal displacement and the displacement in the axial direction of the equivalent cement and the equivalent mud layer. The overall settings of the model boundary conditions are shown in Figure 10.

(4) Submission of calculation and analysis

Finally, the calculation was submitted, and the simulation results were analyzed.

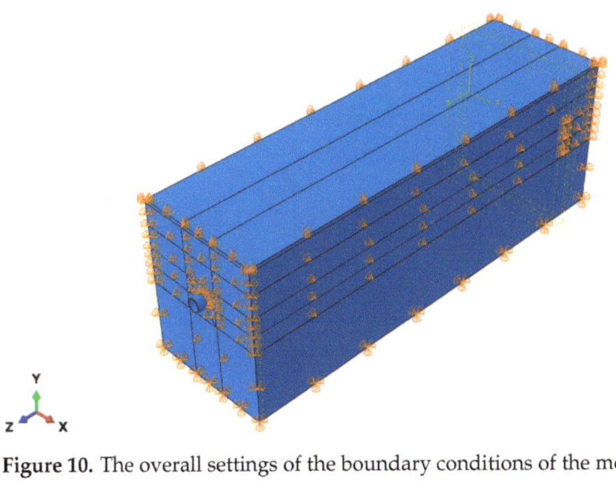

Figure 10. The overall settings of the boundary conditions of the model.

4.3. Analysis of Results

The full-contact pipe-sticking state of the pipe is shown in Figure 11. The outer side of this pipe is wrapped with the equivalent cement. When the friction coefficient with the cement is 0.5 [18], the jacking is simulated. The axial displacement distribution of the pipe is shown in Figure 12. From Figure 12, it can be found that in the full-contact state, the transmission law of the axial displacement of the pipe is that the axial displacement of the upper part of the pipe is greater than the axial displacement of the lower part of the pipe. The reason for this phenomenon is that the contact pressure at the bottom of the pipe is greater than that in other directions due to the self-weight of the pipe, resulting in greater friction resistance at the bottom of the pipe.

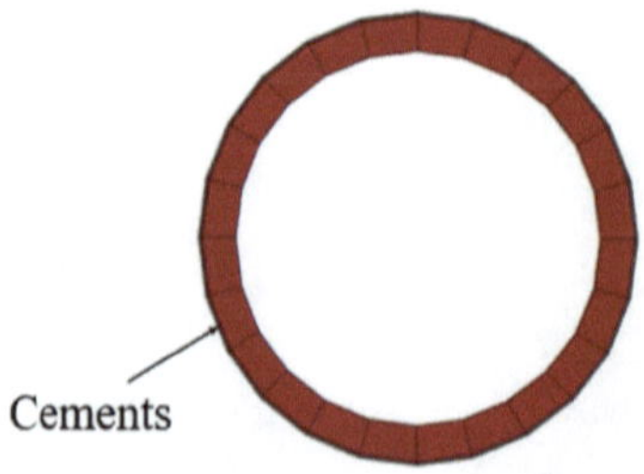

Figure 11. The full-contact state of pipe sticking.

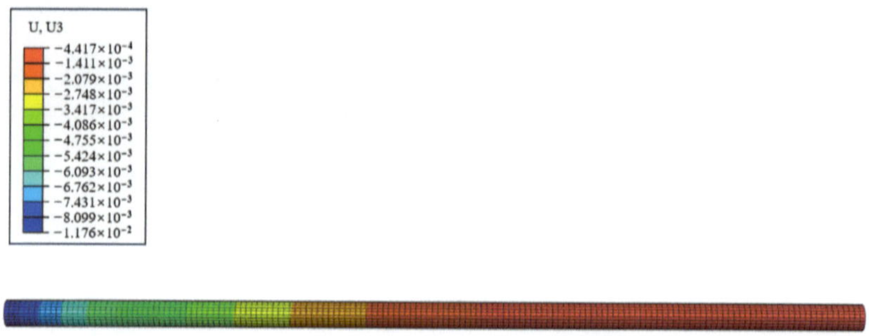

Figure 12. Axial displacement of pipe under full-contact condition when jacking force reaches 3500 t.

The simulated value of the axial displacement of the pipe in the full-contact state of pipe sticking was compared with the actual monitoring value of the displacement in Section 3.1. The comparison results are shown in Table 5. The results show that the simulated value of the pipe is essentially consistent with the actual monitoring data, which proves that the model established in this study is accurate and can effectively simulate the stress state of the pipe section on-site.

Table 5. The comparison between the measured value and the simulated value of the axial displacement of the pipe at 3500 t jacking force.

Pipe Number	The Distance Between the Pipe and the No. 2 Intermediate Jacking Station (m)	The Measured Value (mm)	The Simulated Value (mm)
60	30	0.4	0.4417
66	45	0.4	0.4417
82	85	4	4.086
92	110	11.8	11.76

The axial stress distribution of the pipe is shown in Figure 13. It shows that under the condition of full contact, the transmission of axial jacking force in the upper part of the pipe is greater than that of the lower part of the pipe. This is consistent with the distribution law of the axial displacement of the pipe.

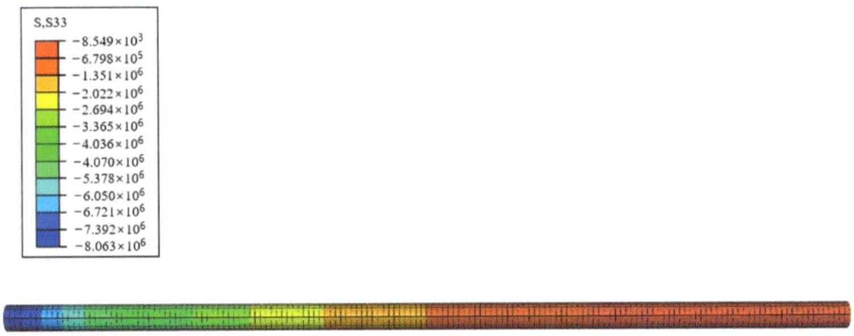

Figure 13. Axial stress distribution of pipe under full-contact condition when jacking force reaches 3500 t.

Then, the contact pressure around the pipe in the simulation results was extracted, as shown in Figure 14. It can be seen from Figure 14 that in the full-contact state, the contact pressure at the bottom of the pipe is the largest. With the increase in the height of the pipe, the contact pressure gradually decreases, and the contact pressure at the top of the pipe is the smallest. In addition, the contact pressure of the left part and the right part of the pipe is equal. The contact pressure of the pipe was converted into the unit friction resistance of the pipe section in the current contact state, and then the maximum friction coefficient that can push the pipe forward was calculated by means of the maximum allowable jacking force and the length of the pipe. The average value of the contact pressure around the pipe is 130 kPa according to these calculations and the length of the pipe affected by pipe sticking is 137.5 m. The calculation results are shown in Table 6.

Table 6. The calculated values in the full-contact state.

Maximum Allowable Jacking Force (kN)	Unit Friction Resistance (kN/m)	The Jacking Length Limit (m)	Coefficient of Friction
35,000	254.55	137.5	0.15

Note: 1. Unit friction resistance (kN/m) = Contact friction coefficient × Average contact stress (kPa) × Pipe circumference (m); 2. Ultimate jacking length = Ultimate bearing capacity of the pipe (kN)/Unit friction resistance (kN/m).

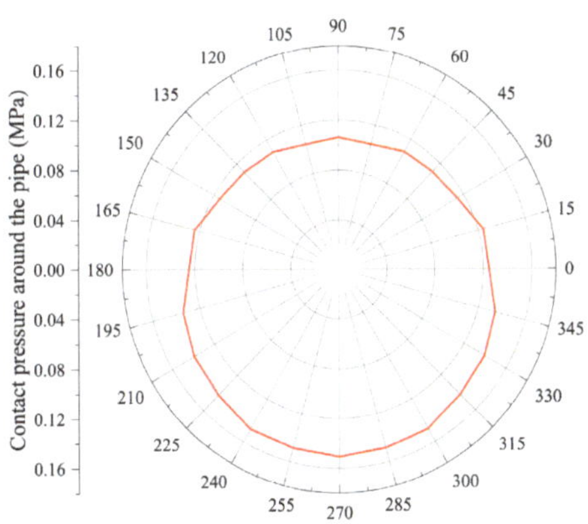

Figure 14. Contact pressure around the pipe.

After calculation, at the maximum allowable jacking force, the contact friction coefficient needs to be reduced to 0.15 for the forward jacking of the pipe affected by sticking. Based on the inversion of the field jacking force data, Stein et al. [19] determined the friction coefficient of the pipe and soil to be 0.1~0.3 under good lubrication conditions. Therefore, in theory, the sticking can be resolved if measures are taken to adequately lubricate the affected pipe.

It should be noted that the safest full-contact state was selected for this research, and the results are conservative. In the follow-up study, if the contact state of the pipe can be accurately judged and used for numerical simulation, the accuracy of the numerical simulation results will be greatly improved, which will provide more accurate data support for the optimization of the intermediate jacking station and the resolution of the pipe sticking state.

5. Technical Measures for Pipe Jacking Sticking

5.1. Add a Temporary Intermediate Jacking Station at the Location of the Pipe Jacking

Under pipe sticking conditions, the reserve jacking force cannot push all the pipe sections at the interval. Therefore, shortening the jacking interval should be considered, along with calculating the number of pipe sections that are convenient to push in combination with the friction resistance calculation model of the stuck pipe, completely breaking and dismantling a complete pipe section behind it, and then cleaning up the filling of the annular air and carrying out over-excavation of the surrounding rock. Afterward, the installation of the relay cylinder is carried out here according to the calculated value, and after the installation is completed, it is used as a temporary intermediate jacking station to push the pipe section. The rear intermediate jacking station and the main top oil cylinder become operational after the pipe section of the front section is pushed so as to advance all pipe sections. The temporary intermediate jacking station should be dismantled as soon as possible after successfully passing through the stuck pipe section in order to form a good mud sleeve in the annular space. The adjacent pipe section should then be closed, and then the new pipe section head should be sealed so as to restore the normal working state at the small cost of losing a pipe section, achieving smooth penetration.

5.2. Instantaneously Increasing the Jacking Force

First of all, under the premise of ensuring safety, it is necessary to determine the timing for increasing the jacking force according to the pipe section jamming situation and pipe operation parameters. Then, by gradually increasing the jacking force, an instantaneous increase in the jacking force is applied to the pipe to push the stuck pipe. In the process of implementation, it is necessary to strictly monitor the operation status of the pipe and the changes in the surrounding environment, adjust the jacking force in time, and take necessary safety measures, such as setting warning signs and blocking dangerous areas, to ensure the safety of operation. After the pipe section jamming is successfully solved, the pipe needs to be inspected and repaired, and the related equipment and tools should be inspected and maintained at the same time to ensure the safety and stability of the project. The entire process requires a high degree of expertise and experience to ensure that the increase in jacking force is smooth and controllable while not causing unnecessary risks and losses.

5.3. Loose Drilling Method

The purpose of the loose drilling method is to loosen the filler on both sides of the clamping tube position by drilling in this location so as to reduce the contact pressure between the tube and the filler and thereby ameliorate the pipe sticking phenomenon. According to the actual situation of the site, the location of each well is accurately determined through measurement and marked. A JKS400 percussion drilling rig is used to drill holes, and the hole diameter is 300 mm. A steel pipe casing is used to prevent the orifice from collapsing. After the drilling is completed, the steel pipe can be retained in the tube well according to the actual situation so as to be used as an anti-slide pile in the subsequent construction using the internal flushing method of pipe jacking. A row of five on the right side is drilled first, and then the jacking attempt is carried out. If the trap is not released, the drilling is continued, and, if necessary, the left side of the pipe section is also drilled. In order to prevent contact with the pipe section, the well side is 0.5 m away from the pipe section side, and five pieces are applied in each jacking attempt. One side is drilled according to three cycles, and the whole line is subjected to grouting before jacking. After 2 h, the jacking starts and the jacking force varies from small to large.

5.4. In-Tube Flushing Method

This method is usually used for steel pipe clamping after the pipe jacking is clamped. After roughly determining the position of the stuck pipe, the flushing section is arranged at a certain distance at the clamping position according to the transmission law of jacking force, and the bottom raft plate is installed and equipped with a ball valve after the opening of the inner wall of the steel pipe. A certain number of cleaning ball valves are installed in other positions, and the circulating mud pipeline is temporarily changed. The closure is formed in the adjacent two sections, the rear section is flushed with bentonite mud or clean water, and the front section continuously sucks out the filling of the annular air washed to the front by the mud. In this way, the outside of the pipe wall can be cleaned, as shown in Figure 15. After the cleaning is completed, when it is confirmed that there is no filler in the empty ring or that only a small amount of filler remains, all ball valves are closed, and the bottom raft is well sealed, and the grouting process is resumed.

Figure 15. In-tube flushing method.

In this project, the jacking force between the No. 3 and No. 2 intermediate jacking stations was first selected to be increased, and repeated jacking was carried out in the positive and negative directions. Thick mud was injected at the same time. The jacking construction can be resumed if this process is successful; otherwise, the loose drilling method is adopted, and the loose well is drilled along the left and right sides of the stuck pipe section. The technical measures of the first method were operated simultaneously to try to resolve the problem: If the first two methods were not successful, the pipe flushing method was used for construction; that is, the hole was opened at the appropriate position through a water grinding drill in the pipe wall, and the external sedimentation or wrapped formation of the pipe section was washed with high-pressure water to reduce the frictional resistance. If this method was not feasible, the tensile steel sheet pile + excavation method was used for treatment. Due to the deep burial depth and considering the construction risk and other issues, the external vibration method and internal vibration method of the pipeline were not considered at this stage. Ultimately, the project was successfully extricated from the difficulties associated with the measures of flushing and slag discharge in the pipe, the loose drilling method, and the instantaneous increase in jacking force.

6. Conclusions

(1) In this study, a total station was used to measure the displacement of a pipe in the limited jacking state to initially determine the interval of pipe sticking. Then, a strain gauge was arranged on the inner surface of the pipe at the interval of pipe sticking, and the displacement of the pipe in the limited jacking state was measured again. This approach can be used to quickly and accurately determine the pipe sticking position and provide a technical reference for related engineering cases.

(2) In the pipe sticking state, the transmission law of the axial displacement of the pipe is that the axial displacement of the upper part of the pipe is greater than the axial displacement of the lower part of the pipe, and the transmission distance of the axial jacking force in the upper part of the pipe is also greater than that in the lower part of the pipe. For the contact pressure around the pipe, the contact pressure at the bottom of the pipe is the largest. With the increase in the height of the pipe, the contact pressure gradually decreases, and the contact pressure at the top of the pipe is the smallest. In addition, the contact pressure of the left part and the right part of the pipe is equal.

(3) According to the numerical simulation results in the limited jacking force state, in the full-contact state, the maximum friction coefficient of the contact interface between the pipe and the cuttings that enables the pipe sticking interval to advance forward is 0.15. These findings provide data support for the optimization of the intermediate jacking station and the resolution of the pipe sticking state.

(4) Four measures were proposed to resolve the sticking phenomenon after pipe jacking and clamping, and they were successfully applied to the project, providing a reference method for subsequent pipe jacking sticking.

Author Contributions: Methodology, T.S.; Validation, P.Z.; Data curation, X.X.; Writing—original draft, S.Z.; Writing—review & editing, T.X.; Project administration, F.L. All authors have read and agreed to the published version of the manuscript.

Funding: This research received no external funding.

Data Availability Statement: The original contributions presented in the study are included in the article, further inquiries can be directed to the corresponding author.

Conflicts of Interest: Authors Shilei Zhang and Xiaodong Xu were employed by the company Inner Mongolia Yin Chao Ji Liao Water Supply Co., Ltd. Author Feilun Luo was employed by the company Sinohydro Bureau 5 Co., Ltd. The remaining authors declare that the research was conducted in the absence of any commercial or financial relationships that could be construed as a potential conflict of interest.

References

1. Liu, K.X.; Xiao, A.F.; Zhang, P.; Zhou, H.; Chen, Z.; Xu, T.; Ma, B.; Ai, H.; Wang, Q. Study on mechanical response of steel pipe jacking considering the effect of pipe sticking. *Tunn. Undergr. Space Technol.* **2022**, *127*, 104617. [CrossRef]
2. Shou, K.J.; Hsieh, C.Y. On the behavior of the overcut and stuck effects during pipejacking. *Procedia Eng.* **2016**, *165*, 593–601. [CrossRef]
3. Li, C.; Zhong, Z.L.; Bie, C.Y.; Liu, X. Field performance of large section concrete pipes cracking during jacking in Chongqing—A case study. *Tunn. Undergr. Space Technol.* **2018**, *82*, 568–583. [CrossRef]
4. Li, C.; Zhong, Z.; Liu, X.; Tu, Y.; He, G. Numerical simulation for an estimation of the jacking force of ultra-long-distance pipe jacking with frictional property testing at the rock mass-pipe interface. *Tunn. Undergr. Space Technol.* **2019**, *89*, 205–221. [CrossRef]
5. Li, C. Experimental and Analytical Study of Construction Mechanics Effect of an Extra-Long Distance Rock Pipe Jacking Through Rock Strata Under Complex Contact Condition. Ph.D. Thesis, Chongqing University, Chongqing, China, 2020.
6. Li, C.; Zhong, Z.; Liu, X.; Deng, Z.; He, G. The investigation of ultra-long-distance concrete pipe stuck in quartz sandstone formation using numerical simulation. *Arab. J. Geosci.* **2018**, *11*, 678. [CrossRef]
7. Zhong, Z.L.; Li, C.; Liu, X.R.; Xiong, Y.; Fan, Y.; Liang, N. Assessment of experimental friction parameters and contact property of pipe string for the estimation and verification of a solution forpipe string for the estimation and cerification of a solution for pipe stuck in the China's first rock pipe jacking. *Tunn. Undergr. Space Technol.* **2021**, *107*, 103671. [CrossRef]
8. Deng, Z.Y. Mechanical Properties of Fiber-Reinforced Concrete Pipe and Its Application in Crack Control of Long-Distance Rock Pipe Jacking. Ph.D. Thesis, Chongqing University, Chongqing, China, 2020.

9. Deng, Z.Y.; Liang, N.H.; Liu, X.R.; de la Fuente, A.; Lin, P.; Peng, H. Analysis and application of friction calculation model for long-distance rock pipe jacking engineering. *Tunn. Undergr. Space Technol.* **2021**, *115*, 104063. [CrossRef]
10. Deng, Z.Y.; Liu, X.; Zhou, X.; Yang, Q.; Chen, P.; Fuente, A.; Ren, L.; Du, L.; Han, Y.; Xiong, F.; et al. Main engineering problems and countermeasures in ultra long-distance rock pipe jacking project: Water pipeline case study in Chongqing. *Tunn. Undergr. Space Technol.* **2022**, *123*, 104420. [CrossRef]
11. Yang, Q.H. Research on technique of reliving the stuck pipe of super-long distance rock jacking pipe project. *Chin. J. Undergr. Space Eng.* **2019**, *15*, 766–772.
12. Yang, Q.H. Studies and discussion on key technology of long distance rock pipe jacking in construction of the Guanjingkou multi-purpose project. *China Water Resour.* **2021**, 37–40. [CrossRef]
13. Zhu, F.Y.; Sun, Y.; Wang, Y.S. Key technologies and measures for rock pipe jacking tunnel of the Guanjingkou multi-purpose project. *China Water Resour.* **2021**, *19*, 50–52.
14. Na, Y. Study on the removal of stuck pipe (inner diameter: 4 m) during jacking in fine silt liquefaction layer. *Archit. Technol.* **2024**, *55*, 2385–2388.
15. He, Z.; Chen, J. Experimental Study on the Complex Contact Frictional Property of an Ultralong Distance Large-Section Concrete Pipe Jacking and Prediction of Pipe String Stuck. *Adv. Mater. Sci. Eng.* **2019**, *2019*, 4353520. [CrossRef]
16. Kamran, M.; Hu, X.W.; Hussain, M.; Sanaullah, M.; Ali, R.; He, K. Dynamic Response and Deformation Behavior of Kadui-2 Landslide Influenced by Reservoir Impoundment and Rainfall, Baoxing, China. *J. Earth Sci.* **2023**, *34*, 911–923. [CrossRef]
17. Tong, D.F.; Su, A.J.; Tan, F.; Tang, J.D.; Yi, X.W. Genetic Mechanism of Water-Rich Landslide Considering Antecedent Rainfalls: A Case Study of Pingyikou Landslide in Three Gorges Reservoir Area. *J. Earth Sci.* **2023**, *34*, 1878–1891. [CrossRef]
18. Xu, T.S. Research on the Mechanical Properties of Large-Diameter Reinforced Concrete Pipe Jacking in Complex Strata and the Mechanism of Pipe Sticking. Ph.D. Thesis, China University of Geosciences, Wuhan, China, 2024.
19. Stein, D. *Microtunnelling: Installation and Renewal of Nonmansize Supply and Sewage Lines by the Trenchless Construction Method;* Ernst & Sohn Verlag fur Architektur und Technische Wissenschaften: Berlin, Germany, 1989.

buildings

Article

Numerical Study on Mechanical Characteristics of Tower Sections with Main Member Disconnection

Hengwei Zheng *, Changli Wu, Jinhong Liu, Lang Zhong, Kai Li and Zhitao Yan

Department of Theoretical and Applied Mechanics, Chongqing University of Science and Technology, Chongqing 401331, China; 2021206013@cqust.edu.cn (C.W.); 2021206054@cqust.edu.cn (J.L.); 2021206046@cqust.edu.cn (L.Z.); 2023206099@cqust.edu.cn (K.L.); 2016023@cqust.edu.cn (Z.Y.)

* Correspondence: zhenghw@cqust.edu.cn

Abstract: Restricted by the existing construction technology, there are a lot of disconnections in the angle steel components of transmission towers. At present, there are more studies on single angle steel or cross bracing, but less on the main member containing disconnection joints. According to the disconnection position in the main member, an upper end disconnection, middle end disconnection, and lower end disconnection were designed in this paper. At the same time, a tower section model of a connected main member and a tower section model of a disconnected main member were established and analyzed by finite element analysis. Considering the loadings acting on transmission towers, the two load conditions of axial loading and tension–compression coupling are set. Considering the loadings acting on transmission towers, the influences of the combination form of the inner and outer steel cladding and the steel cladding area ratio on the ultimate bearing capacity of the main member were studied under 72 groups of different tower sections applied with axial loadings. The influence of the disconnection joints on the stability of the tower section was studied under 24 groups of different tower sections applied with tension and compression coupled loading conditions. Referring to the specifications, the slenderness ratio correction factor formula of the disconnect main member can be derived. The results indicate that when designing the disconnection joint in the main member, it is recommended to choose a middle end disconnection with a steel cladding area ratio of 1.0.

Keywords: disconnection joint; tower section; ultimate bearing capacity; slenderness ratio correction factor

1. Introduction

Due to the advantages of low cost and easy installation, angle steel transmission towers are widely used in practical engineering. A transmission tower is composed of a tower top, a tower body, and tower legs. In this paper, a tower section was selected from the tower body as the object in order to analyze the influence of main member disconnection joints on its mechanical properties. A transmission tower section contains two main members, one cross bracing, and several auxiliary support members. The stable bearing capacity of equilateral angle steel as the main load-bearing component of transmission towers is always a research focus. Experts and scholars usually treat equilateral angle steel main members as axially compressive bars and have studied their stable bearing capacity by using axial compression tests, finite element simulation analysis, or theoretical calculations [1–7]. Some experts also analyzed the mechanical characteristics of the main members by studying the entire tower [8–12].

Restricted by the existing construction technology, there are a lot of disconnections in the angle steel components of transmission towers, and their location is usually in the main member at the bottom of the tower section. The disconnected joints are connected by bolts and steel cladding. Currently, the relevant codes in the field of transmission towers only specify requirements for the length of disconnecting joints, the number of bolts, and the steel cladding area, which are considered relatively conservative [13,14]. Therefore, there is a necessity to conduct further research on disconnecting joints in main members. Few studies have focused on this area. According to Gao [15], the steel cladding area in disconnecting joints has an impact on joint stiffness, and he also studied the layout position of disconnecting joints. Xue [16] examined disconnecting joints of various sizes in main members. However, while many scholars have studied the mechanical properties of bolted connection nodes in transmission towers [17–22], there is relatively little research on the main member disconnection joints. The disconnection of the main member will influence the stability of the tower section. Scholars have conducted extensive research on the out-of-plane instability and the corresponding laws of the intact tower section's cross bracings through experiments and finite element simulations [23–26], but there is relatively little research on tower sections with main member disconnection.

Although domestic and foreign scholars have conducted extensive research on the bearing capacity of main members and the stability of cross bracing, when considering the interaction between the main member, cross bracing, and auxiliary support member, there is relatively little research on the mechanical responses of a tower section with main member disconnection from the perspective of the overall structure. In this paper, INVENTOR was utilized to establish tower section models of both an intact main member and a main member with disconnection. Then, the mechanical responses of these models were studied using a numerical method. The slenderness ratio correction factor formula of the disconnected main members was derived. Therefore, there is significant reference value in optimizing the design of main member disconnection joints in practical engineering.

2. Materials and Methods

2.1. Principle of Slenderness Ratio Correction of Main Members

The slenderness ratio is an important parameter for describing the stability of a member. For the main member, the disconnection joint will directly affect its stable bearing capacity. The slenderness ratio of the main member containing the disconnection joint can be obtained by reverse calculation of its ultimate bearing capacity. In this paper, the ultimate bearing capacity of the disconnected main member was obtained through finite element calculation, and then the slenderness ratio of the disconnected main member was calculated by combining it with the stable bearing capacity calculation method for axial compression components [23] (derived from: Technical code for the design of tower and pole structure of overhead transmission lines DL/T 5154-2012). The slenderness ratio correction factor is derived by comparing the slenderness ratio of disconnected main members with that of connected main members. The following formula for calculating the bearing capacity of axially compressed components has been provided in standard DL/T 5154-2012 [23]:

$$N/(\varphi \cdot A) \leq m_N \cdot f \tag{1}$$

Here, N is axial tension or axial pressure; φ is the stability coefficient of axially compressed components; A is the gross cross-sectional area of the component; m_N is the reduction coefficient of the stable strength of the compression bar; and F is the design value of the steel strength of the component.

The calculation steps for the slenderness ratio of the disconnected main member are as follows:

(1) Substitute the ultimate bearing capacity of each disconnected main member obtained from the finite element calculation into formula (1) to deduce the stability coefficient of the compression φ.
(2) Calculate the corrected slenderness ratio based on the stability coefficient φ value table provided in reference to DL/T 5154-2012.

(3) The slenderness ratio correction coefficient K_λ of the main member can be obtained by dividing the corrected slenderness ratio λ_n derived from step (2) by the nominal slenderness ratio, as shown in Equation (2).

$$K_\lambda = \lambda_n / \lambda \tag{2}$$

2.2. Tower Section Model

2.2.1. Tower Section Structure and Disconnection Design

In this paper, a tower section of an equilateral angle steel transmission tower was selected in a certain high-voltage transmission project as the object with which to study a main member disconnection joint. The structure of the tower section and the dimensions of the components in the structure are shown in Figure 1. The main member and the joint steel cladding are made of Q420 (nominal yield strength = 420 MPa), and the cross bracing and auxiliary support members are made of Q235 (nominal yield strength = 235 MPa). According to the position of the main member disconnection joint within the tower section, three types of main member disconnection joints were designed: upper end disconnection, middle end disconnection, and lower end disconnection. At the same time, a model of the connected main member was designed as the control group, as shown in Figure 2.

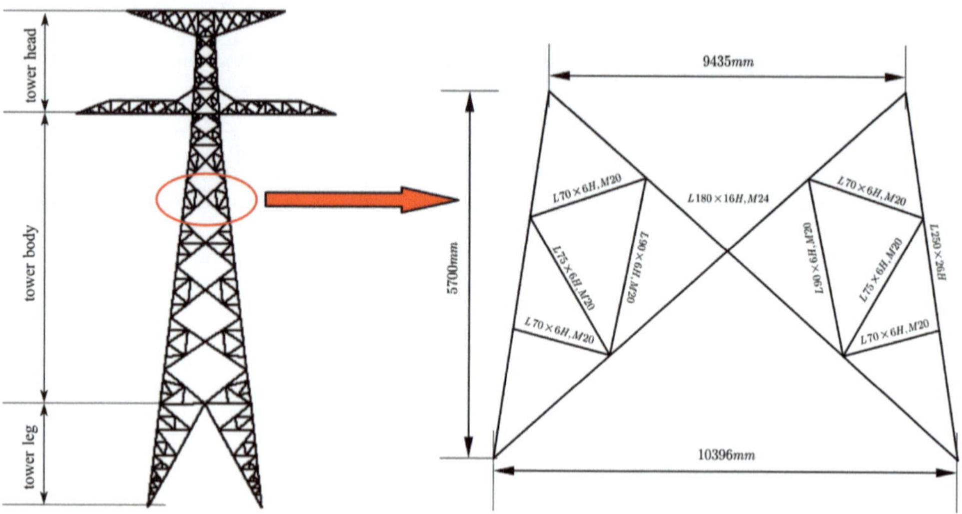

Figure 1. Tower section structure diagram.

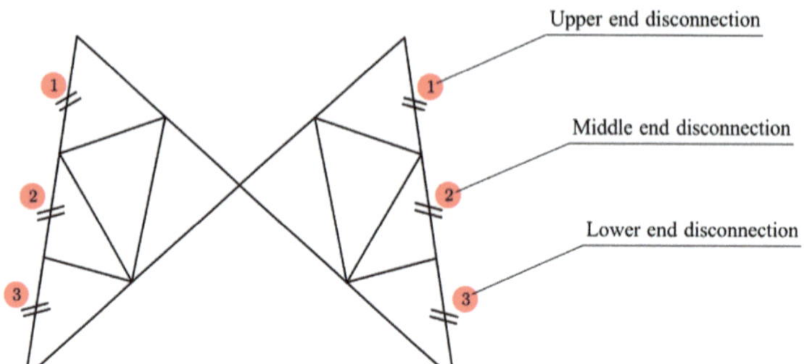

Figure 2. Schematic diagram of tower section disconnection design.

2.2.2. Main Member Disconnection Joints

The external pasting and the internal cladding bolt connection are applied to the main member disconnection joints. The bolts on the joint are uniformly M24 bolts and are arranged in a staggered manner according to the requirements of GB50017-2017 "Code for Design of Steel Structures" [24], as shown below in Figure 3. Here, in Figure 3b, A is the internal steel cladding of the joint, B is the main member, and C is the external pasting plate of the joint. The internal steel cladding and the external pasting plate are collectively referred to as the steel cladding. The ratio of the steel cladding area is defined based on the ratio of the cross-sectional area of the steel cladding to the cross-sectional area of the main member, and the calculation formula is as follows:

$$a = \frac{A_b}{A_0} \tag{3}$$

Here, A_b is the gross cross-sectional area of the steel cladding and A_0 is the gross cross-sectional area of the main member.

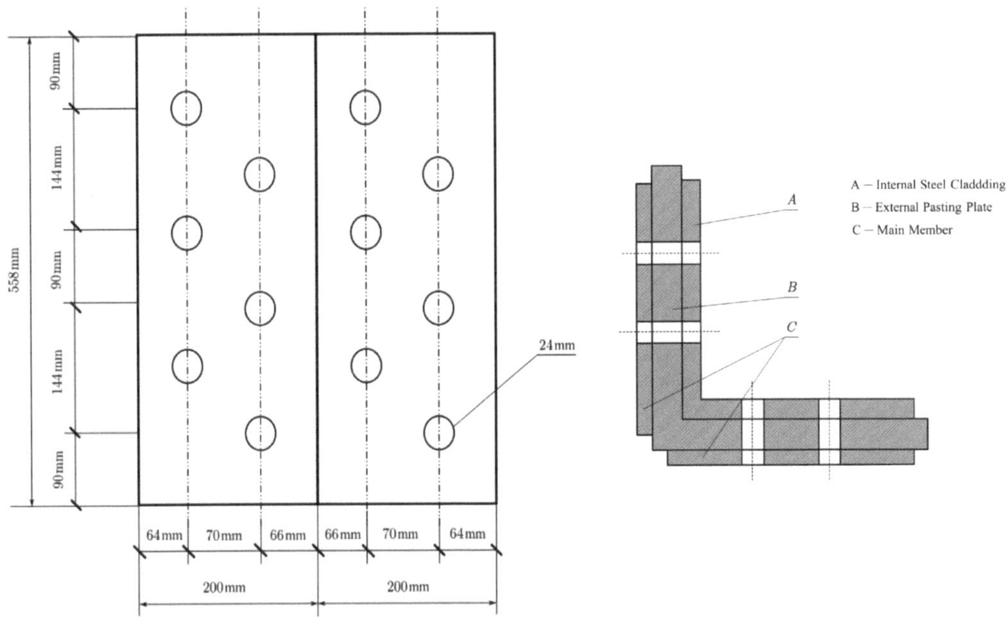

(**a**) Internal steel cladding joint bolt layout diagram. (**b**) Joint cross-section diagram.

Figure 3. Main member disconnection joint.

In this paper, the disconnection joint was taken as an important research object and the area ratio of the steel cladding was studied as the main control variable. Six different steel cladding area ratios were set: 0.8, 0.9, 1.0, 1.1, 1.2, 1.3. Four different combinations of inner and outer steel cladding were given by comparing the steel cladding area ratios of each group, as shown in Table 1. Due to the characteristic of the steel cladding area ratio, the actual calculated steel cladding area ratio cannot be accurately determined as the design value of the steel cladding area ratio. Therefore, an approximate actual steel cladding area ratio is selected for calculation during the calculation process.

Table 1. Steel cladding design for disconnection joints.

Design Steel Cladding Area Ratio	Number	Internal Steel Cladding (mm)	External Pasting Plate (mm)	Main Members (mm)	Actual Steel Cladding Ratio
0.8	A1	L200 × 11	198 × 15	L250 × 26	0.83
	A2	L200 × 13	198 × 13	L250 × 26	0.83
	A3	L200 × 15	198 × 11	L250 × 26	0.82
	A4	L200 × 18	198 × 8	L250 × 26	0.82
0.9	B1	L200 × 14	198 × 15	L250 × 26	0.92
	B2	L200 × 15	198 × 14	L250 × 26	0.92
	B3	L200 × 16	198 × 13	L250 × 26	0.92
	B4	L200 × 17	198 × 12	L250 × 26	0.91
1.0	C1	L200 × 15	198 × 17	L250 × 26	1.01
	C2	L200 × 16	198 × 16	L250 × 26	1.01
	C3	L200 × 17	198 × 15	L250 × 26	1.01
	C4	L200 × 18	198 × 14	L250 × 26	1.01
1.1	D1	L200 × 15	198 × 20	L250 × 26	1.14
	D2	L200 × 16	198 × 19	L250 × 26	1.11
	D3	L200 × 17	198 × 18	L250 × 26	1.11
	D4	L200 × 18	198 × 17	L250 × 26	1.10
1.2	E1	L200 × 13	198 × 25	L250 × 26	1.21
	E2	L200 × 14	198 × 24	L250 × 26	1.21
	E3	L200 × 16	198 × 22	L250 × 26	1.21
	E4	L200 × 17	198 × 21	L250 × 26	1.20
1.3	F1	L200 × 14	198 × 28	L250 × 26	1.34
	F2	L200 × 16	198 × 26	L250 × 26	1.33
	F3	L200 × 18	198 × 24	L250 × 26	1.33
	F4	L200 × 21	198 × 21	L250 × 26	1.32

2.2.3. Load Conditions

Considering the mechanical characteristics of transmission towers, two tower section load conditions were designed. The first load condition is shown in Figure 4a, as the main load-bearing component of the transmission tower; the axial loading applied to the main member of the tower section serves to simulate the vertical loading of the transmission tower. The second load condition is shown in Figure 4b. The most important role of the transmission tower is to support the overhead transmission line; therefore, the application of tension–compression coupling loads to the tower section serves to simulate the transverse load of the transmission tower.

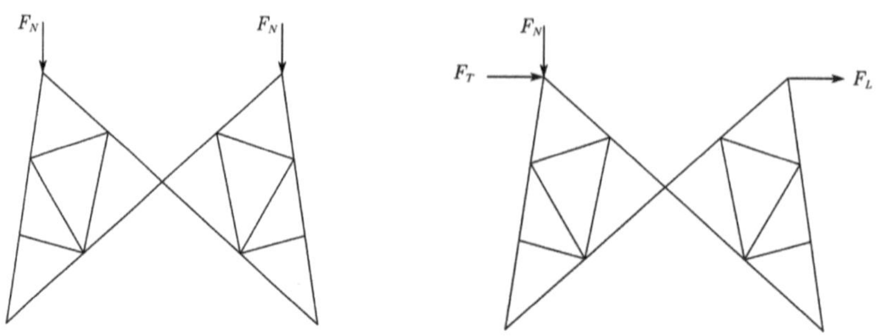

(**a**) Condition1: Axial load. (**b**) Condition2: tension and compression coupling load.

Figure 4. Two load conditions.

2.2.4. Finite Element Model

Combining 3D modeling and finite element analysis in this study, the tower sections were modeled and conducted by finite element simulation analysis. The 8-node 3-dimensional solid element can accurately describe the deformation of the structure and possesses the advantages of good convergence performance and computational stability. Therefore, the 8-node 3-dimensional solid element is used to mesh the tower section models, with a 20 mm swept grid size applied to the models, and the grid is refined at the bolt positions. To enhance the precision and accuracy of the finite element analysis, the mesh at the bolt-hole positions and breaking node locations is refined. Since bolt connections are not the focus of this paper, we utilize the multipoint restraint and beam element simulation bolts to establish the connections between members. The yield strength of the Q420 steel used is 420 MPa, and the elastic modulus and Poisson's ratio are 2.06×10^5 MPa and 0.28, respectively. The yield strength of the Q235 steel used is 235 MPa, and the elastic modulus and Poisson's ratio are 2.1×10^5 MPa and 0.27, respectively. As the angle steel components of the transmission tower are made of steel, the ideal elastoplastic model is selected as the constitutive model for the steel according to the material properties test in Reference [27], as shown in Figure 5. According to the tower section model test device described in Reference [27], the boundary conditions are constrained in terms of both the translational degrees of freedom (in the X, Y, and Z directions) and the rotational degrees of freedom (UR1 and UR3) at the non-loading end at the bottom of the model. The tower section is loaded according to the specified loading conditions, and the loading adopts a gradual incremental method. The load continues to increase until the stiffness matrix of the entire structure becomes singular and can no longer converge iteratively, at which point the load is determined to be the ultimate bearing capacity. In the nonlinearity settings, geometric nonlinearity is enabled. The connection tower section finite element models are shown in Figure 6.

The ultimate bearing capacity of the connection main member will be calculated using two methods. One method was to use formula (1) to calculate the ultimate bearing capacity as a theoretical value of 4052.9 kN; the other method was to calculate the ultimate bearing capacity as a numerical value of 4669.5 kN by finite element analysis. Analysis of the error between the theoretical and finite element results shows that the relative error is 12.21%, which is considered relatively small. Therefore, it is feasible to use the finite element method to simulate the loadings acting on the main member. The finite element model of the tower section with main member disconnection is presented in Figure 7, where the grid refinement at the connection joint positions has been implemented.

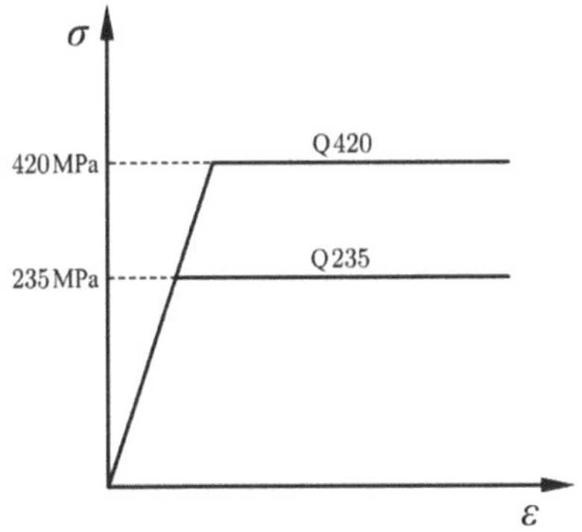

Figure 5. The ideal elastoplastic model.

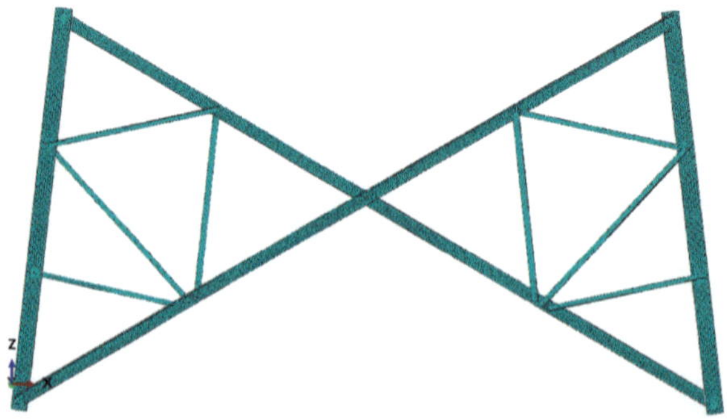

Figure 6. Connection tower section finite element model.

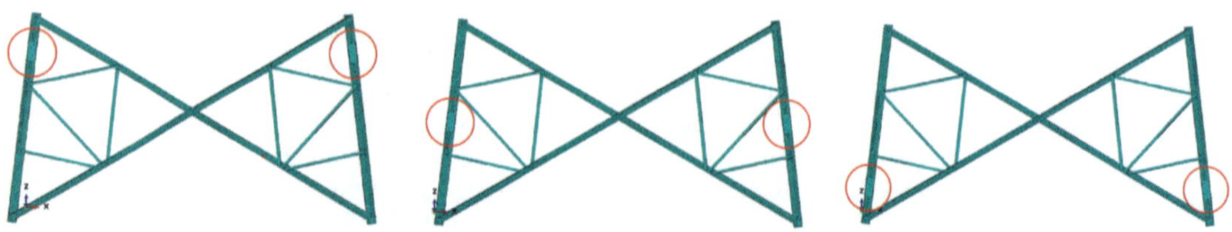

(a) Upper end disconnection. (b) Middle end disconnection. (c) Lower end disconnection.

Figure 7. Disconnection tower section finite element model. (The disconnection joint is in the red circle).

3. Results and Discussion

3.1. Analysis of Finite Element Results for Axial Loading of Tower Section

3.1.1. Combination form of Inner and Outer Steel Cladding

In order to quantify the degree of impact of the combination of inner and outer steel cladding on the ultimate bearing capacity of the main member, the difference between the maximum and minimum ultimate loads of the main member corresponding to different combinations of inner and outer steel cladding was calculated for each steel cladding area ratio under different disconnection joint designs. The results are shown in Figure 8. From the figure, it can be seen that the area ratio of the steel cladding is smaller, and the impact of the combination of inner and outer steel cladding on the ultimate bearing capacity of the disconnected main member is greater. In addition to this, when the area ratios of the steel cladding are 1.0 and 1.1, the ultimate bearing capacity of the main member at the middle end disconnection is most affected by the combination form of the inner and outer steel cladding. When the area ratios of the steel cladding are any other values, the ultimate bearing capacity of the upper end disconnection main member is most affected by the combination form of the inner and outer steel cladding. Therefore, it is necessary to study the combination form of the inner and outer steel cladding when designing the steel cladding area ratio.

In this paper, the ultimate loads of three sets of disconnected main members under different combinations of inner and outer steel cladding with different steel cladding area ratios are given. The results are shown in Figure 9. Comparing the ultimate loads of main members under different combinations of inner and outer steel cladding, the optimal working condition is the maximum ultimate load, with a steel cladding area ratio of 0.8 to 1.3. The optimal working conditions, corresponding to the upper end of the main member are A3, B3, C2, D2, E3, and F2, respectively. The optimal working conditions corresponding to the middle end of the main member are A3, B3, C2, D2, E2, and F2, and the optimal working conditions corresponding to the lower end of the main member are A3, B4, C4, D4, E4, and F4, respectively. All of these optimal operating conditions are explained in Table 1.

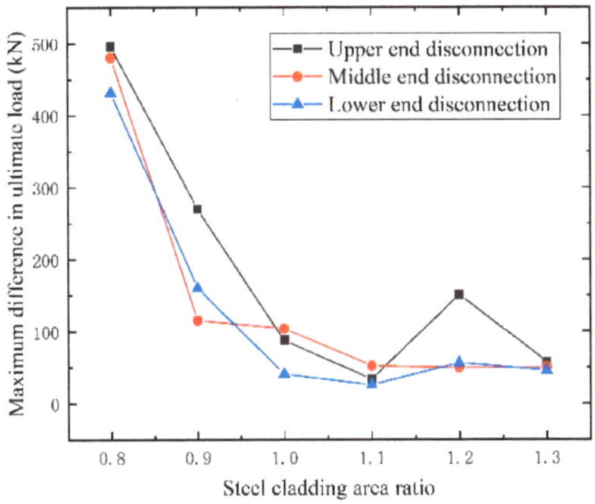

Figure 8. Maximum difference value in ultimate load of different steel cladding combinations under different steel cladding area ratios.

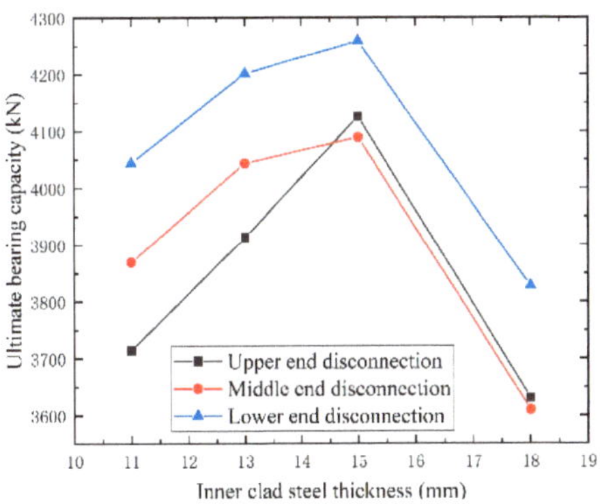

(**a**) Steel cladding area ratio 0.8.

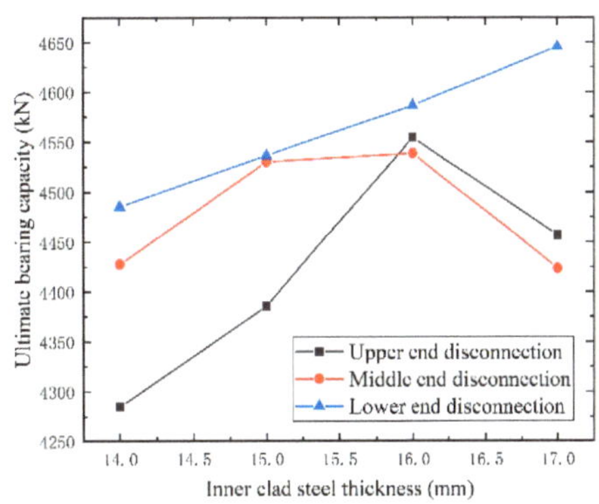

(**b**) Steel cladding area ratio 0.9.

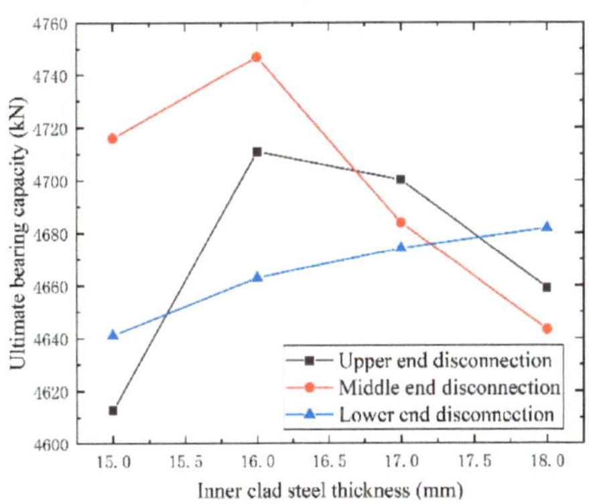

(**c**) Steel cladding area ratio 1.0.

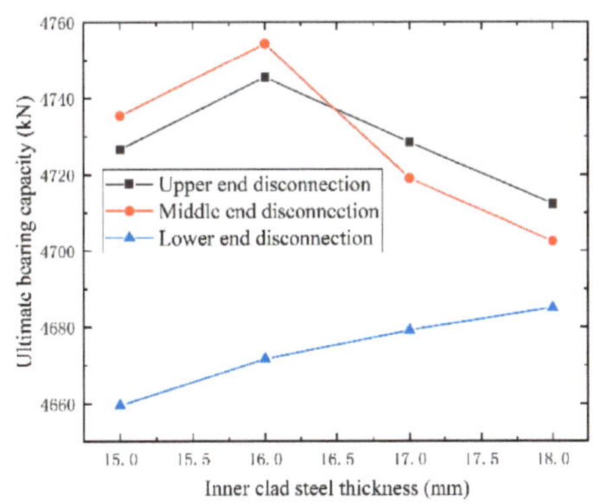

(**d**) Steel cladding area ratio 1.1.

Figure 9. *Cont.*

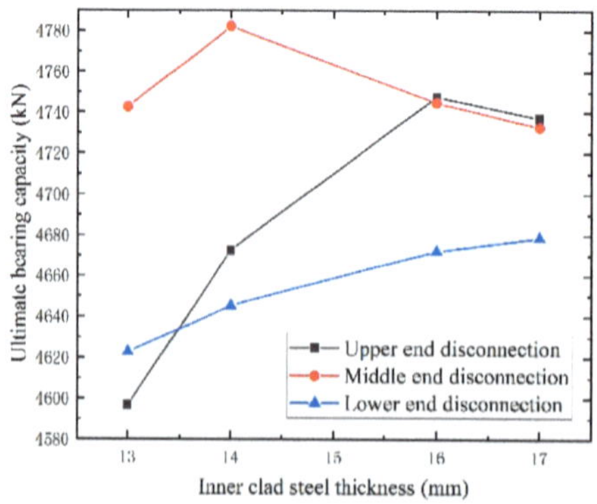

(**e**) Steel cladding area ratio 1.2.

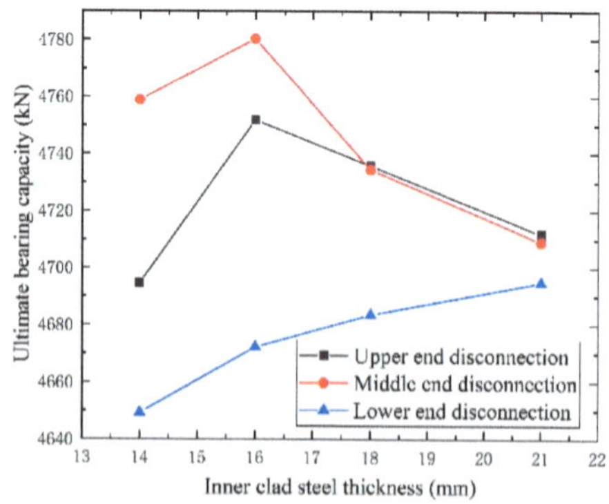

(**f**) Steel cladding area ratio 1.3.

Figure 9. Ultimate load of disconnected main members for different steel cladding combination forms.

As shown in Figure 9, the trend in the ultimate loads of the three sets of disconnected main members as the inner steel cladding thickness changes can be observed. It can be found that, considering the same steel cladding area ratio, there are two distinct trends in the ultimate load of the main member: it first increases and then decreases with the increase in inner steel cladding thickness; or it continuously increases with the increase in inner steel cladding thickness. It is observed that the ultimate bearing capacity of the main member can be increased due to a certain degree of increase in the thickness of the inner steel cladding.

3.1.2. Steel Cladding Area Ratio

As shown in Figure 10, it can be seen that the ultimate load of the disconnected main member under the optimal working condition varies with the steel cladding area ratio. Here, the horizontal reference line is the ultimate bearing capacity of the connected main member. As the steel cladding area ratio increases, the ultimate bearing capacity of the main member increases significantly at first, and then slowly, with the steel cladding area ratio to 1.0 as the boundary. Comparing this with the ultimate bearing capacity of the connected main member, it can be found that when the steel cladding area ratio is less than 1.0, the ultimate bearing capacity of the three sets of disconnected main members significantly will decrease. Conversely, when the steel cladding area ratio is greater than or equal to 1.0, the ultimate bearing capacities of the three sets of disconnected main members significantly will increase, and then the ultimate bearing capacities of the middle end disconnected main member will increase to the maximum. Therefore, it is recommended that the middle end of the main member can be chosen as the disconnect position, and the steel cladding area ratio at the disconnection joint can be set at 1.0.

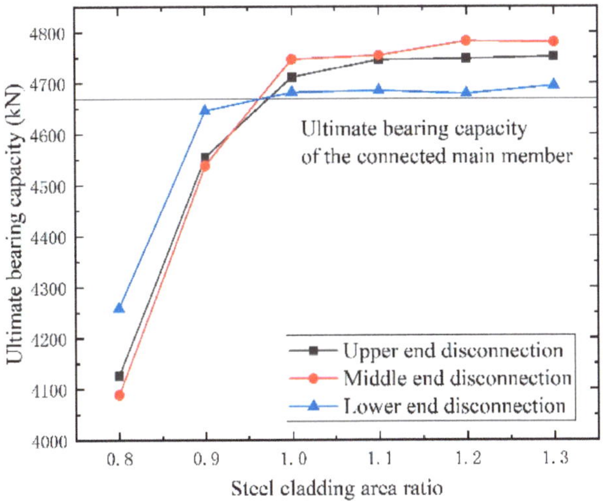

Figure 10. Comparison of ultimate load of main members under different steel cladding area ratios.

3.2. Analysis of Finite Element Results for Tension and Compression Coupled Loading of Tower Section

The out-of-plane displacement (displacement perpendicular to the structural plane) and ultimate thrust at the intersection point of the cross bracing by selecting the optimal working condition and applying tension and compression to the tower section were obtained, as shown in Table 2. As shown in the table, it can be seen that the presence of a disconnecting joint will reduce the stability of the tower section. Furthermore, when there is a disconnection joint in the main member, the layout position of the disconnection joint will have little impact on the stability of the tower section.

Table 2. Out-of-plane displacement of cross bracing.

Type	Steel Cladding Area Ratio	Ultimate Thrust (kN)	Out-of-Plane Displacement (mm)
No disconnection	non	463.6	141.5
Upper end disconnection	0.8	395.4	167.8
	0.9	395.4	167.5
	1.0	395.4	167.4
	1.1	401.2	167.8
	1.2	395.4	168.7
	1.3	395.4	171.0
Mid end disconnection	0.8	401.0	167.7
	0.9	401.0	167.7
	1.0	401.0	167.7
	1.1	401.0	167.7
	1.2	401.0	167.4
	1.3	401.0	167.2
Lower end disconnection	0.8	396.2	165.6
	0.9	396.3	165.6
	1.0	396.3	165.6
	1.1	396.8	165.6
	1.2	396.2	165.6
	1.3	396.4	165.7

3.3. *Slenderness Ratio Correction Factor for Disconnection Main Members*

The slenderness ratio correction factors for the three types of disconnection main members under various optimal working conditions are calculated based on the principle of slenderness ratio correction, as shown in Figure 11.

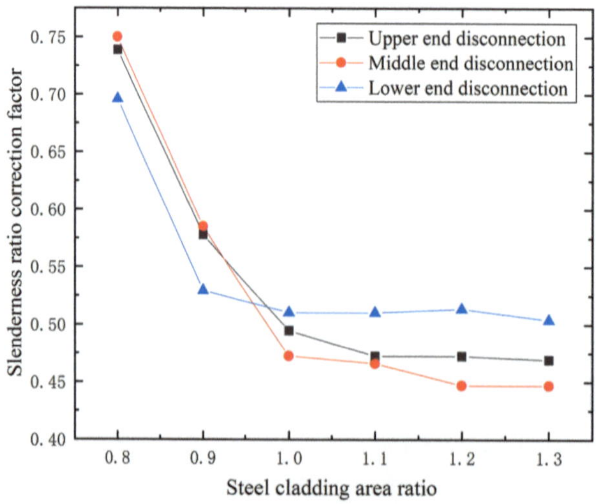

Figure 11. Scatter plot of the slenderness ratio correction factor of the disconnect main member.

From Figure 11, it can be observed that the correction factor decreases with the increase in the steel cladding area ratio, demonstrating a clear inverse proportional relationship. Therefore, the least squares method can be employed to fit the data and obtain the following formula for the slenderness ratio correction coefficient for the disconnected main member at the upper end.

$$K_{\lambda} = \frac{0.5408}{a} + 0.0082 \tag{4}$$

The slenderness ratio correction coefficient for the disconnected main member at the middle end will be given by Equation (4).

$$K_{\lambda} = \frac{0.6168}{a} - 0.0758 \tag{5}$$

The slenderness ratio correction coefficient for the disconnected main member at the lower end will be given by Equation (5).

$$K_{\lambda} = \frac{0.3326}{a} + 0.2185 \tag{6}$$

4. Conclusions

Based on finite element simulation analysis for the four groups of the tower section, the following conclusions were obtained:

(1) Considering the same steel cladding area ratio, the ultimate bearing capacity of the main member will be affected by the combination of internal and external steel cladding on the disconnected joint. Increasing the thickness of the inner cladding will increase the stiffness at the disconnection joint position to a certain extent, thereby enhancing the ultimate bearing capacity of the main member. Therefore, the influence of the combination form of the inner and outer steel cladding on the joint should be considered when designing the disconnection joint in the main member of the transmission tower.

(2) The ultimate bearing capacity of the main member increases significantly at first and then slowly with the steel cladding area ratio set to 1.0 as the boundary. When the area ratio of steel cladding is greater than or equal to 1.0, the disconnection joint can improve the ultimate bearing capacity of the main member. Especially, the most significant improvement in the ultimate bearing capacity of the main member is achieved when the disconnection joint is located at the middle end of the main member. It is recommended that the middle end of the main member can be chosen as the disconnection position, and the steel cladding area ratio at the disconnection joint can be set to 1.0.

(3) The mode of tower section instability is out-of-plane deformation, and the disconnection joint in the main member reduces the stability of the tower section significantly. When there is a disconnection joint in the main member, the stability of the tower section is not affected by the location of the disconnection joint in the main member.

(4) Because the ultimate bearing capacity of the main member is affected by the disconnection joint in the main member, the slenderness ratio correction coefficient formulas have been derived in this paper for the upper end disconnected main member, the middle end disconnected main member, and the lower end disconnected main member.

Author Contributions: Conceptualization, H.Z. and Z.Y.; methodology, H.Z.; software, C.W.; validation, C.W., J.L. and L.Z.; formal analysis, C.W.; investigation, C.W.; resources, C.W.; data curation, C.W.; writing—original draft preparation, H.Z. and C.W.; writing—review and editing, H.Z. and Z.Y.; visualization, C.W., J.L., L.Z. and K.L.; supervision, H.Z. and Z.Y.; project administration, H.Z. and Z.Y.; funding acquisition, Z.Y. All authors have read and agreed to the published version of the manuscript.

Funding: This research was funded by the National Natural Science Foundation of China (Grant No. 11602042 and 11802047), Foundation and Frontier Projects in Chongqing City (Grant Nos. cstc2019jcyj-msxmX0317 and cstc2015jcyjA50010).

Data Availability Statement: The data used to support the findings of this study are included in the article.

Conflicts of Interest: The authors declare no conflicts of interest.

References

1. Shanley, F.R. Inelastic column theory. *J. Aeronaut Sci.* **1947**, *14*, 261–267. [CrossRef]
2. Mengelkoch, N.S.; Yura, J.A. Single-angle compression members loaded through one leg. *Gainesville SSRC* **2002**, 212–218.
3. Liu, Y.; Hui, L. Behavior of steel single angles subjected to eccentric axial loads. *Can. J. Civ. Eng.* **2010**, *37*, 887–896. [CrossRef]
4. Dobri, J.; Filipovi, A.; Baddoo, N. Design procedures for cold-formed stainless steel equal-leg angle columns. *Thin-Walled Struct.* **2020**, *159*, 107210. [CrossRef]
5. Fasoulakis, Z.C.; Lignos, X.A.; Avraam, T.P.; Katsatsidis, S.P. Investigation on single-bolted cold-formed steel angles with geometric imperfections under compression. *J. Constr. Steel Res.* **2019**, *162*, 265–278. [CrossRef]
6. Iftesham, B.; Mahmud, K.A. Comparison of Codes for Axial Compression Capacity of Eccentrically Loaded Single Angles. *J. Constr. Steel Res.* **2021**, *185*, 106829.
7. Giacomo, V.; Zongchen, L.; Christian, A. Evaluation of the Ultimate Collapse Load of a High-Voltage Transmission Tower under Excessive Wind Loads. *Buildings* **2023**, *13*, 513. [CrossRef]
8. Wang, J.; Qi, H.; Zhang, Z.; Zhang, B.; Sun, Q. Failure mechanism of a Q690 steel tubular transmission Tower: Full-Scale experiment and numerical simulation. *Eng. Struct.* **2023**, *295*, 116881. [CrossRef]
9. Shen, G.; Yu, H.; Yu, L.; Li, B.; Yao, J. Seismic responses of a long-span transmission tower and tower-line system. *J. Phys. Conf. Ser.* **2022**, *2351*, 012023. [CrossRef]
10. Deng, Z.H.; Huang, B. Study on ultimate bearing capacity of main member in transmission tubular tower leg. *Thin-Walled Struct.* **2018**, *127*, 51–61. [CrossRef]
11. Li, R.; Qi, L.; Dong, Y.-R.; Wang, H. Nonlinear Performance of Steel Tube Tower in Ultra-High Voltage Transmission Lines under Wind Loads. *Buildings* **2024**, *14*, 140. [CrossRef]
12. Kim, P.; Han, W.S.; Kim, J.H.; Lee, J.; Kang, Y.J.; Kim, S. Analytical Investigation of the Effects of Secondary Structural Members on the Structural Behaviors of Transmission Towers. *Buildings* **2023**, *13*, 223. [CrossRef]
13. *DL/T 5154-2012*; Technical Code for the Design of Tower and Pole Structures of Overhead Transmission Line. China Planning Press: Beijing, China, 2012. (In Chinese)

14. *GB50017-2017*; Steel Structure Design Standards. China Construction Industry Press: Beijing, China, 2017. (In Chinese)
15. Fan, G. Research on the Stability Capacity of the Equal Angle Steel Main Member in Transmission Tower Considering Middl Disconnected Joint. Master's Thesis, Chongqing University, Chongqing, China, 2021. (In Chinese)
16. Xue, S.; Wang, L.; Li, X.; Zhou, Y. Finite element analysis of axial pressure stability bearing capacity of angle steel such a transmission tower. *Build. Struct.* **2023**, *53* (Suppl. 1), 1431–1435. (In Chinese)
17. Ma, L.; Bocchini, P. Hysteretic Model of Single-Bolted Angle Connections for Lattice Steel Towers. *J. Eng. Mech.* **2019**, *145*, 0401905 [CrossRef]
18. Mahmoudi, M.; Kosari, M.; Lorestani, M.; Abad, M.J. Effect of contact surface type on the slip resistance in bolted connections. *Constr. Steel Res.* **2020**, *166*, 105943. [CrossRef]
19. Poovakaud, V.V.; Jiménez-Peña, C.; Talemi, R.; Coppieters, S.; Debruyne, D. Assessment of fretting fatigue in high strength stee bolted connections with simplified Fe modelling techniques. *Tribol. Int.* **2020**, *143*, 106083. [CrossRef]
20. Abdelrahman, A.H.A.; Liu, Y.; Chan, S. Advanced joint slip model for single-angle bolted connections considering various effect *Adv. Struct. Eng.* **2020**, *23*, 2121–2135. [CrossRef]
21. de Souza, R.R.; Miguel, L.F.; Kaminski, J., Jr.; Lopez, R.H.; Torii, A.J. Optimization of transmission towers considering the bol slippage effect. *Eng. Struct.* **2020**, *211*, 110436. [CrossRef]
22. Abad, J.; Franco, J.M.; Celorrio, R.; Lezáun, L. Design of experiments and energy dissipation analysis for a contact mechanics 3D model of frictional bolted lap joints. *Adv. Eng. Softw.* **2012**, *45*, 42–53. [CrossRef]
23. Huang, Z.; Liu, H.; Zhang, J.; Li, Z.; Ohsaki, M.; Bai, Q. Experimental and numerical study on the effective length of tower cros bracing. *Thin-Walled Struct.* **2023**, *182*, 110296. [CrossRef]
24. Li, J.; Li, Z.; Huang, X.; Liu, H.; Wang, W. Study on the influence of restraint stiffness of main material on the compression stabilit bearing capacity of cross-bracing in transmission tower. *J. Phys. Conf. Ser.* **2022**, *2260*, 012004. [CrossRef]
25. Tang, Z.; Li, Z.; Wang, T. Probabilistic bearing capacity assessment for unequal-leg angle cross-bracings in transmission towers. *Constr. Steel Res.* **2023**, *200*, 107672. [CrossRef]
26. Li, Y.; Yan, Z.; Zhou, D.; Ren, Q.; Hua, C.; Zhong, Y. Analysis of the stability behavior of cross bracings in transmission tower based on experiments and numerical simulations. *Thin-Walled Struct.* **2023**, *185*, 110554. [CrossRef]
27. Li, Y. Theoretical and Experimental Study on Bearing Capacity of Unequal Angle Steel in Transmission Tower. Ph.D. Thesis Chongqing University, Chongqing, China, 2017. (In Chinese)

buildings

MDPI

Article

Experimental Study on the Influence of Sidewall Excavation Width and Rock Wall Slope on the Stability of the Surrounding Rock in Hanging Tunnels

Hao Zhang [1], Tianyu Luo [2], Zhao Xiang [1,3], Zhiwei Cai [1,*], Tongqing Wu [4,*], Dong Zhang [5], Bing Liu [2] and Hu Feng [6]

1 CCTEG Chongqing Engineering (GROUP) Co., Ltd., Chongqing 400016, China
2 School of Architecture and Luqiao, Chongqing Vocational College of Transportation, Chongqing 402260, China
3 Guang'an Vocational & Technical College, Guang'an 638000, China
4 School of Civil and Hydraulic Engineering, Chongqing University of Science and Technology, Chongqing 401331, China
5 Qinyuan Technology Co., Ltd., Chengdu 610213, China
6 Chongqing Mas Sci. & Tech. Co., Ltd., Chongqing 400052, China
* Correspondence: caizhiwei@cqmsy.com (Z.C.); wutongq@cqust.edu.cn (T.W.)

Abstract: Hanging tunnels are a unique type of highway constructed on hard cliffs and towering mountains, renowned for their steep and distinctive characteristics. Compared to traditional full tunnels or open excavations, hanging tunnels offer significant advantages in terms of cost and construction time. However, the engineering design and construction cases of such tunnels are rarely reported, and concerns about construction safety and surrounding rock stability have become focal points. Taking the Shibanhe hanging tunnel as a case study, this paper focuses on the stability of the surrounding rock during the excavation of limestone hanging tunnels using physical analog model (PAM) experiments and numerical calculation. Firstly, based on the similarity principle and orthogonal experiments, river sand, bentonite, gypsum and P.O42.5 ordinary Portland cement were selected as the raw materials to configure similar materials from limestone. Secondly, according to the characteristics of hanging tunnels, geological models were designed, and excavation experiments with three different sidewall excavation widths and rock wall slopes were carried out. The effects of these variables on the stress and displacement behavior of the surrounding rock were analyzed, and the laws of their influence on the stability of the surrounding rock were explored. Finally, numerical simulations were employed to simulate the tunnel excavation, and the results of the numerical simulations and PAM experiments were compared and analyzed to verify the reliability of the PAM experiment. The results showed that the vertical stress on the rock pillars was significantly affected by the sidewall excavation widths, with a maximum increase rate of 53.8%. The displacement of the sidewall opening top was greatly influenced by the sidewall excavation widths, while the displacement of the sidewalls was more influenced by the rock wall slope. The experimental results of the PAM are consistent with the displacement and stress trends observed in the numerical simulation results, verifying their reliability. These findings can provide valuable guidance and reference for the design and construction of hanging tunnels.

Keywords: hanging tunnel; physical analog model; tunnel excavation; rock wall slope; sidewall excavation width

1. Introduction

In recent years, China has undergone a period of accelerating highway, railway, and urban rail transit construction and development due to the implementation of the rural revitalization strategy, in particular in mountainous areas, which required building numerous tunnels. In 2022, China's rural highway mileage exceeded 4.53 million km with prefectural, township, and village mileages of 0.699, 1.243 and 2.589 million km, respectively. By the end of 2022, China had built 24,850 highway tunnels, exhibiting a 6% increase compared with 2021 [1]. In remote areas in southwest China, highways are usually built near steep cliffs, which requires the additional construction of bridges, tunnels, or high slopes. In these cases, highway construction is complicated by the steepness and instability of slopes. In addition to high costs and severe environmental damage, large-scale tunnel excavation can trigger slope instability. Hanging tunnels offer a good solution to the above problems. As a special type of road, hanging tunnels are carved out in hard cliffs and high steep mountains.

Compared with conventional tunnels or open excavation, hanging tunnels are dug out by making openings in the sidewalls of the cliffs. This method solves various problems in tunnel construction, such as ventilation, illumination, and waste slag at the working face. In addition, favorable ventilation and illumination conditions are generally ensured in hanging tunnels, reducing energy consumption during tunnel operation. Hanging tunnels are aligned with the initial terrain and merge into the mountains, preventing large-scale sloping and mitigating damage to the surrounding environment. Furthermore, hanging tunnels fully leverage the stability of the rock masses and require no special protection or support, such as linings or local reinforcement, thereby involving much lower cost and less time [2,3]. In a sense, hanging tunnels have the dual attributes of tunnels and slopes. It can be imagined that the construction safety of a hanging tunnel is directly determined by the sidewall excavation width and rock wall slope. Therefore, it is necessary to study the influence of these two factors on the stability of the surrounding rock of the hanging tunnel.

According to incomplete statistics, there are currently more than 10 hanging tunnels in China, with a total length of about 30 km, most of which were spontaneously built by villagers from the 1990s to the early 21st century. Some hanging tunnels operate normally in China's mountainous areas where limestone is predominant, such as in the Guizhou, Henan, and Shanxi Provinces, see Figure 1c–f. Hanging tunnels have also been found in the United States, Pakistan, Greece, Saudi Arabia, and the Philippines, see Figure 1a,b. These hanging tunnels provide greater travel convenience for local residents and become famous tourist attractions to boost economic growth in mountainous areas. How to ensure the safe operation of a hanging tunnel, promote the construction of hanging tunnels under appropriate and necessary conditions, and choose the most reasonable and safe construction techniques with the highest economic efficiency in specific situations has become an urgent problem.

(a)

(b)

Figure 1. *Cont.*

Figure 1. The hanging tunnels. (**a**) The half tunnel T-7 located in the Sutlej Valley of the western Himalayas, "Reprinted/adapted with permission from Ref. [2]. Copyright 2003, Anbalagan, R"; (**b**) the Ganji half tunnel located on Skardu Road near the Karakoram Highway, "Reprinted/adapted with permission from Ref. [3]. Copyright 2022, Emad, M."; (**c**–**f**) Shibanhe hanging tunnel in Guizhou Province, "Reprinted/adapted with permission from Refs. [4–6]. Copyright 2019, 2020, Xianpu Han".

In the case of hanging tunnels, the rock masses that undergo stress redistribution after excavation are considered in the surrounding rock, and their stability and safety form the foundation for the tunnel's engineering safety throughout its life cycle. The safety of the surrounding rock is an important indicator of underground engineering stability and depends on various factors, including rock mass strength, the geometric shape of the excavation, the stress induced around the tunnel entrance, construction methods, groundwater, and weathering processes [7]. The commonly used methods for stability analysis of the surrounding rocks in the tunnels include the following:

(1) Engineering analogy method [8–11]. The empirical approach, particularly the engineering analogy method, is the mainstream in surrounding rock and support structure design for tunnels. This method allows for the rapid formulation of preliminary support design schemes based on rock mass rating. However, it is impossible to obtain the stress and displacement distributions of surrounding rocks or perform a quantitative analysis of the mechanical performance of the surrounding rocks and support structures using this method.

(2) Theoretical analysis. Theoretical analysis centers on mechanical calculation, especially estimating surrounding rock stress. To be specific, the analytical approach is founded upon a series of empirical formulae derived from engineering practice [12–15] and those from mechanical theories [16,17]. These theoretical formulae are modified based on geological engineering conditions, rock mass conditions, excavation methods, and the characteristics of the tunnel itself [18–20]. The above references are all research results on the theoretical analysis of the stability of the rock surrounding the tunnel

under specific conditions. Considering that hanging tunnels have both tunnel and slope properties, it is difficult for us to directly apply them. However, the above references provide us with ideas and methods for the theoretical analysis of the stability of hanging tunnel-surrounding rock in the next step.

(3) Numerical simulation. The numerical approach usually yields intuitive results and involves the consideration of many influencing factors. Along with the rapid progress of computer technology, the finite element method [3,21–23], finite difference method [24–26], discrete element method [27–33], discontinuous deformation analysis method [34], and boundary element method [35] have been applied to predict surrounding rock stability. For example, Schlotfeldt et al. [36] and Tokashiki et al. [37] investigate the geotechnical aspects and stability of overhanging cliffs using numerical modeling. R. Anbalagan et al. [2] employed numerical analysis methods to investigate the distribution of stresses in the surrounding rock of half tunnels. Emad et al. [3] established a two-dimensional computational model of the Ganji half tunnel. They conducted a comparative analysis of the yielded volumes for tunnel heights ranging from 4.7 to 14.1 m while varying the tunnel span from 7 to 13 m, thereby exploring the stability of the surrounding rock of the half tunnel. This method only considers the effects of span size and tunnel height (i.e., the magnitude of the overlying load), and does not investigate the impact of different sidewall excavation widths and rock wall slope on the stability of the surrounding rock.

(4) Case analysis of field measurements. Various types of sensors are deployed in the construction areas to acquire data and observations, combined with the construction experience in other projects under similar conditions and the engineer's own experience to predict engineering safety [38–40]. To ensure the safety of tunnel construction, excavation methods or shield tunneling parameters can be adjusted based on the field-monitoring data. Although field tests can capture the deformation of surrounding rock and actual stress changes, installing the necessary testing components for a hanging tunnel connecting rural areas can be quite costly.

(5) Physical analog model (PAM) tests. It is still difficult to obtain field test measurements directly when performing a stability analysis for confining pressure in tunnels due to the complexity of the tunnel engineering itself. For this reason, numerical simulations are usually conducted instead. The numerical simulation has the advantages of clarified working principles, ease of use, and strong applicability. However, given the complexity of the tunnel-surrounding rocks, some mechanisms of mechanical response are as yet unclear, restricting the applicability of numerical simulation techniques. PAM tests can offer a full picture of the stress and deformation damage mechanisms, failure morphology, and instability phase under the interaction of engineering structures and rock and soil masses involved. PAM tests provide a convenient pathway to grasp mechanical properties, deformation and failure characteristics, as well as the stability of the surrounding rocks in the tunnels. Properly selecting analog materials and designing tunnel model test systems are vital for successful PAM tests. Analog materials are usually selected according to the similarity principle. For example, sand, quartz sand, and barite powder are used as aggregate [41–43], while cement, clay, and gypsum are used as cementing materials [44–46]. Various combinations of analog materials for tunnel-surrounding rocks are assessed via the orthogonal design. Next, a laboratory model test is conducted to thoroughly investigate various aspects and parameters of the tunnel, including the buried depth, weak intercalated layer, rupture failure mechanism, span, excavation method, and cross-sectional shape [47–50].

Wedge theory and numerical simulation are the mainstream methods of studying hanging tunnel stability. To clarify the effects of the opening width in the sidewalls and rock wall slope on the surrounding rock stability of a hanging tunnel, this study constructed a 1:20 scale indoor test PAM simulating the particular hanging tunnel in Shibanhe Village in Hezhang County of the Guizhou Province of China. PAM tests were conducted for different combinations of three sidewall excavation widths and three rock wall slopes, which affect the hanging tunnel stress and displacement variations, as well as the surrounding rock stability. These were experimentally estimated, providing a deeper insight into the surrounding rock stability assessment in similar tunnel engineering scenarios.

2. Engineering Background

The Hanging Highway Tunnel in Shibanhe Village, Hezhang County, Guizhou Province of China, was taken as a case study. As shown in Figure 1d–f, this hanging tunnel is located in the northern part of the Yangtze/Pre-Yangtze Platform, with parallel fold bundles as the main structure, trending NE–SW. The folds are mostly elongated in shape, with the dip angle mainly gentle to moderate. The site has a middle mountain topography, with large elevation differences ranging from 1700 to 2100 m. The area has strong tectonic and erosion effects, with river valleys mostly in canyon form and the valley slopes mainly steep cliffs. Bedrock is exposed on the cliffs, with well-developed vegetation at the top and debris accumulation at the foot of the slope. The rock types in the area are mainly Permian Maokou Formation limestone, mainly light gray dolomite limestone, dolomite limestone, and marl, with some distribution of dolomite. The rock layers are mainly thick to very thick, with intact rock mass, no developed fractures, no groundwater, high rock strength, no support measures, and good slope and tunnel rock mass self-stability.

3. Methods

3.1. Similarity Principle

3.1.1. Basic Principles of Similarity Principle

Similarity principle provides a theoretical basis for physical analog model (PAM) design and testing. Its core idea is to establish a similarity relationship between the PAM and prototype via various similarity criteria concerning their geometry, dimensions, physical quantities, etc., so that the PAM test would reflect the objective laws of the prototype. The ratio of the same dimensional physical quantities in the prototype (P) and the model (M) is called the similarity scale, denoted by C. Geometric similarity is a prerequisite for other phenomena to be similar, and based on geometric similarity, other physical quantities can be derived through dimensional analysis, combined with geometric equations, physical equations, boundary conditions, etc.

According to the physics equation, the similarity relationship between displacement (C_δ), geometry (C_L), and strain (C_ε) similarity coefficients is as follows [44,46,50]:

$$C_\delta = C_L C_\varepsilon \tag{1}$$

Based on geometric and physical equations, as well as stress and displacement boundary conditions, the following similarity relationships can be derived [47,48]:

$$C_\sigma = C_\gamma C_L \tag{2}$$

$$C_\sigma = C_\varepsilon C_E \tag{3}$$

where C_σ and C_γ are stress and bulk density similarity coefficients, respectively.

According to dimensional analysis, if the similar scales of physical quantities with the same dimensions are equal, then the similarity ratio of dimensionless physical quantities is 1.

3.1.2. Determine the Similarity Coefficient

Hereinafter, the following designations are used: L is length, ρ is density, u is displacement, E is elastic modulus, σ is stress, σ_c is compressive strength, σ_t is tensile strength, ε is strain, f is internal friction coefficient, c is cohesive force, φ is internal friction angle, μ is Poisson's ratio, and k is permeability coefficient.

The respective similarity coefficients are as follows: C_μ is the Poisson's ratio similarity scale, C_ε is the strain similarity scale, C_f is the friction coefficient similarity scale, C_c is the cohesion similarity scale, C_φ is the internal friction angle similarity scale, C_σ is the stress similarity scale, $C_{\sigma c}$ is the compressive strength similarity scale, $C_{\sigma t}$ is the tensile strength similarity scale, C_E is the elastic modulus similarity scale, C_L is the geometric similarity scale, and C_ρ is the density similarity scale.

According to similarity theory and the actual engineering background, the related design principles of the model hanging tunnels are presented in Table 1. In this test, based on the geometric dimensions of the prototype tunnel, as shown in Figure 3, and taking into account experimental conditions, testing costs, etc., the geometry ratio was determined to be $C_L = 20$. The other ratios can be obtained by solving Equations (1)–(3), and the results are shown in Table 1 [41,46,50].

Table 1. Similarity coefficients of key physical quantities of physical analog model (PAM) materials.

Key Physical Quantities	Dimension	Similarity Relation	Similarity Coefficient (Control Amount *)
Density (ρ)	ML^{-3}	C_ρ	1 *
Elastic modulus (E)	$ML^{-1}T^{-2}$	C_E	20 *
Poisson's ratio (μ)	Dimensionless	C_μ	1
Cohesion (c)	$ML^{-1}T^{-2}$	$C_c = C_E C_\varepsilon$	20
Internal friction angle (φ)	Dimensionless	C_φ	1
Stress (σ)	$ML^{-1}T^{-2}$	$C_\sigma = C_E C_\varepsilon$	20 *
Strain (ε)	Dimensionless	$C_\varepsilon = C_\rho C_g C_L C_E{}^{-1}$	1
Displacement (u)	L	C_L	20 *

Note: Parameters with "*" represent the key control quantity in each physical parameter, which strongly affect other parameters.

3.2. Similar Materials of PAM

3.2.1. Selection of PAM Materials

The similarity of model materials is an important prerequisite to ensure the accuracy of physical simulation results, and it significantly impacts the physical and mechanical properties of similar models. It is unrealistic to expect all physical quantities of similar materials to remain highly similar to the prototype. This article mainly studies the stress and displacement field changes of the surrounding rock in hanging tunnels under different widths of sidewall openings and slopes of rock walls. Therefore, it is only necessary to control the similarity of the main physical indicators of similar materials. Based on a large number of research results [41–46] and combined with engineering practice, orthogonal experiments were conducted to determine the optimal mix ratio using fine river sand as aggregate, gypsum powder, bentonite, and P.O42.5 ordinary Portland cement as binders. The raw materials are shown in Figure 2a.

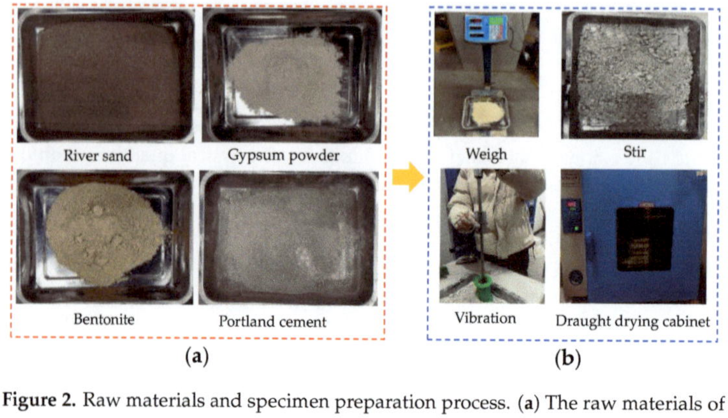

Figure 2. Raw materials and specimen preparation process. (**a**) The raw materials of surrounding rock; (**b**) steps in specimen preparation and maintenance.

3.2.2. Preparation of Samples

PAM materials were mixed with a water-to-cement ratio of 1:7 and a sand-to-cement ratio of 3:1; mix ratios of bentonite, gypsum powder, and cement in the binding materials were 2:7:1, 3:6:1, 4:5:1, 4:4:2, 5:3:2, and 5:4:1, respectively. Comprehensive PAM tests were conducted on samples with the above ratios. The raw materials and preparation steps of the specimens are as shown in Figure 2.

3.2.3. Testing of Basic Mechanical Parameters

Physical and mechanical parameters of the prepared PAM samples were determined via orthogonal tests, as shown in Table 2. Similarity coefficients obtained for various mix ratios are listed in Table 3. The model-to-prototype similarity degrees of groups with various mix ratios are summarized in Table 4. According to field investigation and laboratory mechanics experiment, the physical and mechanical parameters of the surrounding rocks are shown in Table 5. The similarity coefficients of elastic modulus, cohesion, and compressive strength are close to 20, and the similarity coefficients of weight, Poisson's ratio, and internal friction angle are close to 1. Therefore, the mix ratio of 451 sets between bentonite, gypsum powder, and ordinary Portland cement was selected as the optimal mix ratio.

Table 2. Physical and mechanical parameters of sample groups with various mix ratios.

Group	ρ/g/cm^3	σ_t/kPa	σ_c/kPa	μ	E/MPa	c/kPa	φ/°
271	1.96	185	1750	0.16	345	121.3	29
361	1.96	135	1417	0.16	294	125.4	32
451	1.93	126	1240	0.17	230	123.8	31
442	1.96	120	1182	0.17	195	121.8	33
532	1.93	119	1176	0.18	170	122.6	27
541	1.96	95	1257	0.17	152	124.6	32

Table 3. Similarity coefficients of sample groups with various mix ratios.

Group	C_γ	$C_{\sigma t}$	$C_{\sigma c}$	C_μ	C_E	C_c	C_φ
271	1.20	13.78	14.29	1.56	14.49	5.36	1.03
361	1.20	18.89	17.64	1.56	17.01	5.18	0.94
451	1.19	20.24	20.16	1.47	21.74	5.25	0.97
442	1.20	21.25	21.15	1.47	25.64	5.34	0.91
532	1.22	21.43	21.26	1.39	29.41	5.30	1.11
541	1.20	26.84	19.89	1.47	32.89	5.22	0.94

Table 4. Model-to-prototype similarity of each group.

Group	271	361	451	442	532	541
$C_\sigma/(C_L C_\gamma)$	0.60	0.74	0.85	0.88	0.87	0.83
C_E/C_σ	1.01	0.96	1.08	1.21	1.38	1.65

Table 5. Physical and mechanical parameters of the surrounding rocks.

Type	Young's Modulus E/MPa	Bulk Density γ/kN/m^3	Poisson's Ratio μ	Internal Friction Angle φ/°	Cohesion Force c/kPa	Compressive Strength σ_c/kPa	Tensile Strength σ_t/kPa
Model	230	19.26	0.17	31	123.8	1240	126
Prototype	5000	23.5	0.25	30	650	25,000	2550

3.3. Design and Testing of the Hanging Tunnel

3.3.1. Model Test Platform

The hanging tunnel under study mainly comprised low-grade mountain roads, with lower road grades and mostly two-way two-lane roads. Taking the actual situation of the Shibanhe hanging tunnel as the background, referring to the "Specifications for Design of Highway Tunnels Section 1 Civil Engineering" (JTG 3370.1-2018) [15], the hanging tunnel was designed as a circular arch structure with a radius of 4.4 m. The tunnel is 6.55 m high and 8.8 m wide, as shown in Figure 3a. The size of the sidewall openings was derived based on numerous engineering precedents with mostly irregular rectangular shapes, with a height of 55 cm below the tunnel vault, as shown in Figure 3b. The effect of the width of sidewall openings on the stability of the surrounding rock was analyzed; therefore, the height remains unchanged, while three widths of 500, 1000, and 4000 cm were used, as shown in Figure 3b. The tunnel section and the size of the sidewall openings are shown in Figure 3.

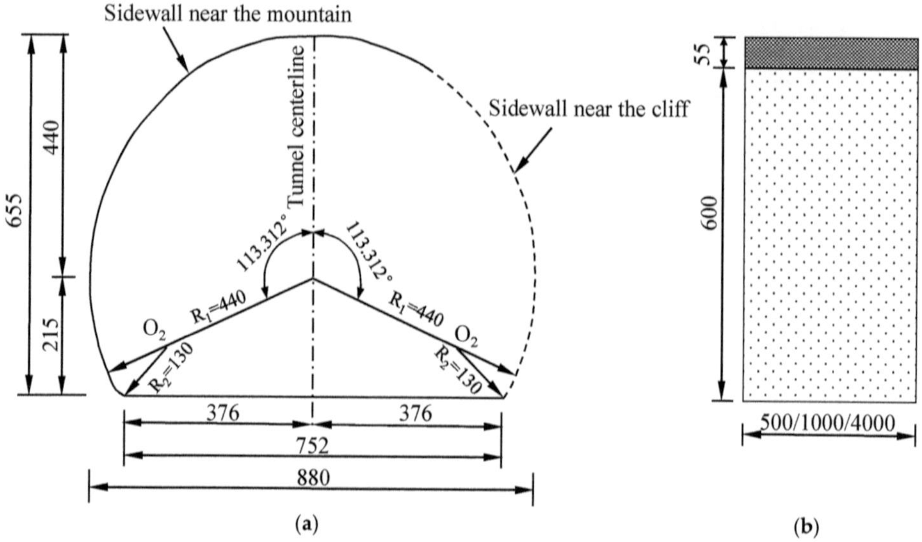

Figure 3. Tunnel cross-section and sidewall opening dimensions of the prototype hanging tunnel (unit: cm). (**a**) The sectional dimensions of the prototype hanging tunnel; (**b**) the dimensions of the sidewall openings.

Based on the geometric dimensions of the prototype tunnel, and taking into account experimental conditions, testing costs, etc., the geometric similarity ratio of the selected model was 1:20. Considering the tunnel's maximum burial depth, excavation width, length, etc., the height, width, and thickness of the PAM box were as follows: 2 m × 2 m × 2 m. Its main frame was welded from square steel, and 4 mm thick steel plates were used to enclosure the four sides. The entire model box was subdivided into faces 1–5, with face 1 and face 3 facing each other, where face 1 was movable to simulate differently inclined overburden faces in the test; face 2 and face 4 were opposite, representing tunnel excavation faces, as shown in Figure 4.

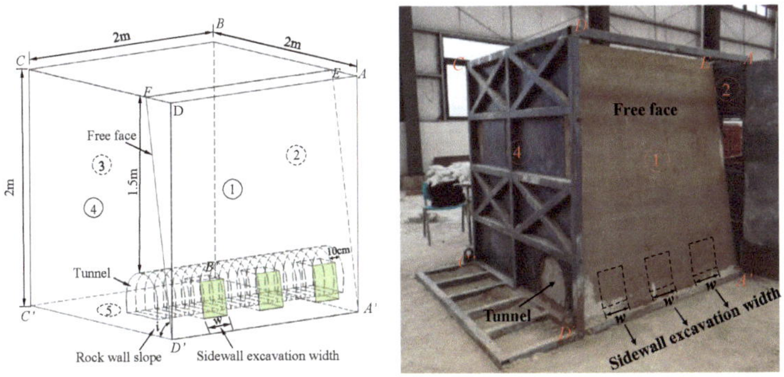

(a) (b)

Figure 4. Model test platform. (**a**) Schematic diagram of hanging tunnel model; (**b**) the PAM of hanging tunnel.

3.3.2. PAM Test Monitoring Scheme

The main purpose of this experiment was to monitor variations in internal rock stresses and displacements during the excavation process. In PAM experiments, the stability of tunnel-surrounding rock was analyzed by obtaining the stress and displacement of the surrounding rock. By burying soil pressure gauges, the stress changes in the surrounding rock at the monitoring points could be obtained, and by burying displacement sensors, the displacement changes in the monitoring points during excavation could be obtained. The measurement system included a miniature DMTY soil pressure gauge (diameter of 10 mm, thickness of 10 mm, and range of 0–100 kPa), a DMTY-type full-bridge embedded single-point displacement meter (with a 0.003 mm accuracy), and a DM-YB1820 dynamic and static strain measurement system (Nanjing Danmo Electronic Technology Co., Ltd., Nanjing, China), as shown in Figure 5.

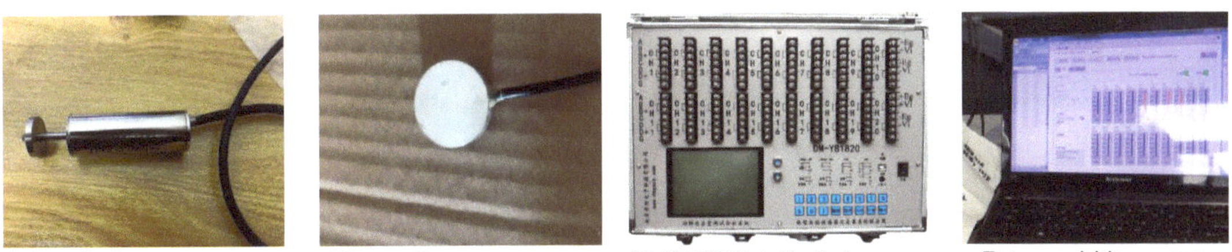

Micro displacement sensors Micro soil pressure gauges DM-YB1820 static strain gauge Data acquisition system

Figure 5. PAM test measuring equipment.

In this study, one stress monitoring section was set at the rock pillar near the longitudinal middle of the tunnel; five soil pressure gauges were set at the tunnel sidewall near the mountain (1), the tunnel vault (2), the tunnel sidewall near the cliff (3), the tunnel floor (4) and the rock pillar (5). The soil pressure gauge on the side wall was vertically arranged to obtain its radial stress value, while other positions were horizontally arranged, as shown in Figure 6a. One displacement monitoring section was set along the longitudinal middle of the tunnel; four displacement meters were set at the tunnel sidewall near the mountain (1), the tunnel hance near the mountain (2), the tunnel vault (3) and the tunnel sidewall opening top (4), to collect radial displacement, as shown in Figure 6c. Layout diagram of measuring test components is shown in Figure 6.

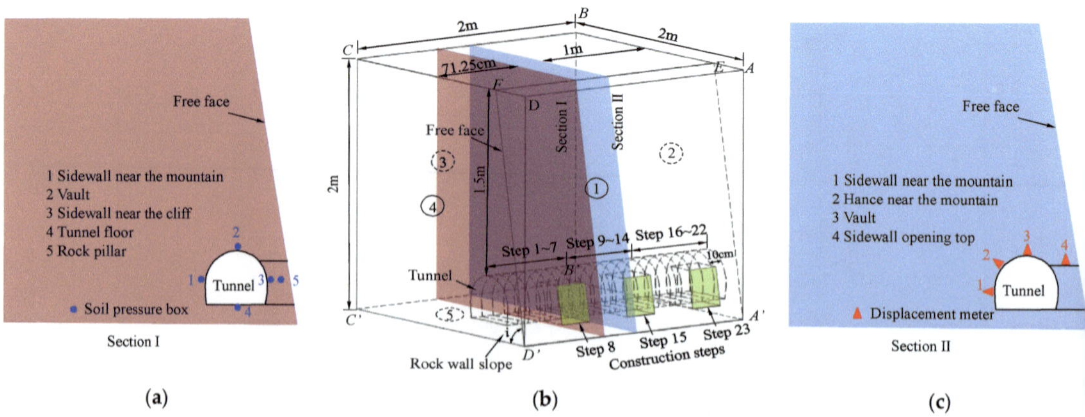

Figure 6. Layout diagram of measuring test components. (**a**) Schematic diagram of stress monitoring layout section; (**b**) schematic diagram of monitoring layout section; (**c**) schematic diagram of displacement monitoring layout section.

3.3.3. PAM Test Program

As already mentioned, the geometric scale of this indoor large-scale physical model test was 1:20. Based on the actual engineering situation and considering the cost issue of model tests, the working conditions of the model tests were simplified. Three sidewall widths and three rock wall slopes were used to provide nine working conditions for the tunnel-surrounding rock-stability-model test. The study investigated the effects of different rock wall slopes and sidewall excavation widths on the stability of hanging tunnel-surrounding rock. The test conditions are presented in Table 6.

Table 6. Test conditions.

Working Condition	1	2	3	4	5	6	7	8	9
Rock wall slope/°		80			85			90	
Sidewall excavation width/cm	25	50	200	25	50	200	25	50	200

3.3.4. PAM Elaboration Steps

Using the optimal mix ratio in the PAM test, the calculated amounts of filling materials were evenly mixed via a vertical mixer. Since the excavation process and surface of PAM did not reflect those of the actual hanging tunnel, 10 cm thick tunnel blocks made of steel plates were prefabricated according to the excavation tunnel profile via the layered filling method. To ensure smoothness and compaction of each layer, a 5 cm layer was filled each time, first leveled with wooden boards, then compacted with a self-made compactor. After filling to the tunnel position, the prefabricated tunnel blocks were placed in the predetermined position, filled as tightly as possible. After filling at the specified position, sensors were inserted/buried, recording the instrument number and position for later data collection. After filling was completed, the PAM top and open face were covered with plastic films for curing, and excavation was carried out one week later. This process is illustrated in Figure 7.

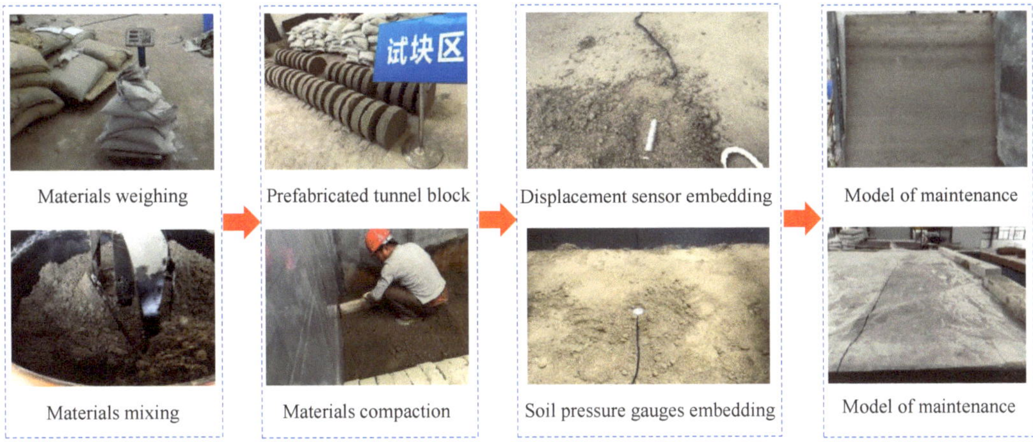

Figure 7. PAM elaboration scheme.

3.3.5. Model Test Steps

The PAM longitudinal length was 2 m, simulating full-section excavation tunneling. Each excavation (i.e., prefabricated tunnel block thickness) was 10 cm at a time. When reaching the sidewall opening, its excavation was carried out and repeated until completion of the hanging tunnel excavation process. After excavation, no support was provided in the physical model. Micro soil pressure gauges and displacement meters embedded in the model were used for stress and displacement monitoring, respectively. Before excavation, the sensor connector was wired to the stress–strain data acquisition instrument to collect the initial values. Data collection was immediately carried out after each excavation step to monitor variations in the surrounding rock stresses and strains during the excavation process. The PAM test steps and excavation process are depicted in Figure 8.

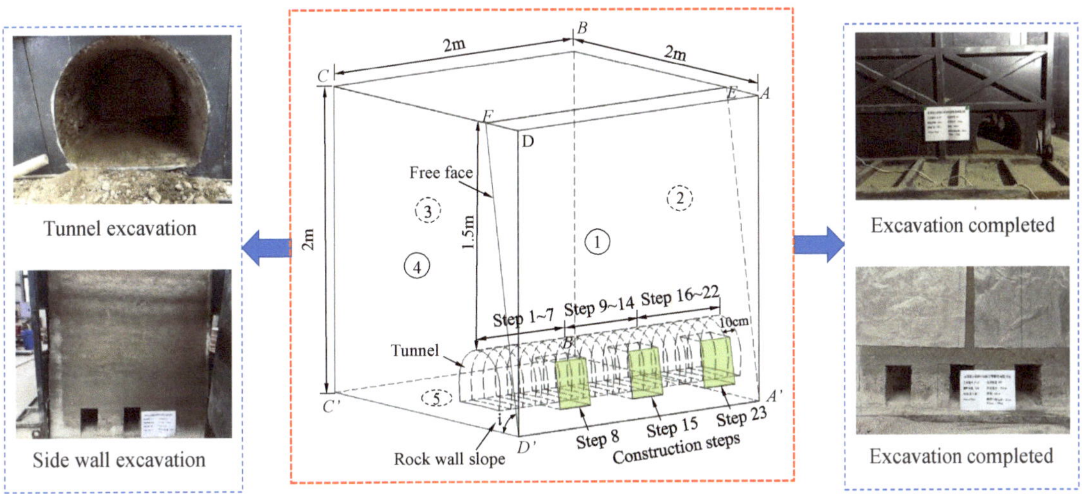

Figure 8. PAM test steps and excavation process.

4. Experimental Results and Analysis

4.1. Surrounding Rock Stress Distribution

4.1.1. Tests with Varying Sidewall Excavation Widths

The experimental assessment of the surrounding rock pressure used relative pressure values, i.e., pressure fluctuations after tunnel excavation, with positive values indicating a decrease in pressure and negative values indicating an increase in pressure. Figure 9 shows the excavation of different sidewall widths (25, 50, and 200 cm) under a rock wall slope of 80°, the stress curves of the sidewalls near the mountain, vaults, tunnel floors, sidewalls near the cliff, and rock pillars of the hanging tunnel model.

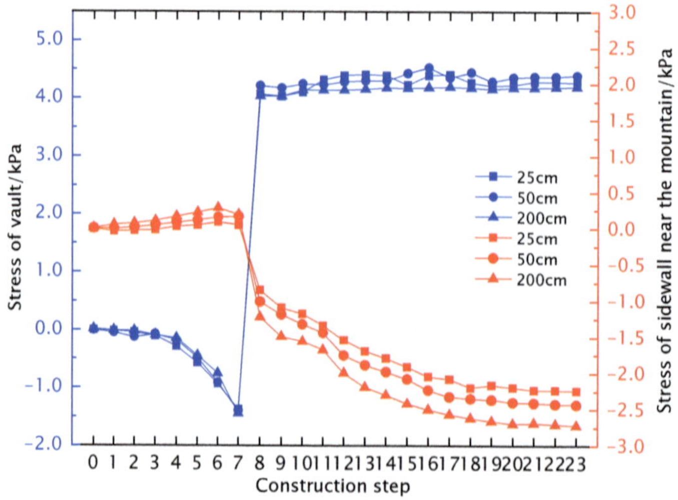

(a)

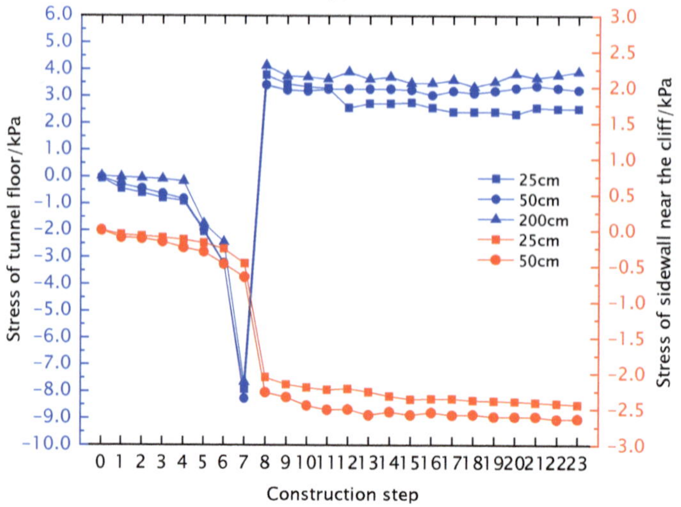

(b)

Figure 9. *Cont.*

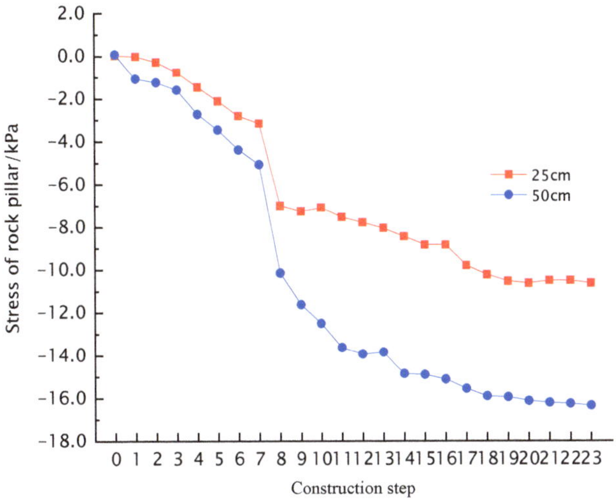

(c)

Figure 9. Stress variation curves of surrounding rock under different sidewall excavation widths and different construction steps at a rock wall slope of 80°. (**a**) Stress of the vault and sidewall near the mountain; (**b**) stress of tunnel floor and sidewall near the cliff; (**c**) stress of rock pillar.

As shown in Figure 9, the radial compressive stress of the vault gradually increased with the advancement of the excavation, with a maximum increase of 1.5 kPa. After excavation of step 7 in the monitoring section, the compressive stress decreased by 5.5 kPa due to load release and stabilization. The difference in stress changes under various conditions was small, indicating that the width of the opening on the sidewall had little effect on the radial compressive stress of the vault. The trend in radial compressive stress on the tunnel floor was similar to that in the vault, with a maximum increase of 8 kPa, and unloading during excavation reduced the compressive stress by 12 kPa. The radial compressive stress on both sidewalls significantly increased after excavation to the monitoring section. On the sidewall near the mountain, at w = 25 cm, the radial compressive stress increased by 2.25 kPa, and at w = 200 cm, it increased by 2.61 kPa, increasing by 16% compared to the former. On the sidewall near the cliff, at w = 25 cm, the radial compressive stress grew by 2.4 kPa, and at w = 50 cm, it increased by 2.63 kPa, i.e., by 9.6%. Due to the sidewall excavation, its stress could not be measured at w = 200 cm. The tunnel sidewall near the cliff was closer to the free surface, and its stress variation exceeded that of the sidewall near the mountain. The vertical compressive stress of the rock pillar was most affected by the width w of the sidewall opening. The compressive stress of the rock pillar at w = 25 cm increased by 10.6 kPa, while at w = 50 cm, it increased by 16.3 kPa, i.e., by 53.8% compared to the former. This indicates that the vertical compressive stress of the rock pillar was sensitive to the sidewall excavation width.

4.1.2. Tests with Varying Rock Wall Slopes

Stress analysis revealed the stress changes at each monitoring point during the excavation process compared with the initial state before excavation, as shown in Figure 10.

As shown in Figure 10, the tunnel vault was affected by excavation disturbance, and the radial stress increased as the excavation progressed. After excavation reached the monitoring section, the unloading stress dropped sharply, and the stress stabilized. This trend was consistent under slopes i of 80°, 85°, and 90°, with maximum changes of 5.8, 6.0, and 6.3 kPa, respectively; the latter two exceeding the former by 3.45% and 5%. The tunnel floor stress trend was consistent with that of the tunnel vault stress, but the change was larger, with maximum changes of 12, 12.4, and 13 kPa, respectively; the latter two exceeding the former by 3.3% and 4.8%. The radial stress of the sidewall near the mountain decreased first and then increased as excavation progressed, with incremental changes of 2.25, 2.3, and 2.38 kPa under various working conditions, with increments of 2.2% and 3.4%. The stress increments of the tunnel sidewall near the cliff were 2.42, 2.52, and 2.67 kPa, with increments of 4.1% and 5.6%, with a larger change than the tunnel sidewall near the cliff due to the existence of side wall openings and free surfaces; the pressure on both sides of the tunnel is asymmetric. The rock pillar was the main load-bearing element of the open-side rock mass, with a relatively large vertical stress change compared to other monitoring points, with changes of 10.6, 11.1, and 11.7 kPa under the varying rock wall slopes, with increments of 4.7% and 5.4%. The stress variations at all positions were about 4%, indicating a slight effect of the rock wall slope variation on the tunnel's stress states.

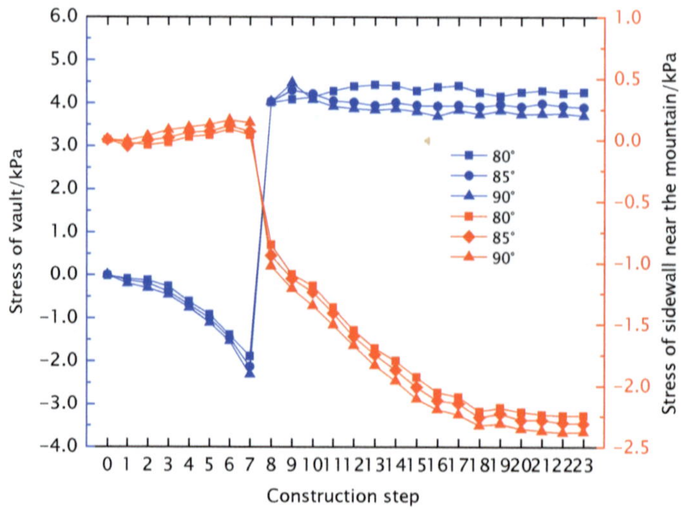

(**a**)

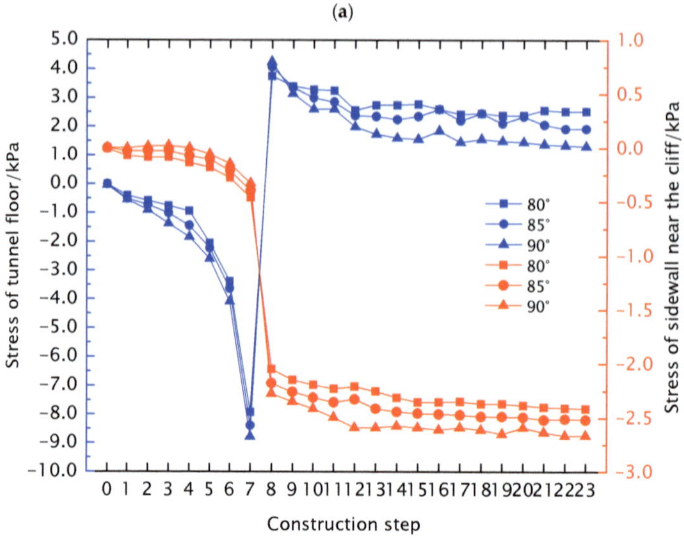

(**b**)

Figure 10. *Cont.*

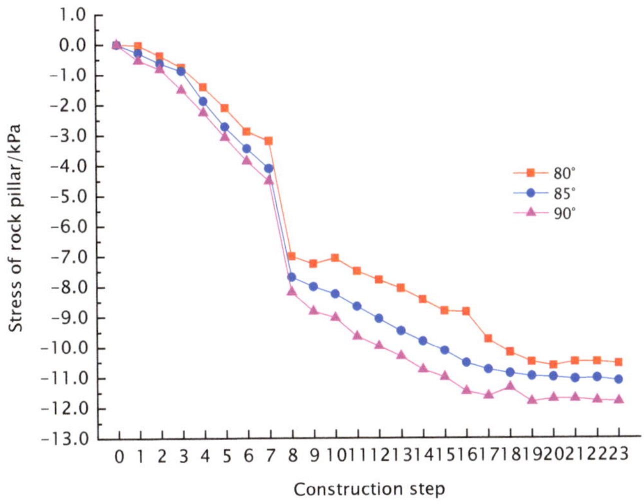

(c)

Figure 10. Stress variation curves of surrounding rock under rock wall slopes and different construction steps. (**a**) Stress of the vault and sidewall near the mountain; (**b**) stress of the tunnel floor and sidewall near the cliff; (**c**) stress of the rock pillar.

4.2. Surrounding Rock Displacement Distribution

4.2.1. Tests with Varying Sidewall Excavation Widths

During the excavation process of the tunnel, data collection was carried out through displacement meters embedded in the box body, monitoring the displacement evolution in the surrounding rock. Under different sidewall excavation widths w, the surrounding rock displacements were measured at four points, corresponding to the vault, the sidewall near the mountain, the sidewall opening top, and the hance near the mountain. Displacement probe elongation and contraction increased or decreased the initial micro-strain values, respectively. Therefore, positive values appear on the vault and sidewall opening tops, and negative values appear on the sidewalls and hances near the mountain. The radial displacement variation diagram of the surrounding rock under sidewall excavation widths is shown in Figure 11.

As shown in Figure 11, the displacement of each monitoring point positively correlated with the sidewall excavation width, and the variation was greater within a range of 1.5 times the diameter of the monitoring section. At the sidewall excavation widths w = 25 cm, 50 cm, and 200 cm, the displacements of the vault were 0.14, 0.17, and 0.21 mm, respectively; the latter two exceeding the former by 21.4% and 23.5%. The settlement of the sidewall opening top was larger than that of the vault, with displacements of 0.15, 0.2 and 0.25 mm, respectively, with increments of 33.3% and 31.6%. It can be seen that the sidewall excavation width had a greater impact on the displacement of the sidewall opening top. The displacement of the sidewall near the mountain was larger than that of the hance near the mountain, with displacement values of −0.053, −0.06 mm, and −0.072 mm, respectively, with the width of the sidewall opening with increments of 12.3% and 20.8%. The displacement values of the hance near the mountain were −0.045, −0.052, and −0.062 mm, respectively, with increments of 14.5% and 19.0%.

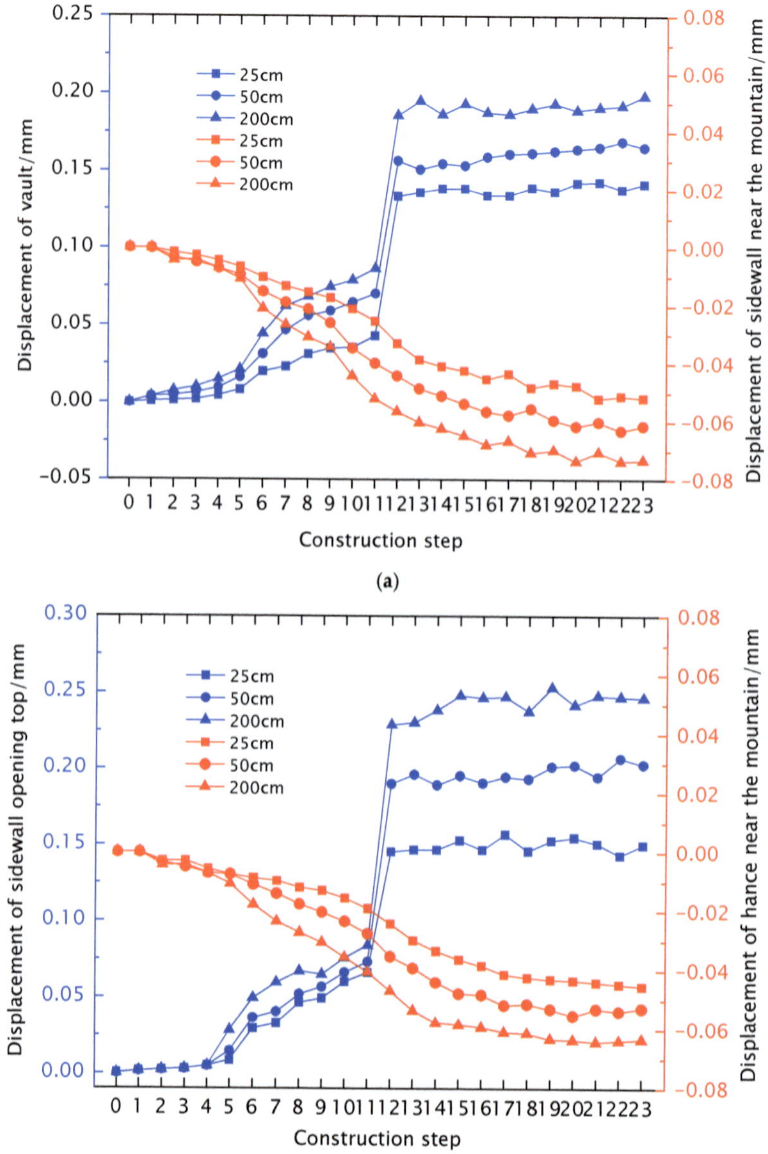

(**a**)

(**b**)

Figure 11. Radial displacement variation diagrams of the surrounding rock under sidewall excavation widths. (**a**) Displacement of the vault and sidewall near the mountain; (**b**) displacement of the sidewall opening top and hance near the mountain.

The displacement results can be ranked in decreasing order as follows: sidewall opening top > tunnel vault > sidewall near the mountain > hance near the mountain, due to the sidewall opening also forming a lateral "arch structure", similar to the deformation of the main tunnel; in addition, with its closer distance to the free face, the displacement of the sidewall opening top is more sensitive to the width of the sidewall opening compared to the tunnel.

4.2.2. Tests with Varying Rock Wall Slopes

The effect of different rock wall slopes on the displacement around the tunnel was studied to analyze the variation law, and collect the displacement changes in the vault, sidewall near the mountain, hance near the mountain, and sidewall opening top during excavation under three different rock wall slopes. The displacement curves of each measuring point during excavation are shown in Figure 12.

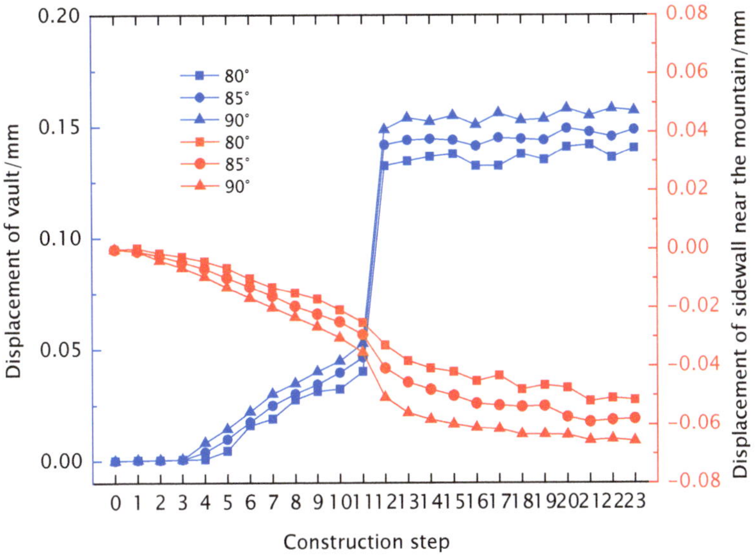

(a)

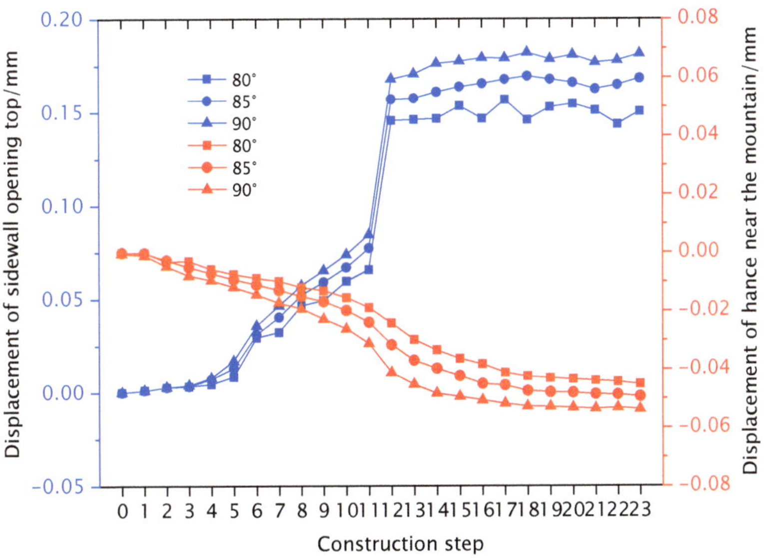

(b)

Figure 12. Radial displacement variation diagrams of the surrounding rock under different rock wall slopes. (**a**) Displacement of the vault and sidewall near the mountain; (**b**) displacement of the sidewall opening top and hance near the mountain.

As shown in Figure 12, the displacement of the tunnel vault changed significantly within 1.5 times the diameter of the monitoring section. At the rock wall slopes of $i = 80°$, 85°, and 90°, the displacements in the tunnel vault were 0.14, 0.15, and 0.16 mm, respectively, with increments of 7.1% and 6.7%. The displacement in the sidewall opening top was larger than that of the tunnel vault, reaching values of 0.152, 0.162, and 0.173 mm, with increments of 6.6% and 6.2%. Based on the relative displacement variations in the tunnel vault and sidewall opening top, it can be concluded that the change in rock wall slope had similar impacts on the two monitoring points. The radial displacement of the sidewall near the mountain increased continuously with excavation depth, with a steep increase in displacement when excavating in the monitored section, followed by a gradual decrease in the rate of change. The final displacements under each condition were −0.053, −0.058, and −0.066 mm, with increments of 9.4% and 12.1%. The displacement of the hance near the mountain was similar to that of the sidewall near the mountain, but with smaller changes in displacement, with values of −0.045, −0.049, and −0.054 mm, with increments of 8.8%, 10.2%. The rock wall slope effect on the displacement of the sidewall near the mountain was quite strong.

5. Discussions

5.1. Analysis of Surrounding Rock Displacement Law

The above PAM tests revealed that displacements of the sidewall opening top and the sidewall near the mountain were mainly affected by the sidewall excavation width and rock wall slope, respectively. To explore the law of surrounding rock displacement evolution with changes in sidewall excavation width and rock wall slope, the displacement values of the vaults and sidewall opening tops at w = 25~200 cm and those of the sidewall near the mountain at i = 80~90° were experimentally determined and converted to the prototype, according to the geometric similarity ratio of 1:20, as shown in Table 7. Due to the opening on the sidewall, the displacement data of the sidewall near the cliff could not be measured.

Table 7. Sidewall excavation width, rock wall slope, and surrounding rock displacement.

Sidewall Excavation Width/m	**5**	**10**	**40**
Displacement of sidewall opening top/mm	3.01	4.01	4.94
Displacement of vault/mm	2.83	3.31	3.97
Rock wall slope/°	**80**	**85**	**90**
Displacement of sidewall near the mountain/mm	1.06	1.16	1.32

From Table 7, it can be seen that the displacement of the tunnel vault and sidewall opening top increased with the sidewall excavation width, and that of the sidewall near the mountain increased with the rock wall slope, both exhibiting positive correlation. To quantitatively analyze the relationship between them, considering that the experiment with w = 40 m contained no sidewall, the opening width in the first few working conditions was that of a single opening. When the opening width in the sidewall condition was adjusted to w = 13.3 m for fitting, the displacement curves were constructed for the sidewall opening top under different sidewall excavation widths and for the sidewall near the mountain under different rock wall slopes, as shown in Figure 13. From Figure 13, it can be seen that the width of the sidewall opening was practically exponential with the vault displacement value, the latter sharply increasing with the former growth. The width of the sidewall opening were nearly linearly distributed with the sidewall opening top and sidewall near the mountain displacement value, the latter increasing with the former growth.

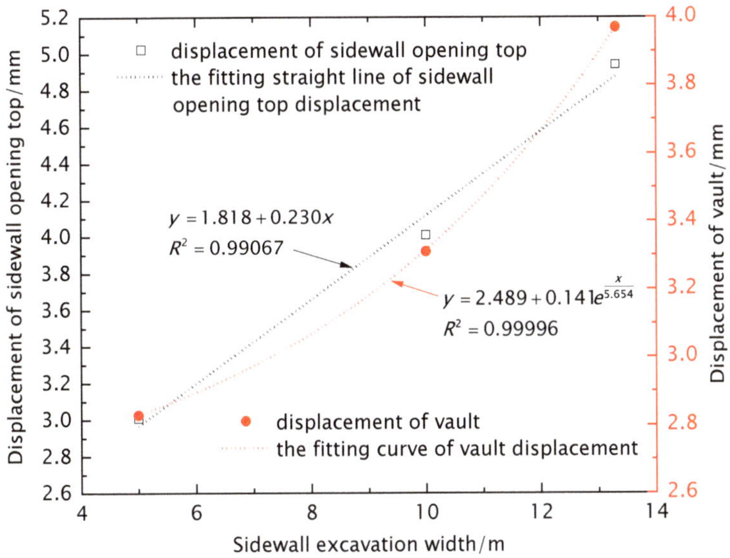

(a)

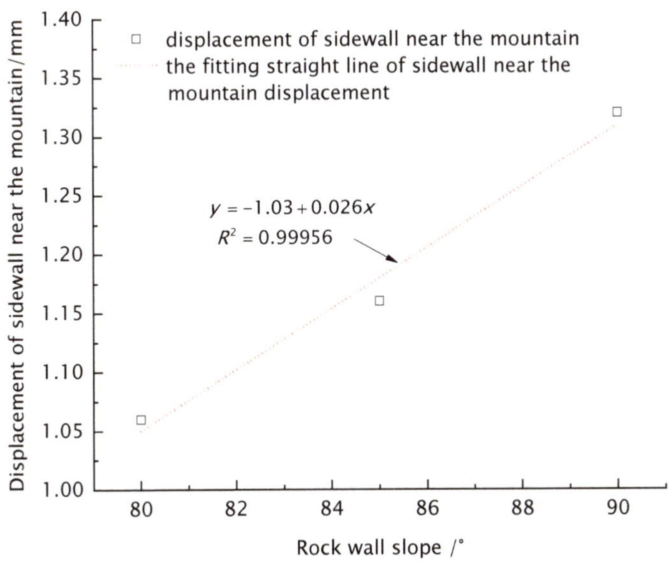

(b)

Figure 13. Relationship between surrounding rock displacement and sidewall excavation width, and rock wall slope. (**a**) Relationship diagram between the displacement of sidewall opening top, vault and the sidewall excavation width; (**b**) relationship diagram between the displacement of sidewall near mountain and rock wall slope.

5.2. Comparison of Numerical Simulation and PAM Test Results

Using the finite difference software FLAC3D 6.0 for numerical simulation calculations, the Drucker–Prager model was adopted as the constitutive model, with the physical and mechanical parameters of the surrounding rock listed in Table 5. The coordinate system implied the X-axis perpendicular to the tunnel axis, with the direction to the right as positive; the Y-axis was parallel to the tunnel axis, with the direction inward as positive; and the Z-axis coincided with the vertical direction, with upward as positive. The boundary range was as follows: in the X direction, four times the diameter of the tunnel was taken to the right from the tunnel boundary, and the left, the tunnel was taken to the rock wall with a clearance face of 3 m, which was the sidewall thickness based on the supporting engineering; in the Y direction, four times the diameter of the tunnel was taken. In the Z direction, from the upper boundary of the tunnel to the ground surface, the burial depth was 31 m, and the lower boundary was taken as four times the diameter of the tunnel, with an overall model size of 47 m (X-direction) × 35.2 m (Y-direction) × 72.75 m (Z-direction). The surrounding rock was simulated using hexahedral solid elements, with a total of 182,351 nodes and 176,710 elements in the model. Z-direction displacement constraints were applied to the bottom of the model, X-axis positive direction displacement constraints were applied to the right side of the model, Y-direction displacement constraints were applied to the front and back sides of the model, and the top surface and X-axis negative direction were free surfaces. The model was subdivided into grids before and after excavation, as shown in Figure 14.

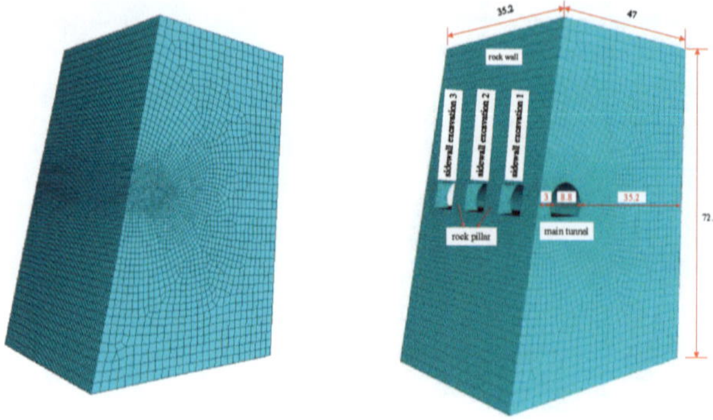

(a) (b)

Figure 14. Overall and grid map after excavation (unit: m). (**a**) The configuration of the numerical model before excavation; (**b**) the configuration of the numerical model after excavation.

A numerical simulation of hanging tunnel excavation was performed via FLAC3D 6.0, computing displacement and stress variations in the surrounding rocks for different sidewall excavation widths and rock wall slopes. When performing numerical simulation, no external load was applied and only the weight of the surrounding rock itself was considered. The numerical results were compared with the PAM test results. The displacement of the middle section of the hance near the mountain, the tunnel vault, and the sidewall opening top with a rock wall slope $i = 80°$ and a sidewall excavation width of 500 cm were selected for comparative analysis, as shown in Figure 15.

As can be seen in Figure 15, when the PAM results were converted to the prototype according to the similarity coefficients, the PAM-derived displacements in the hance near the mountain, vault, and sidewall opening top near the mountain generally exceeded the respective numerical results. It can be seen that the final displacement convergence value of the PAM is approximately twice that of the numerical simulation method. But both methods showed a sudden displacement change during excavation steps 11 to 12, and the trend in the change was consistent, indicating that the disturbance patterns during the excavation process were consistent for both methods. The observed discrepancies between numerical and PAM test results can be attributed to simplified boundary conditions, selection of analog material parameters, similarity coefficients and excavation simulation. Since PAM test and numerical curves exhibited the same trend, the numerical verification proved the proposed PAM model efficiency.

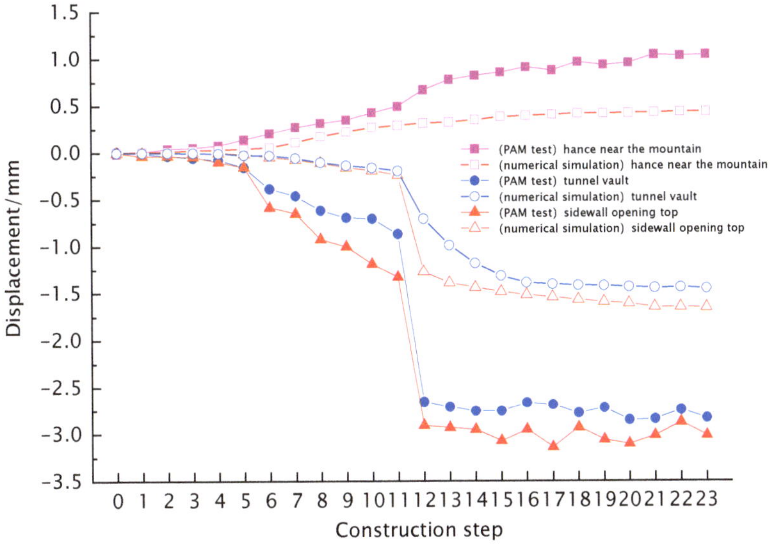

Figure 15. Comparison of displacements obtained by numerical simulation and PAM tests.

When conducting numerical simulation calculations, they are based on the actual parameters of the engineering prototype. In PAM experiments, the selection of similar materials cannot be completely consistent with the prototype, coupled with the influence of factors such as tunnel model construction and excavation process simulation, resulting in certain differences in displacement values between the two methods. However, their changing trends are consistent, achieving the purpose of mutual verification. Therefore, numerical simulation model can be used to predict and analyze the behavior of the hanging tunnels.

6. Conclusions

The stress and displacement variations during the excavation of the hanging tunnel with three different rock wall slopes (80°, 85°, 90°) and three sidewall excavation widths (25, 50, and 200 cm) were studied by conducting physical analog model (PAM) tests. The variation patterns were analyzed, and the following conclusions were drawn:

(1) The radial stress of the vault and floor of the tunnel increased gradually as the excavation face advanced. After excavation to the monitored section, the stress dropped sharply due to unloading. The radial stress on the two sidewalls changed inconsistently, and the vertical stress in the rock pillar gradually grew during the excavation process. The maximum increment in the vertical stress of the rock pillar due to the sidewall excavation width variation was 53.8%, exceeding those of other positions in the hanging tunnel; the rock wall slope variation changed stresses of the monitoring points around the tunnel only by 4%, indicating its slight effect on the hanging tunnel stressed state.

(2) The hanging tunnel excavation caused the vault to sink, the tunnel floor to rise, and the sidewall opening top to sink. Due to the lack of support near the rock wall, the sidewall and the hance near the mountain deformed toward the mountain side. The sidewall opening top displacement positively correlated with the sidewall excavation width and the rock wall slope. The former parameter had a stronger effect on the

sidewall opening top displacement, while the latter on the sidewall displacement, providing a certain reference for similar engineering projects.

(3) The surrounding rock displacement and stress variation trends obtained via PAM tests were numerically verified, yielding similar curves for various sidewall excavation widths and rock wall slopes, proving the proposed model feasibility.

(4) Although there is no on-site measurement data in this article, two methods, PAM experiments and numerical simulation, were used for mutual verification, and the results are reliable. In the future, if conditions are suitable, on-site monitoring of actual projects can be carried out to further verify the reliability of the tests.

(5) This article assumes that the surrounding rock of the tunnel is a homogeneous body, but in actual engineering, the surrounding rock is a non-homogeneous body with groundwater present, as well as faults and joints. In the future, we will conduct analysis and research on the impact of faults, joints, and other factors on the stability of tunnel-surrounding rock.

Author Contributions: Conceptualization, Z.C., T.W. and Z.X.; methodology, Z.C., H.Z. and T.W.; software, Z.C., Z.X. and D.Z.; validation, H.Z., Z.C. and T.W.; formal analysis, H.Z. and Z.C.; investigation, Z.C., T.W. and D.Z.; resources, Z.C., H.Z. and T.L.; data curation, Z.X., D.Z., B.L. and H.F.; writing—original draft preparation, T.W., Z.C., Z.X. and D.Z.; writing—review and editing, T.W. and Z.C.; visualization, T.W., Z.C. and H.F.; supervision, Z.C., T.W. and T.L.; project administration, Z.C., T.L. and B.L.; funding acquisition, H.Z. and Z.C. All authors have read and agreed to the published version of the manuscript.

Funding: This research was funded by the Scientific and Technological Research Program of Chongqing Municipal Education Commission (Grant No. KJQN202005703) and the Natural Science Foundation of Chongqing (Grant No. cstc2021jcyj-msxmX1048).

Data Availability Statement: The original contributions presented in the study are included in the article, further inquiries can be directed to the corresponding authors.

Acknowledgments: We extend our sincere gratitude to Xianpu Han of Bijie Daily in Guizhou Province for his invaluable contribution to this research by providing onsite photographs of the Shibanhe hanging tunnel.

Conflicts of Interest: Authors Hao Zhang, Zhao Xiang and Zhiwei Cai were employed by the company CCTEG Chongqing Engineering (GROUP) Co., Ltd. Author Dong Zhang was employed by the company Qinyuan Technology Co., Ltd. Author Hu Feng was employed by the company Chongqing Mas Sci. & Tech. Co., Ltd. The remaining authors declare that the research was conducted in the absence of any commercial or financial relationships that could be construed as a potential conflict of interest.

References

1. Available online: https://www.gov.cn/lianbo/bumen/202306/content_6887539.htm (accessed on 10 August 2024).
2. Anbalagan, R.; Singh, B.; Bhargava, P. Half tunnels along hill roads of Himalaya—An innovative approach. *Tunn. Undergr. Space Technol.* **2003**, *18*, 411–419. [CrossRef]
3. Emad, M.; Khan, M.; Raza, M.; Saki, S.; Aijaz, M.; Tahir, M. Optimum design of half tunnels for transportation in the Himalayas. *Transp. Infrastruct. Geotechnol.* **2022**, *9*, 101–116. [CrossRef]
4. Available online: https://h.xinhuaxmt.com/vh512/share/6244427?d=13415a0&channel (accessed on 10 August 2024).
5. Available online: https://h.xinhuaxmt.com/vh512/share/9650463?d=1343f0c&channel (accessed on 10 August 2024).
6. Available online: https://www.chinadaily.com.cn/a/202110/22/WS61721009a310cdd39bc707b2.html (accessed on 10 August 2024).
7. Hoek, E.; Brown, E.T. *Underground Excavations in Rock*; Institution of Mining and Metallurgy: London, UK, 1980.
8. Carranza-Torres, C.; Fairhurst, C. Application of the Convergence-Confinement method of tunnel design to rock masses that satisfy the Hoek-Brown failure criterion. *Tunn. Undergr. Space Technol.* **2000**, *15*, 187–213. [CrossRef]
9. Deere Don, U.; Deere Don, W. Rock quality designation (RQD) after twenty years. In *U.S. Army Corps of Engineers Contract Report GL-89-1*; Waterways Experiment Station: Vicksburg, MI, USA, 1989.
10. Hoek, E.; Diederichs, M.S. Empirical estimation of rock mass modulus. *Int. J. Rock Mech. Min.* **2006**, *43*, 203–215. [CrossRef]
11. Cai, M.F.; Kaiser, P.K.; Tasaka, Y.; Minami, M. Determination of residual strength parameters of jointed rock masses using the GSI system. *Int. J. Rock Mech. Min.* **2007**, *44*, 247–265. [CrossRef]

12. Singh, B.; Goel, R.K.; Jethwa, J.L.; Dube, A.K. Support pressure assessment in arched underground openings through poor rock masses. *Eng. Geol.* **1997**, *48*, 59–81. [CrossRef]
13. Terzaghi, K. *Theoretical Soil Mechanics*; Wiley and Sons: Hoboken, NJ, USA, 1943; pp. 195–202.
14. Barton, N.; Lien, R.; Lunde, J. Engineering classification of rock masses for the design of tunnel support. *Rock Mesh.* **1974**, *6*, 189–236. [CrossRef]
15. *JTG 3370.1-2018*; Specifications for Design of Highway Tunnels Section 1 Civil Engineering. China Merchants Chongqing Communications Technology Research and Design Institute Co., Ltd.: Beijing, China; China Communications Press: Beijing, China, 2018.
16. Fraldi, M.; Guarracino, F. Limit analysis of progressive tunnel failure of tunnels in Hoek-Brown rock masses. *Int. J. Rock Mech. Min.* **2012**, *50*, 170–173. [CrossRef]
17. Cai, W.Q.; Zhu, H.H.; Liang, W.H. Three-dimensional tunnel face extrusion and reinforcement effects of underground excavations in deep rock masses. *Int. J. Rock Mech. Min.* **2022**, *150*, 104999. [CrossRef]
18. Perazzelli, P.; Leone, T.; Anagnostou, G. Tunnel face stability under seepage flow conditions. *Tunn. Undergr. Space Technol.* **2014**, *43*, 459–469. [CrossRef]
19. Liang, Q.; Xu, J.J.; Wei, Y.G. Three-dimensional stability analysis of tunnel face based on unified strength theory. *Sci. Rep.* **2023**, *13*, 12326. [CrossRef] [PubMed]
20. Ukritchon, B.; Keawsawasvong, S. Stability of unlined square tunnels in Hoek-Brown rock masses based on lower bound analysis. *Comput. Geotech.* **2019**, *105*, 249–264. [CrossRef]
21. Wang, Y.C.; Jing, H.W.; Su, H.J.; Xie, J.Y. Effect of a fault fracture zone on the stability of tunnel-surrounding rock. *Int. J. Geomech.* **2017**, *17*, 837. [CrossRef]
22. Zhang, Z.Q.; Chen, F.F.; Li, N.; Swoboda, G.; Liu, N.F. Influence of fault on the surrounding rock stability of a tunnel: Location and thickness. *Tunn. Undergr. Space Technol.* **2017**, *61*, 1–11. [CrossRef]
23. Ding, Z.D.; Ji, X.F.; Li, X.Q.; Ren, Z.H.; Zhang, S. Influence of symmetric and asymmetric voids on mechanical behaviors of tunnel linings: Model tests and numerical simulations. *Symmetry* **2019**, *11*, 802. [CrossRef]
24. Zheng, H.B.; Li, P.F.; Ma, G.W. Stability analysis of the middle soil pillar for asymmetric parallel tunnels by using model testing and numerical simulations. *Tunn. Undergr. Space Technol.* **2020**, *108*, 103686. [CrossRef]
25. Jiang, Q.; Song, S.G.; Li, T.; Wang, K.; Gu, R.H. Study on surrounding rock stability of small clear-distance twin highway tunnel with eight lanes. *Geotech. Geol. Eng.* **2019**, *37*, 593–598. [CrossRef]
26. Zhen, C.; Li, J.H.; Fu, X.W.; Sheng, Q.; Zhou, G.X.; Ma, Y.L.N.; Wang, T.Q. Evaluating the response of a tunnel subjected to strike-slip fault rupture in conjunction with model test and hybrid discrete-continuous numerical modeling. *Rock Mech. Rock Eng.* **2022**, *55*, 4743–4764. [CrossRef]
27. He, P.; Li, S.C.; Li, L.P.; Zhang, Q.Q.; Xu, F.; Chen, Y.J. Discontinuous deformation analysis of super section tunnel surrounding rock stability based on joint distribution simulation. *Comput. Geotech.* **2017**, *91*, 218–229. [CrossRef]
28. Xiang, Y.Z.; Zeng, Z.K.; Xiang, Y.J.; Abi, E.D.; Zheng, Y.R.; Yuan, H.C. Tunnel failure mechanism during loading and unloading processes through physical model testing and DEM simulation. *Sci. Rep.* **2021**, *11*, 16753. [CrossRef]
29. Li, C.C.; Mo, Z.G.; Jiang, H.B.; Yang, F.C. Analysis of the stability and mechanical characteristics of the jointed surrounding rock and lining structure of a deeply buried hydraulic tunnel. *Fluid Dyn. Mater. Proc.* **2022**, *18*, 29–39. [CrossRef]
30. Chen, H.M.; Yu, H.S.; Smith, M.J. Physical model tests and numerical simulation for assessing the stability of brick-lined tunnels. *Tunn. Undergr. Space Technol.* **2016**, *53*, 109–119. [CrossRef]
31. Vazaios, I.; Diederichs, M.S.; Vlachopoulos, N. Assessment of strain bursting in deep tunnelling by using the finite-discrete element method. *J. Rock Mech. Geotech. Eng.* **2018**, *11*, 12–37. [CrossRef]
32. Fu, H.L.; Deng, H.S.; Zhao, Y.B.; Chang, X.B.; Yi, H.D. Study on the disturbance of existing subway tunnels by foundation sloping excavation. *Appl. Sci.* **2023**, *13*, 948. [CrossRef]
33. Wang, L.X.; Zhu, Z.D.; Zhu, S.; Wu, J.Y. A case study on tunnel excavation stability of columnar jointed rock masses with different dip angles in the Baihetan diversion tunnel. *Symmetry* **2023**, *15*, 1232. [CrossRef]
34. Ma, H.P.; Daud, N.N.N.; Yusof, Z.M.; Yaacob, W.Z.; He, H.J. Stability analysis of surrounding rock of an underground cavern group and excavation scheme optimization: Based on an optimized DDARF method. *Appl. Sci.* **2023**, *13*, 2152. [CrossRef]
35. Panji, M.; Koohsari, H.; Adampira, M.; Alielahi, H.; Asgari Marnani, J. Stability analysis of shallow tunnels subjected to eccentric loads by a boundary element method. *J. Rock Mech. Geotech. Eng.* **2016**, *8*, 480–488. [CrossRef]
36. Schlotfeldt, P.; Elmo, D.; Panton, B. Overhanging rock slope by design: An integrated approach using rock mass strength characterisation, large-scale numerical modelling and limit equilibrium methods. *J. Rock Mech. Geotech. Eng.* **2018**, *10*, 72–90. [CrossRef]
37. Tokashiki, N.; Aydan, Ö. The stability assessment of overhanging Ryuku limestone cliffs with an emphasis on the evaluation of tensile strength of rock mass. *Doboku Gakkai Ronbunshuu C* **2010**, *66*, 397–406. [CrossRef]
38. Yuan, B.X.; Liang, J.K.; Lin, H.Z.; Wang, W.Y.; Xiao, Y. Experimental Study on Influencing Factors Associated with a New Tunnel Waterproofing for Improved Impermeability. *J. Test. Eval.* **2024**, *52*, 344–363.
39. Fang, Q.A.; Zhang, D.L.; Li, Q.Q.; Wong, L.N.Y. Effects of twin tunnels construction beneath existing shield-driven twin tunnels. *Tunn. Undergr. Space Technol.* **2015**, *45*, 128–137. [CrossRef]

40. Xue, Y.D.; Zhao, F.; Zhao, H.X.; Li, X.; Diao, Z.X. A new method for selecting hard rock TBM tunnelling parameters using optimum energy: A case study. *Tunn. Undergr. Space Technol.* **2018**, *78*, 64–75. [CrossRef]
41. Tian, Q.Y.; Zhang, J.T.; Zhang, Y.L. Similar simulation experiment of expressway tunnel in karst area. *Constr. Build. Mater.* **2018**, *176*, 1–13. [CrossRef]
42. Huang, F.; Zhu, H.H.; Xu, Q.W.; Cai, Y.C.; Zhuang, X.Y. The effect of weak interlayer on the failure pattern of rock mass around tunnel–Scaled model tests and numerical analysis. *Tunn. Undergr. Space Technol.* **2013**, *35*, 207–218. [CrossRef]
43. Wu, J.; Fu, H.; Zhang, L.W.; Zhang, X.Y.; Guo, D.Y. Stability analysis of surrounding rock in underground chamber excavation of coral reef limestone. *Rock Mech. Rock Eng.* **2022**, *55*, 4717–4742. [CrossRef]
44. Xu, Z.L.; Luo, Y.B.; Chen, J.X.; Su, Z.M.; Zhu, T.T.; Yuan, J.P. Mechanical properties and reasonable proportioning of similar materials in physical model test of tunnel lining cracking. *Constr. Build. Mater.* **2021**, *300*, 123960. [CrossRef]
45. Zhao, W.S.; Gao, H.; Chen, W.Z.; Liu, J.G.; Peng, W.B.; Zhou, S. Experimental study on similar materials for tunnel lining concrete in geomechanical model tests. *Eng. Fail. Anal.* **2023**, *152*, 107456. [CrossRef]
46. Jiang, N.; Guo, P.P.; Zhang, H.; Lu, F.Q.; Liu, J. Mechanical behavior of a new similar material for weathered limestone in karst area: An experimental investigation. *Appl. Rheol.* **2023**, *33*, 20220154. [CrossRef]
47. Lei, M.F.; Peng, L.M.; Shi, C.H. Model test to investigate the failure mechanisms and lining stress characteristics of shallow buried tunnels under unsymmetrical loading. *Tunn. Undergr. Space Technol.* **2015**, *46*, 64–75. [CrossRef]
48. Lin, P.; Liu, H.Y.; Zhou, W.Y. Experimental study on failure behaviour of deep tunnels under high in-situ stresses. *Tunn. Undergr. Space Technol.* **2015**, *46*, 28–45. [CrossRef]
49. Xu, S.S.; Lei, H.; Li, C.; Liu, H.Q.; Lai, J.X.; Liu, T. Model test on mechanical characteristics of shallow tunnel excavation failure in gully topography. *Eng. Fail. Anal.* **2021**, *119*, 104978. [CrossRef]
50. Peng, L.F.; Hu, X.C.; Su, G.S.; Qin, Z.H.; Lu, H.; He, B.Y. Cracking characteristics of the surrounding rocks of a hydraulic tunnel under high geothermal conditions: A model test. *Rock Mech. Rock Eng.* **2021**, *54*, 1369–1390. [CrossRef]

buildings

MDPI

Article

Settlement of a Pile Foundation Considering Linear and Rheological Properties of Soils

Armen Z. Ter-Martirosyan *, Lyubov Yu. Ermoshina * and George O. Anzhelo

Department of Soil Mechanics and Geotechnical Engineering, National Research Moscow State Civil Engineering University, 26, Yaroslavskoye Shosse, 129337 Moscow, Russia; nocgeo@mail.ru
* Correspondence: gic-mgsu@mail.ru (A.Z.T.-M.); lyubov.ermoshina1801@mail.ru (L.Y.E.)

Abstract: Despite numerous studies of single piles and practical experience with their application, methods for calculating settlements of pile foundations remain limited. The existing objective need for specialized methods of pile foundation settlement calculation that take into account the rheological properties of the base soils is becoming more and more important, especially in the construction of unique objects in complex ground conditions. When predicting the stress–strain state of the pile–raft-surrounding soil mass system, it is allowed to consider not the entire pile foundation as a whole, but only a part of it—the computational cell. In the present work, we have solved the problems of determining the strains of the computational cell consisting of the pile, the raft and the surrounding soil according to the column pile scheme and hanging pile scheme, on the basis of the Kelvin–Voigt rheological model, which is a model of a viscoelastic body consisting of parallel connected elements: Hooke's elastic spring and Newtonian fluid. According to our results, we obtained graphs of the dependence of strains of the computational cell on time at different pile spacing and different values of coefficients of viscosity of the surrounding soil, and a formula for calculating the reduced modulus of deformation of the pile. The results of the present study can significantly improve the accuracy of calculations during construction on clayey soils with pronounced rheological properties and, as a result, increase the reliability of pile structures in general.

Keywords: pile foundation; computational cell; column pile; hanging pile; rheologic properties of soil; coefficient of viscosity; reduced modulus of deformation

1. Introduction

For many years, the pile foundation has been one of the most popular types of foundations used in practice due to its reliability, speed of construction and the possibility of construction in almost any engineering–geological and climatic conditions. At the same time, the widespread use of pile foundations is due to the increasing number of stories and height of buildings, as well as the increasing weight of technological equipment.

When an external load is applied to a pile foundation, in addition to the settlements of the piles themselves, there is settlement of the soil in the inter-pile space, causing the friction forces developed on the lateral surfaces of the piles to not fully develop, unlike in single piles. When increasing the spacing between piles to 3–4 pile diameters, the development of friction forces on the lateral surface will only be partially realized.

It should be noted that the stress–strain state of the active zone in pile foundations significantly differs from the stress–strain state of the soil mass around a single pile, as in pile foundations there is a mutual influence of piles on each other. Therefore, it is not permissible to use the results of static tests of single piles and analytical methods to determine their settlements when calculating the settlements of a pile foundation.

Despite a large number of scientific works devoted to the study of single piles (Luga A.A. [1,2], Ogranovich A.B. [3], Golubkov V.N. [4,5], etc.), and the availability of many years of experience in the use of piles in the construction of various buildings and structures, for quite a long time there was no single engineering method for calculating the settlement of pile foundations and pile groups.

In the 1960s, the method by Egorov K.E. began to be used to calculate the settlement of pile groups [6], based on the calculation of a rectangular foundation placed on a natural base. The specific feature of this calculation method was that the pile foundation was considered as a massive foundation, and the place of application of the calculated pressure on the soil under the foundation bottom was dependent on the ratio of the length of the piles to the width of the foundation. If this ratio was less than or equal to one, then the point of application of the calculated pressure was taken as the surface located at the tip of the pile. If the ratio of pile length to foundation width exceeded unity, the design pressure was distributed to the surface at a depth equal to the foundation width [7].

At present, the prediction of the rate of settlement of foundations over time, formed mainly by weak clay soils, is quite relevant in the design of various types of foundations, including pile foundations. Since most settlements are uneven and can develop over years and even decades, it is necessary to know the rate of development of these settlements over time to ensure the normal operation of buildings and structures erected on weak clay soils.

Foundation settlement over time can be predicted by taking into account the long-term complex physical processes that occur during load transfer to soils that have rheological properties (creep properties). To describe the stress–strain state of the base foundation system, taking into account the viscoelastic and viscoelastic–plastic properties of the base soils, it is necessary to take into account many parameters, the values of which may change during the operation of the structure [8]. Based on the above, it is important to note that the time factor must be taken into account when calculating foundations on bases formed by soils possessing rheological properties.

Numerous experimental and theoretical studies aimed at studying the rheological properties of clay soils, presented in the works of Tertsagi K. [9], Gersevanov N.M. [10], Florin V.A. [11,12], Cytovich N.A. [13], Meschyan S.R. [14], Vyalov S.S. [15], Maslov N.N. [16], Ter-Martirosyan Z.G. [17] and many others, allowed us to develop methods for calculating foundations on a natural base over time.

Many researchers have studied the interaction of piles with clay soils, including Vyalov S.S. [15], Bartolomei, A.A. [18], Dalmatov B.I. and Lapshin F.K. [19], Ter-Martirosyan Z.G. [17], Omelchak I.M. [8] and many others.

Liu J. et al. [20] proposed a tri-linear softening model to describe the interaction between a single pile and the surrounding soil, and a tri-linear plastic model that describes the interaction between the pile and the underlying soil. Based on the mechanical properties of soils and the theory of elasticity, the authors developed equations that describe the dependence of the settlement of a single pile on the vertical load in layered soils. In performing the calculations, the pile was considered as a flexible shaft with a constant cross-section. The authors of the study have demonstrated the effectiveness of the proposed calculation method for a single pile and provided recommendations on using the proposed models in different soil layers for a correct analysis of the relationship between the pile settlement and the applied vertical load.

The study [21] presents an analytical solution for determining the stress–strain state of a soil mass interacting with underground structures, including with piles.

The interaction of a single long pile with a multilayer soil mass, taking into account the elastic and rheological properties of the soil mass, was investigated in [22,23]. In [22], the solution of the problem by analytical method taking into account linear and nonlinear properties of soils is presented. For describing nonlinear shear deformations, the elasto-plastic Timoshenko S.P. model was used. The authors performed a comparison of the results of the analytical solution to the elastic problem with the results of numerical modeling in the Plaxis 3D software package, and also obtained an expression for determining the reduced shear modulus for a multilayer soil mass. The study [23] considers the solution to the problem of determining the displacements of a long pile in a surrounding multilayer base. It was found that the rate of stress change under the pile heel depends on the viscosity of the soil, while the rheological hardening coefficient significantly influences the time of pressure stabilization under the pile heel and the time of pile settlement stabilization.

Based on the modified Burgers model, in which the spring is replaced by a hyperbolic model, a nonlinear approach to calculate the settlement over time of a vertically loaded single pile as well as a group of piles in a layered soft soil was developed in [24]. In this work, parametric studies were carried out to investigate the influence of the parameters included in the calculation on the time-dependent settlement of a single pile. Based on the conducted research, it was shown that the theoretical results agree well with the measurement results, indicating a sufficiently high accuracy of the approach proposed in the work.

For calculating the long-term settlement of a single pile embedded in viscoelastic base soils, a three-dimensional viscoelastic model with a fractional-order derivative was proposed by Li, Zhang and Liu to describe the rheological behavior of the soils around the pile [25].

A simple approach to analyzing the elasto-plastic behavior of a single pile in layered soils, which allows for a quick assessment of the settlement of a single pile as well as a group of piles, is presented in [26]. The proposed approach is based on the application of two models, one of which describes the nonlinear relationship between the shear stresses occurring on the lateral surface of the pile and the displacement of the soil around the pile. The second model is based on a bilinear hardening model to simulate the relationship between the settlement of the pile base and the resistance of the pile tip.

In [27], a model was proposed for determining the settlement of a vertically loaded single hanging pile embedded in layered base soils.

To evaluate the behavior of a single vertical pile over time, embedded in soil with elasto-plastic properties, the authors of [28] conducted an analysis using the finite element method in a two-dimensional setting. For modeling the soil around the pile, a linear elastic perfectly plastic Mohr–Coulomb model was used, and for modeling the pile material, a linear elastic model was employed. Based on the calculations performed, it was found that a pile in non-cohesive soil has greater resistance under rapid loading than under long-term loading, whereas the opposite results were obtained for a pile in cohesive soil. The geotechnical model presented in this work allows the determination of lateral deformations as well as lateral soil stress and its variation over time.

An analytical method for evaluating the behavior of a single tapered pile and a group of piles from applied static axial compressive loads was proposed in [29]. Based on the proposed method, an iterative computer program was developed to calculate the settlement and bearing capacity of a single tapered pile.

An analytical approach for predicting the settlement of vertically loaded pile and pile groups was proposed by Xia and Zou [30]. A piecewise function was applied to investigate the nonlinear relationship between surface friction (friction on the side surface of the pile) and relative displacement of the pile–soil system. In this paper, a comparative analysis of the results obtained using the proposed calculation methodology with field test data and results obtained by other researchers has been carried out.

In the study by Long Li and Yousheng Deng [31], an algorithm is proposed to calculate the settlement of a pile foundation in a soil base having linear–viscous–elastic properties. The calculation algorithm presented in this paper is based on the analysis of the change in linear viscoelastic settlement of the soil depending on the geometric parameters of the pile (pile spacing, pile length-to-diameter ratio) and the deformation characteristics of the soil (elastic modulus and Poisson's ratio). The authors of the work performed a comparative analysis of the obtained results of the study with the results of field tests of piles.

A methodology based on the use of stress coefficients is presented by Gendy [32] to determine the settlement of hanging piles located in clay soil. The stress coefficients were obtained from the Mindlin equation by eliminating the Poisson's ratio from this equation. Based on the results of the conducted research, equations were derived that allow predicting the settlement of a single pile, a pile group and a pile–raft foundation on clay soils.

It should be emphasized that taking into account the rheological properties of the base soils is important not only in static calculations, but also in dynamic calculations of foundations. This statement is supported by a number of studies and scientific papers that demonstrate the influence of rheological characteristics on the behavior of foundations subjected to dynamic loads [33,34]. The correct interpretation of these properties allows for a more accurate assessment of the stability and durability of structures, especially under variable and dynamic loads such as seismic impacts or vibrations induced by machinery.

Based on the analysis of the scientific and technical literature on the research topic, the following main conclusions can be drawn:

- Most scientific papers focus on determining the settlement and bearing capacity of a single pile, rather than the pile foundation as a whole;
- When determining the settlement of pile foundations, most scientific works do not take into account the rheological properties of the base soils;
- The existing scientific papers present limited data on the influence of one of the most important rheological parameters—the coefficient of soil viscosity—on the settlement of pile foundations in general;
- A number of papers present calculations that take into account the rheological properties of base soils in relation to dynamic calculations.

Based on the above, the purpose of this paper is to determine the settlement of a pile foundation, taking into account the linear and rheological properties of the base soils based on the Kelvin–Voigt viscoelastic rheological model of the viscoelastic body according to the column pile and hanging pile schemes.

2. Materials and Methods

It is known that when a uniformly distributed load is applied to a pile foundation, it will be distributed among the piles, the raft and the surrounding soil mass located in the inter-pile space. Moreover, this distribution significantly depends on the diameter, length and spacing of the piles, as well as on the physical–mechanical and rheological properties of the soils and the pile material. With such a load transfer mechanism (using piles and a raft), a complex, heterogeneous stress–strain state is formed in the soil mass, which can change in space and time, especially when considering the rheological properties of the soils.

The important and determining design parameters when calculating a pile foundation are the settlement and bearing capacity of both the pile foundation as a whole and the individual piles within it (in the case of uneven distribution of piles in the pile foundation).

When predicting the stress–strain state of the pile–raft-surrounding soil mass system, it is allowed to consider not the entire pile foundation as a whole, but only a part of it—the computational cell.

The computational cell is a thick-walled soil cylinder of limited dimensions $(L,\ 2b_1)$, accommodating the pile, the raft and the surrounding soil mass. The geometric dimensions of the computational cell are chosen based on the load on the foundation and the engineering–geological conditions of the construction site.

Figure 1 shows the design scheme of the pile foundation with the delineation of the boundaries of the computational cell $(L,\ 2b_1)$ (dashed line).

When calculating piles as part of a pile foundation, the main thing is the quantitative assessment of the settlement of the pile foundation as a whole, which is largely determined by the magnitude of the load on the raft and the correct assessment of the stress–strain state of the pile–raft-surrounding soil mass system.

Depending on the ratio of the stiffness of the surrounding soil (G_1) and the underlying soil (G_2), the calculation of the stress–strain state of the pile foundation should be carried out either according to the column pile scheme when $G_2 \gg G_1$, or according to the hanging pile scheme when $G_2 > G_1$.

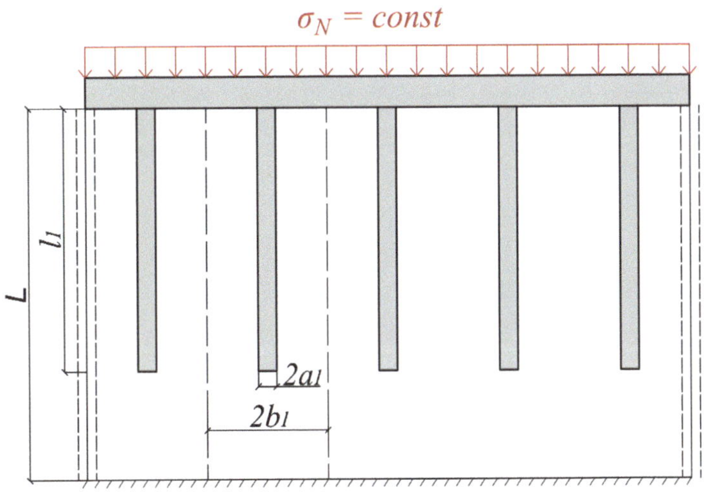

Figure 1. Delineation of the computational cell boundaries in a pile foundation.

According to SP 24.13330.2021 [35], a column pile is a pile resting on rocky or weakly deformed soils and transmitting the load to the base mainly through the heel.

A hanging pile is a pile that transmits the load to the base through the lateral surface and the heel [35].

Figure 2 shows the design scheme of the interaction of the pile as part of the pile foundation with the surrounding and underlying soils and the raft according to two schemes.

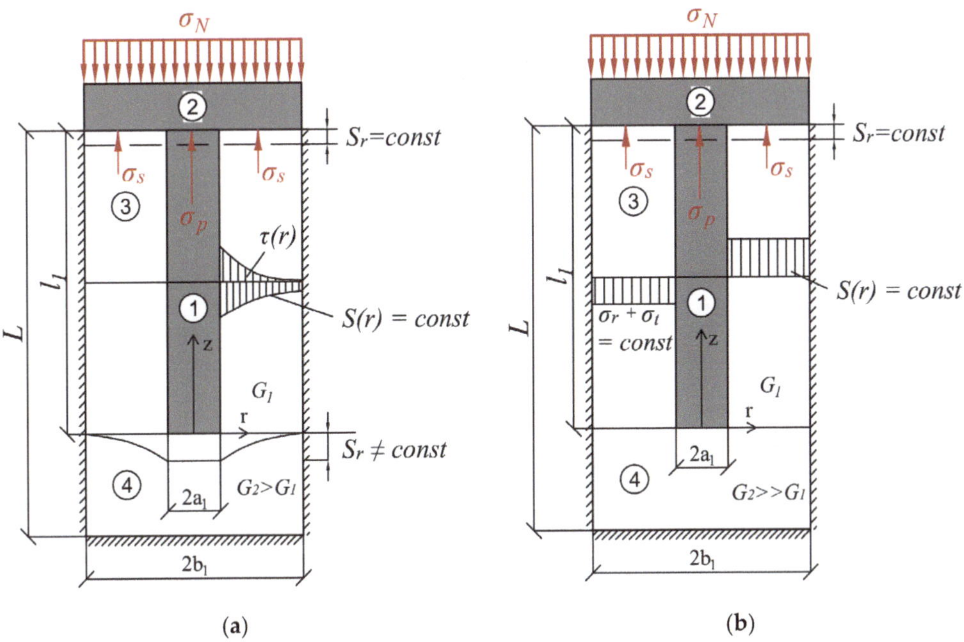

Figure 2. Design scheme of interaction of a pile as part of the pile foundation with the surrounding and underlying soils and the raft according to the following schemes: (**a**) hanging pile; (**b**) column pile: 1—pile; 2—raft; 3—surrounding soil mass; 4—underlying soil mass.

Since, in the following, all main calculations will be based on the Kelvin–Voigt rheological model, the main features of this model are presented below for a better understanding of the results obtained.

The Kelvin–Voigt model is a rheological model of a viscoelastic body consisting of parallel connected elements: an elastic Hooke spring (H) and a Newtonian fluid (N) (Figure 3). When two elements are connected in parallel, the strain of each element will be the same and the total stress will be defined as the sum of the stresses on each element.

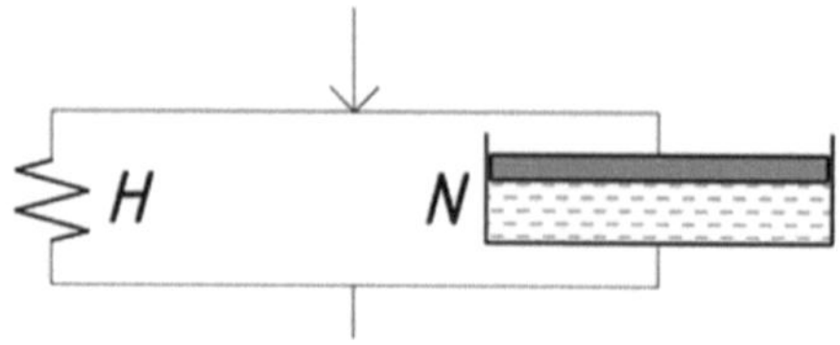

Figure 3. The Kelvin–Voigt rheological model.

The dependence of strains (ε) on stresses (σ) according to the Kelvin–Voigt rheological model will be as follows:

$$\sigma = \varepsilon \cdot E + \dot{\varepsilon} \cdot \eta \tag{1}$$

where E is the modulus of deformation; $\dot{\varepsilon}$ is the strain rate; and η is the coefficient of soil viscosity.

Note that the Kelvin–Voigt model is designed to analyze the mechanical properties of continuous media. In the context of our study, the ground can be classified as a continuous medium, which allows us to use this model to describe its behavior. This is confirmed by the ability of the model to take into account both elastic and viscous properties of the material, which makes it convenient for analyzing the stress–strain state of the soil mass under various loads.

A more detailed description of the Kelvin–Voigt rheological model can be found in the following references: [17,36,37].

3. Results

3.1. Solution of the Problem in a Linear Formulation According to the Column Pile Scheme

Let a long pile with a given diameter of $2a_1$ and length l_1 in a pile foundation be embedded in the soil mass, with its lower end resting on an underlying relatively dense soil layer ($G_2 \gg G_1$). The computational domain of the massif containing the pile and the foundation is a thick-walled soil cylinder of limited dimensions $(L, 2b_1)$.

When a static uniformly distributed load ($\sigma_N = const$) is applied to the raft, it will be distributed between the piles and the surrounding soil within each computational cell. Considering the compressive deformation of the pile shaft and the surrounding soil mass (in accordance with the problem conditions), we can write the following equilibrium equation:

$$\sigma_N = \sigma_p \cdot \omega + \sigma_s \cdot (1 - \omega), \tag{2}$$

where σ_N is the uniformly distributed load applied to the raft; σ_p is the stress acting in the pile shaft at the level of the pile head; σ_s is the stress acting in the soil mass at the contact with the raft; ω is the dimensionless coefficient equal to the ratio of the cross-sectional area of the pile shaft to the cross-sectional area of the entire computational cell, i.e., $\omega = a_1^2/b_1^2$; and a_1 and b_1 are the radius of the pile and the radius of the computational cell, respectively.

Let us also write down the condition of equality of settlements of the raft, the pile and the surrounding soil mass:

$$S_N = S_p = S_s \tag{3}$$

Equation (3) can be represented in the following form:

$$m_r \cdot \sigma_N = m_p \cdot \sigma_p = m_s \cdot \sigma_s \tag{4}$$

where m_p and m_s are the coefficients of relative compressibility of the pile and the soil, respectively.

The joint solution of Equations (2) and (4) will determine σ_p and σ_s, respectively:

$$\sigma_p = \sigma_N \cdot \frac{E_p}{E_p \cdot \omega + E_s \cdot (1-\omega)};\ \sigma_s = \sigma_N \cdot \frac{E_s}{E_p \cdot \omega + E_s \cdot (1-\omega)} \tag{5}$$

The reduced modulus of deformation of the computational soil cell can be found from condition (2), i.e.:

$$E_r = E_p \cdot \omega + E_s \cdot (1-\omega) \tag{6}$$

From Formula (6), it follows that as ω increases, E_r also increases, and conversely. Consequently, other things being equal, by changing the ratio $\omega = a_1^2/b_1^2$, it is possible to adjust both the value of the reduced modulus of deformation of the computational cell E_r and the calculated value of the settlement of the cell as a whole.

Based on the above, the settlement of the pile foundation can be determined according to the following formula:

$$S_{pr} = \frac{\sigma_N}{E_r} \cdot 0.8 \cdot l_1 \tag{7}$$

where l_1 is the length of the pile.

When calculating by Formula (7), the settlement of the underlying layer should be neglected.

Based on the known dependency $S = \varepsilon \cdot l$, the solution to the settlement problem of a pile foundation cell can be obtained when the length (l) and strain (ε) are known. Since length is an input parameter, the following problems will be considered in order to determine the strains of the computational cell.

3.2. The Stress–Strain State of the Cell in an Elasto-Viscous Formulation According to the Column Pile Scheme Based on the Kelvin–Voigt Model

Let us write down the equilibrium condition characterizing the process of development of settlement of the raft over time, assuming in the first approximation that it coincides with the equilibrium Equation (2), then we obtain:

$$\sigma_N = \sigma_p(t) \cdot \omega + \sigma_s(t) \cdot (1-\omega) \tag{8}$$

The stresses acting in the pile shaft at the level of the pile head (σ_p) and the stresses acting in the soil mass at the contact with the raft (σ_s) can be represented by the following relationships:

$$\sigma_p(t) = \varepsilon_p \cdot E_p;\ \sigma_s(t) = \varepsilon_s \cdot E_s + \dot{\varepsilon_s} \cdot \eta_s \tag{9}$$

where ε_p is the strain of the pile; E_p is the modulus of deformation of the pile; ε_s is the strain of the surrounding soil; E_s is the modulus of deformation of the surrounding soil; $\dot{\varepsilon_s}$ is the rate of development of strains of the surrounding soil; and η_s is the coefficient of viscosity of the surrounding soil.

In Figure 4, the calculated model of the interaction of a pile in a pile foundation and the surrounding soil mass under compressive deformation is presented based on the Kelvin–Voigt model.

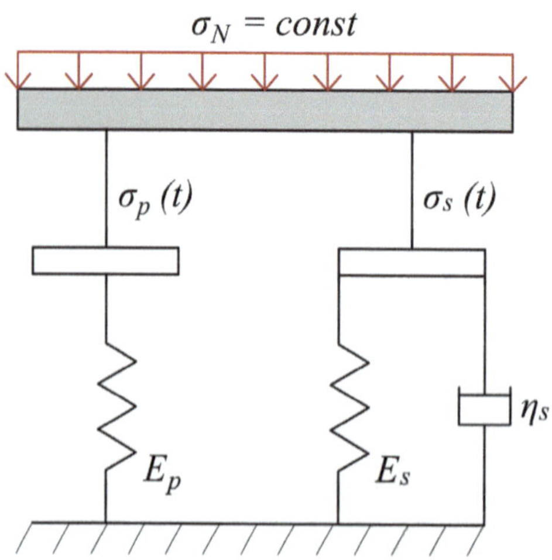

Figure 4. The calculated model of the interaction of a pile in a pile foundation and the surrounding soil mass under compressive deformation based on the Kelvin–Voigt model.

Substituting into the equilibrium condition (8) the dependences obtained in (9), we obtain:

$$\sigma_N = \varepsilon_p \cdot E_p \cdot \omega + (\varepsilon_s \cdot E_s + \dot{\varepsilon_s} \cdot \eta_s) \cdot (1 - \omega) \tag{10}$$

Based on the condition of equality of strains of the pile and the surrounding soil $\varepsilon_p = \varepsilon_s = \varepsilon$, we obtain:

$$\sigma_N = \varepsilon \cdot (E_p \cdot \omega + E_s \cdot (1 - \omega)) + \dot{\varepsilon} \cdot \eta_s \cdot (1 - \omega) \tag{11}$$

where $\varepsilon = \varepsilon_p = \varepsilon_s$, $\dot{\varepsilon} = \dot{\varepsilon_s}$.

Let us represent the differential Equation (11) in the following form:

$$\dot{\varepsilon} + \varepsilon \cdot P = Q \tag{12}$$

$$P = \frac{E_p \cdot \omega + E_s \cdot (1 - \omega)}{\eta_s \cdot (1 - \omega)}; Q = \frac{\sigma_N}{\eta_s \cdot (1 - \omega)} \tag{13}$$

The solution of the differential Equation (12) is known [38] and has the following form:

$$\varepsilon(t) = e^{-\int P dt} \left[\int Q \cdot e^{\int P dt} dt + C \right] \tag{14}$$

By performing the corresponding transformations in (14), we obtain the following equation:

$$\varepsilon(t) = \frac{Q}{P} + C \cdot e^{-P \cdot t} \tag{15}$$

We determine the integration constant from the initial condition when $\varepsilon(0) = 0$ at $t = 0$:

$$C = -\frac{Q}{P} \tag{16}$$

Substituting the integration constant C rom Equation (16) into Equation (15), we obtain:

$$\varepsilon(t) = \frac{Q}{P} \cdot (1 - e^{-P \cdot t}) \tag{17}$$

It follows from Equation (17) that:

- at $t \to 0$ $\varepsilon(0) = 0$;
- at $t \to \infty$ $\varepsilon(\infty) = \frac{Q}{P} = \frac{\sigma_N}{E_p \cdot \omega + E_s \cdot (1-\omega)}$.

Table 1 shows the characteristics of the pile and surrounding soil used in the calculation using Formula (17).

Table 1. Characteristics of the pile and surrounding soil (according to the column pile scheme).

Name of the Parameter	Symbol	Value	Units
Characteristics of the pile			
Modulus of deformation of the pile	E_p	3×10^7	kPa
Radius of the pile	a_1	0.30	m
Pile spacing *	$2b_1$	1.80	m
	$2b_2$	2.40	m
	$2b_3$	3.00	m
	$2b_4$	3.60	m
Characteristics of the surrounding soil			
Modulus of deformation of the surrounding soil	E_s	2×10^4	kPa
Coefficient of viscosity of the surrounding soil	η_{s_1}	1×10^9	Poise
	η_{s_2}	1×10^{11}	Poise
	η_{s_3}	1×10^{13}	Poise

*—in the present study, the pile spacing is equal to the diameter of the computational cell (Figure 1).

$E_p a_1 2b_1 2b_2 2b_3 2b_4 E_s \eta_{s_1} \eta_{s_2} \eta_{s_3}$Figures 5–7 show the graphs of the dependence of strains of the computational cell on time $(\varepsilon - t)$, obtained with different pile spacing $(2b_1 < 2b_2 < 2b_3 < 2b_4)$ and with different coefficients of viscosity of the surrounding soil $(\eta_{r_1} < \eta_{r_2} < \eta_{r_3})$ based on Formula (17).

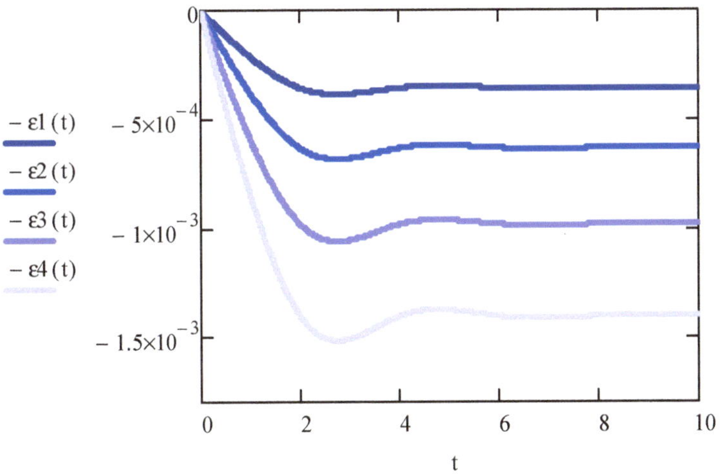

Figure 5. Graphs of the dependence of strains of the computational cell on time $(\varepsilon - t)$, obtained with different pile spacing $(2b_1 < 2b_2 < 2b_3 < 2b_4)$ and with a coefficient of viscosity of the surrounding soil η_{s_1}.

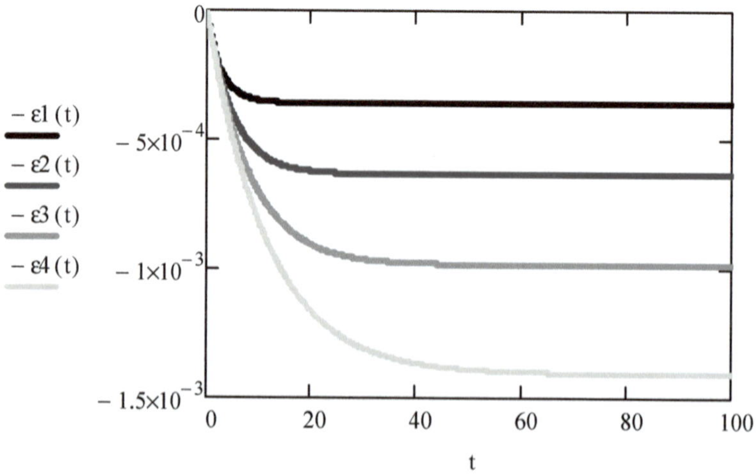

Figure 6. Graphs of the dependence of strains of the computational cell on time $(\varepsilon - t)$, obtained with different pile spacing $(2b_1 < 2b_2 < 2b_3 < 2b_4)$ and with a coefficient of viscosity of the surrounding soil η_{s_2}.

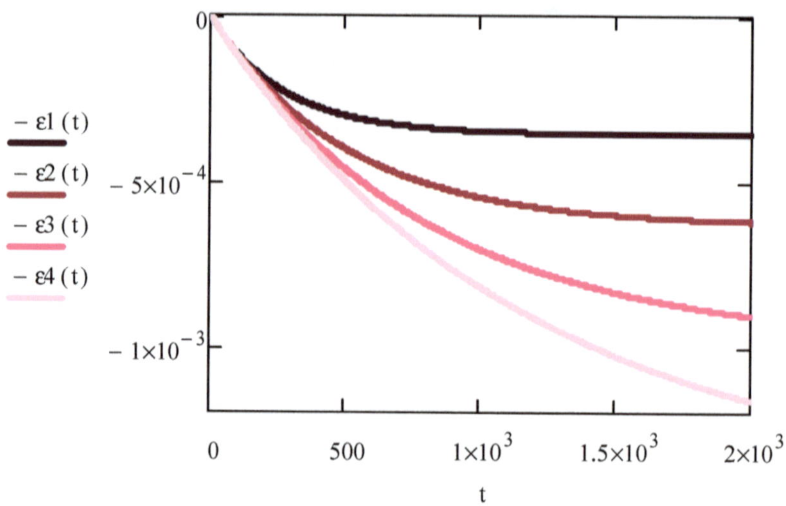

Figure 7. Graphs of the dependence of strains of the computational cell on time $(\varepsilon - t)$, obtained with different pile spacing $(2b_1 < 2b_2 < 2b_3 < 2b_4)$ and with a coefficient of viscosity of the surrounding soil η_{s_3}.

Figure 8 shows the graphs of the dependence of strains of the computational cell on time $(\varepsilon - t)$, obtained with different coefficients of viscosity of the surrounding soil $(\eta_{s1} < \eta_{s2} < \eta_{s3})$ and with pile spacing $2b_2$, based on Formula (17).

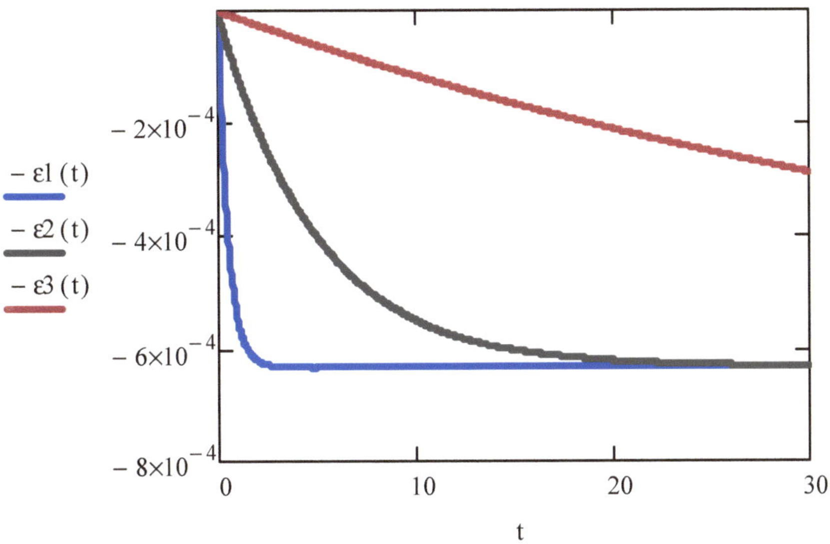

Figure 8. Graphs of the dependence of strains of the computational cell on time $(\varepsilon - t)$, obtained with different coefficients of viscosity of the surrounding soil $(\eta_{s1} < \eta_{s2} < \eta_{s3})$ and with pile spacing $2b_2$.

3.3. The Stress–Strain State of the Cell in an Elasto-Viscous Formulation According to the Hanging Pile Scheme Based on the Kelvin–Voigt Model

The analysis of the solution of the problem in the elastic–viscous formulation according to the column pile scheme showed that if we replace the modulus of deformation of the pile with the reduced modulus of deformation, taking into account the local settlement under the pile heel, we can find the solution of the problem according to the hanging pile scheme. For this purpose, it is sufficient to change the stiffness of the pile in such a way that the following condition is satisfied:

$$\frac{\sigma_p}{E'_p} \cdot l_1 = \frac{\sigma_p}{E_p} \cdot l_1 + \frac{\sigma_p \cdot \pi \cdot a_1 \cdot \left(1 - \nu_2^2\right)}{2 \cdot E_2} \tag{18}$$

$$E'_p = \frac{E_p}{1 + \frac{E_p \cdot \pi \cdot a_1 \cdot \left(1 - \nu_2^2\right)}{2 \cdot E_2 \cdot l_1}} \tag{19}$$

where E'_p is the reduced modulus of deformation of the pile; E_p is the modulus of deformation of the pile; E_2 is the modulus of deformation of the underlying soil mass; ν_2 is the Poisson's ratio of the underlying soil mass; a_1 is the radius of the pile; and l_1 is the length of the pile.

Table 2 shows the characteristics of the pile and surrounding soil used in the calculation using Formula (17), taking into account Formula (19).

Based on the above, we can conclude that by replacing the modulus of deformation of the pile E_p with the reduced modulus of deformation E'_p, in the rheological Equation (11), we obtain the stress–strain state of the cell according to the hanging pile scheme (Figure 9a).

As a result, we can predict the strains of the computational cell both using the column pile scheme $(\varepsilon_1 - t)$ and the hanging pile scheme $(\varepsilon'_2 - t)$ (Figure 10) based on Formula (17), taking into account (19).

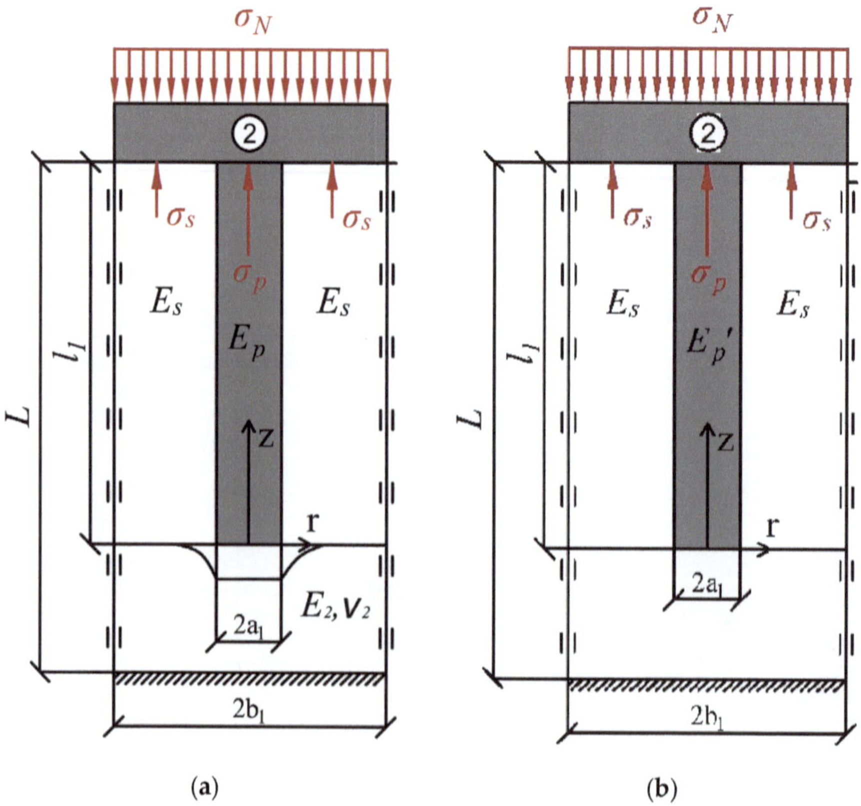

Figure 9. Design diagram for determining the settlement of: (**a**) hanging pile (S_2); (**b**) column pile (S_1).

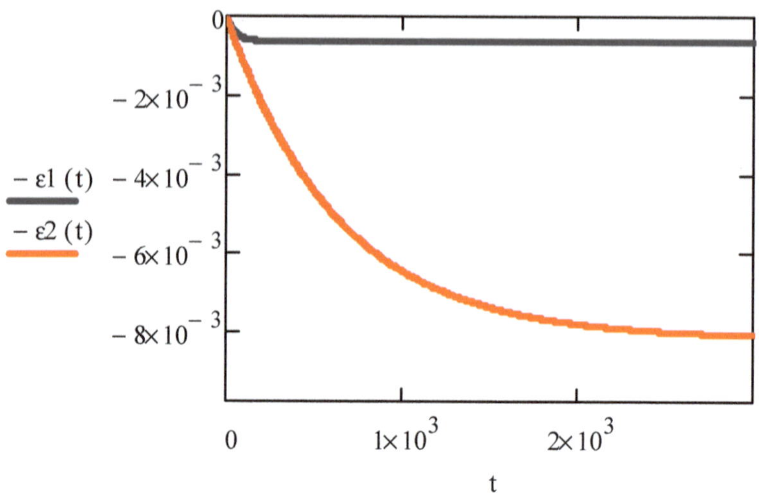

Figure 10. Graphs of dependence of strains of the computational cell on time by the column pile scheme $(\varepsilon_1 - t)$ and by the hanging pile scheme $(\varepsilon_2' - t)$.

Table 2. Characteristics of the pile and surrounding soil.

Name of the Parameter	Symbol	Value	Units
	Characteristics of the pile		
Modulus of deformation of the pile	E_p	3×10^7	kPa
Radius of the pile	a_1	0.30	m
Pile spacing *	$2b_2$	2.40	m
Length of the pile	l_1	30	m
	Characteristics of the soil		
Modulus of deformation of the surrounding soil	E_s	2×10^4	kPa
Modulus of deformation of the underlying soil mass	E_2	3×10^4	kPa
Poisson's ratio of the underlying soil mass	ν_2	0.36	-
Coefficient of viscosity of the surrounding soil	η_s	1×10^{12}	Poise

*—in the present study, the pile spacing is equal to the diameter of the computational cell (Figure 1).

4. Discussion

When the load is transferred to the pile foundation, in addition to the settlement of the piles themselves, the settlement of the surrounding soil located in the inter-pile space occurs. When considering the stress–strain state of the active zone in pile foundations, it should be noted that it differs significantly from the stress–strain state of the surrounding soil mass around a single pile, because in pile foundations there is a mutual influence of the piles on each other, as well as a time distribution of the external load between the raft, piles and the surrounding soil. However, despite this, numerous studies [20–23,25–29] have focused on determining the settlement of a single pile rather than the pile foundation as a whole.

The formation of a complex and heterogeneous stress–strain state of the system pile–raft-surrounding soil mass system depends on:

- geometric characteristics (cross-sectional and longitudinal profile shape, length) and pile material, as well as pile spacing;
- engineering–geological and hydrogeological conditions of the construction site;
- mechanical characteristics of the surrounding soil, which include deformation (E, ν), strength (c, φ) and rheological (η) properties.

Indeed, the factors presented above have a significant influence on the formation of the stress–strain state of the active zone in the calculation of the pile foundation as a whole, which has been reviewed and noted by Long Li and Yousheng Deng in [31]. This study investigated the change in linear viscoelastic settlement of soil depending on the pile spacing, the ratio of pile length to diameter, soil elasticity modulus and Poisson's ratio. However, unfortunately, this study did not pay sufficient attention to the influence of the rheological properties of clay soils (including soil viscosity) on the obtained values of settlement of the pile foundation over time.

Taking into account the rheological properties of clay soils is necessary in order to predict over time the final settlements of pile foundations constructed on clay soils, to prevent in advance the negative phenomena associated with creep of clay soils and, if necessary, to optimize the design solutions of foundations.

In the present study, the problem of the interaction between the pile, the raft and the surrounding soil mass was solved in a linear formulation using the column pile scheme. As a result, Formula (6) was obtained for calculating the reduced deformation modulus of the computational soil cell, as well as Formula (7) for calculating the settlement of the pile foundation.

To take into account the rheological properties of the base soils, a problem was solved using an elastic–viscous formulation in the column pile scheme and in the hanging pile scheme based on the well-known Kelvin–Voigt rheological model. This model allows us to describe creep of clay soils over time quite well because it includes a viscosity parameter.

Let us consider in more detail the main advantages of this model for a better understanding of the results obtained in this study.

1. The structure of the model: as mentioned earlier in this paper, the Kelvin–Voigt model consists of an elastic element (spring) and a viscous element (Newtonian fluid) connected in parallel. This allows this model to describe the processes of elastic deformation and viscous flow, which is characteristic of the behavior of clayey soils under the action of a constant load. We also note that the Kelvin–Voigt model takes into account both the rate and magnitude of deformation, describing the behavior of the soil under the action of constant or temporary loads. For example, when the load is suddenly increased, the elastic deformation will occur immediately, while the viscous part will determine how the soil will continue to deform over time.
2. The ability to take creep into account: in contrast to simple elastic models, the Kelvin–Voigt model allows us to take into account the creep phenomenon inherent in clayey soils. This means that when a constant load is applied, deformations will slowly continue to develop in the clayey soil over a long period of time. These deformations can be quantified using this model, which represents an important aspect in the design of various foundations and enclosing structures, since the final settlement can significantly affect the project and operation of the structure.
3. Taking into account the time factor: the Kelvin–Voigt model makes it possible to estimate deformations of soil with rheological properties both in the short-term (a few days) and long-term (several decades) time range, which is especially important in the design of foundations and various enclosing structures.
4. Flexibility and adaptability: the Kelvin–Voigt model can be used to calculate both single-layer and multilayer bases.
5. Ease of use: the mathematical description of the model is relatively simple, which will allow its future implementation in numerical calculation methods.
6. Empirical validity: the model agrees well with experimental data obtained from tests of clayey soils. This makes it a reliable tool for engineering analysis.

Based on the above, we conclude that the Kelvin–Voigt model is a powerful tool for calculating and predicting long-term deformations of soil bases with rheological properties. The model's ability to take into account both elastic and viscous components allows engineers to more accurately estimate long-term deformations and make informed decisions in the design of various foundations and enclosing structures. Thus, the application of this model provides an opportunity for more reliable and efficient design and minimization of potential risks for the objects under construction.

Based on the solution to the problem using the column pile scheme based on the Kelvin–Voigt model, the following results were obtained:

- It is shown that the vertical strains of the computational cell also increase with increasing pile spacing ($2b_1 < 2b_2 < 2b_3 < 2b_4$) (Figures 5–7), because with increasing pile spacing, the load applied to the raft starts to be supported more by the surrounding soil mass located in the inter-pile space, which has a modulus of deformation much lower than that of the pile itself;
- At low values of the coefficient of viscosity of the surrounding soil ($\eta_{s_1} = 10^9$ Poise) (Figure 5), comparable to the liquid medium, there is a peak value of vertical strains exceeding the residual vertical strains, which is not observed in more viscous media ($\eta_{s_2} = 10^{11}$ Poise, $\eta_{s_3} = 10^{13}$ Poise) (Figures 6 and 7). This phenomenon indicates significant differences in soil behavior depending on its rheological properties;
- In accordance with the character of location of the graphs of dependence of strains of the computational cell on time ($\varepsilon - t$) presented in Figure 8, it can be seen that the time of stabilization of vertical strains of the computational cell is directly proportional to the value of the coefficient of viscosity of the surrounding soil, i.e., the higher the

value of the coefficient of viscosity, the more time will be needed in order to stabilize the vertical strains of the computational cell from the applied load. However, it should be noted that changing the coefficient of viscosity of the surrounding soil, both increasing and decreasing, will affect only the time of stabilization of vertical strains of the computational cell, and will not change their value. The results obtained confirm the importance of considering time factors in assessing the stability and durability of pile structures, especially under variable loads.

Based on the solution to the problem using the hanging pile scheme based on the Kelvin–Voigt model, the following results were obtained:

- Formula (19) for calculating the reduced modulus of deformation of the pile (E'_p) was obtained, which makes it possible to take into account the strains of the underlying soil layer and considerably simplify the procedure for calculating the stress–strain state of the computational cell according to the hanging pile scheme. The resulting equation takes into account both the stiffness of the pile itself (E_p) and the stiffness of the underlying soil mass under the pile heel (E_2);
- Building graphs of dependence of vertical strains of the computational cell over time according to the column pile scheme ($\varepsilon_1 - t$) and according to the hanging pile scheme ($\varepsilon'_2 - t$), presented in Figure 10, showed that they are significantly different. This difference shows that taking into account the local settlement under the pile heel in the design diagram leads to an increase in settlement, which corresponds to the "hanging pile" design diagram (Figure 9a);
- When the modulus of deformation of the surrounding soil (E_s), the modulus of deformation of the underlying soil mass (E_2) and the Poisson's ratio of the underlying soil mass (ν_2) decrease, the vertical strains of the computational cell in the hanging pile scheme will increase, and conversely, when these parameters increase, the vertical strains of the computational cell will decrease;
- Change in the coefficient of viscosity of the surrounding soil (η_s) does not influence the value of vertical strains of the computational cell both for the column pile and hanging pile schemes. However, this parameter significantly influences the time of stabilization of vertical strains, which can be important in the design of foundations, especially under conditions of dynamic loads on building structures.

The analysis of the obtained results allows us to conclude that when designing pile foundations, it is necessary to take an individual approach to each specific construction case, taking into account the type of soil, its rheological properties and its mechanical behavior. The use of the proposed Kelvin–Voigt viscoelastic body model and the obtained formula for calculating the reduced modulus of deformation can significantly increase the accuracy of calculations and, consequently, the reliability of pile structures in general. Importantly, further research in this area will help to optimize existing design standards, leading to safer and more efficient solutions in engineering practice.

5. Conclusions

The acceleration of scientific and technical progress in foundation engineering is inextricably linked to the development of new calculation methods that allow for a more reliable description of the interaction between different types of foundations and base soils during the transfer of loads from buildings, structures and technological equipment [8]. Numerous observations of the settlement of various buildings and structures over the years have shown that a promising direction of research into the interaction of various foundations and soil bases is to take into account the rheological properties of soils (consideration of viscous deformation under the action of constant and variable loads).

Taking into account the fact that many buildings and structures are constructed on bases composed of weak clay soils, the settlement of which, as a rule, is uneven and can develop over years and even decades, it is necessary to consider the rheological properties of clay soils when calculating and designing various types of foundations, including pile foundations. The widespread use of pile foundations is due to their reliability, speed of construction and possibility of construction in almost any engineering–geological and climatic conditions.

The following main results were obtained, meeting the purpose of this study as outlined in the introduction:

- The dependence (17) was obtained, which allows us to determine the vertical strains of the computational cell (and, at known pile length, it is possible to obtain the settlement of the pile foundation as a whole) both for the column pile scheme and for the hanging pile scheme;
- The Formula (19) for calculation of the reduced modulus of deformation of the pile (E'_p) was obtained, which allows us to take into account the local settlement of the underlying soil under the heel of the pile and thus to pass from the solution of the problem according to the column pile scheme to the solution of the problem according to the hanging pile scheme (Figure 10 clearly shows the difference in the values of vertical strains of the computational cell obtained when solving the problem using the column pile scheme ($\varepsilon_1 - t$) and the hanging pile scheme, taking into account the reduced modulus of deformation of the pile ($\varepsilon'_2 - t$));
- It was found that the coefficient of viscosity of the surrounding soil has no influence on the values of vertical strains of the computational cell. This parameter only influences the stabilization time of the vertical strains, i.e., how quickly the vertical strains stabilize. Figure 8 effectively demonstrates how the stabilization time of vertical strains of the computational cell varies significantly with different values of the coefficients of viscosity of the surrounding soil ($\eta_{s1} < \eta_{s2} < \eta_{s3}$).

The practical significance of this study consists in solving the actual problem of determining the settlement of a pile foundation, taking into account the linear and rheological properties of soils under the column pile scheme and under the hanging pile scheme on the basis of the known viscoelastic Kelvin–Voigt model.

As future research, it is planned to further develop methods of quantitative assessment of the stress–strain state of the base-foundation system, taking into account rheological properties of the clay soil in order to exclude excessive development of settlements, tilts and displacements of various buildings and structures, as well as to optimize the adopted design solutions of foundations.

Author Contributions: Conceptualization, methodology, A.Z.T.-M.; software, validation, L.Y.E. and G.O.A.; formal analysis, writing—review and editing, all authors; investigation, A.Z.T.-M., L.Y.E. and G.O.A., supervision, project administration, A.Z.T.-M.; resources, A.Z.T.-M. and G.O.A.; visualization, L.Y.E. All authors have read and agreed to the published version of the manuscript.

Funding: The research was funded by the National Research Moscow State University of Civil Engineering (grant for fundamental and applied scientific research, project No. 02-392/130).

Data Availability Statement: The original contributions presented in this study are included in the article. Further inquiries can be directed to the corresponding author.

Conflicts of Interest: The authors declare no conflicts of interest.

References

1. Luga, A.A. *Methodical Instructions on Calculation of Settlements of Single Piles*; TsNIIS: Moscow, Russia, 1963.
2. Luga, A.A. *Settlement of Pile Foundations Working in Conditions of Absence of Cluster Effect VNITS. Investigations of Bearing Capacity of Foundation Bases*; Transizdat: Moscow, Russia, 1965.
3. Ogranovich, A.B. To the question of determining the settlement of a single pile. *Bases Found. Soil Mech.* **1963**, *1*.
4. Golubkov, V.N.; Tugaenko, Y.F.; Khutoryansky, B.O. Investigation of the process of deformation zone formation in single pile foundations. *Bases Found.* **1971**, *4*, 9–13.
5. Golubkov, V.N. On the nature of joint work of piles and soil. *Bases Found.* **1972**, *5*, 29–35.
6. Egorov, K.E. To the question of permissible settlement of foundations of structures. *Bases Found.* **1952**, 18.
7. Bartolomei, A.A. *Calculation of Settlements of Strip Pile Foundations*; Stroyizdat: Moscow, Russia, 1972; 128p.
8. Omelchak, I.M. Fundamentals of Calculation of Pile Foundations Taking into Account Rheological Properties of Foundation Soils. Ph.D. Thesis, Perm State Technical University, Perm, Russia, 2003.

9. Tertsagi, K. *Theory of Soil Mechanics*; Gostroyizdat: Moscow, Russia, 1961; 507p.
10. Gersevanov, N.M. *Theoretical Foundations of Soil Mechanics and Their Practical Applications*; Strojizdat: Moscow, Russia, 1948; 248p.
11. Florin, V.A. *Fundamentals of Soil Mechanics. Volume 1*; Strojizdat: Leningrad, Russia, 1959; 356p.
12. Florin, V.A. *Fundamentals of Soil Mechanics. Volume 2*; Gosstroyizdat: Leningrad, Russia, 1961; 543p.
13. Cytovich, N.A. *Soil Mechanics*; State Publishing House of Literature on Construction, Architecture, and Building Materials: Moscow, Russia, 1963; 437p.
14. Meschyan, S.R. *Creep of Clay Soils*; Publishing House of the Academy of Sciences of the Armenian SSR: Yerevan, Armenia, 1967; 320p.
15. Vyalov, S.S. *Rheological Foundations of Soil Mechanics: Textbook for Construction Universities*; Vysshaya Shkola: Moscow, Russia, 1978; 447p.
16. Maslov, N.N. *Long-Term Stability and Deformation of Displacement Retaining Walls*; Energiya: Moscow, Russia, 1968; 467p.
17. Ter-Martirosyan, Z.G. *Rheological Parameters of Soils and Foundation Calculations*; Strojizdat: Moscow, Russia, 1990; 200p.
18. Bartolomei, A.A.; Omelchak, I.M.; Yushkov, B.S. *Pile Foundation Settlement Prediction*; Stroyizdat: Moscow, Russia, 1994; 384p.
19. Dalmatov, B.I.; Lapshin, F.K.; Rossikhin, Y.V. *Design of Pile Foundations in Conditions of Weak Soils*; Stroyizdat: Leningrad, Russia, 1975; 240p.
20. Liu, J.; Xiao, H.B.; Tang, J.; Li, Q.S. Analysis of load-transfer of single pile in layered soil. *Comput. Geotech.* **2004**, *31*, 127–135. [CrossRef]
21. Ter-Martirosyan, Z.G.; Ter-Martirosyan, A.Z. Foundation and Pile Interaction with Swelling Soil. *Soil Mech. Found. Eng.* **2018**, *55*, 291–297. [CrossRef]
22. Ter-Martirosyan, Z.G.; Akuletskii, A.S. Interaction between a long pile and multi-layer underlying soils. *Vestn. MGSU* **2021**, *16*, 168–175. [CrossRef]
23. Ter-Martirosyan, Z.G.; Akuletskii, A.S. Interaction between a long pile and the multilayered soil body with account for elastic and rheological properties as well as soil stabilization. *Vestn. MGSU* **2021**, *16*, 608–614. [CrossRef]
24. Feng, S.; Li, X.; Jiang, F.; Lei, L.; Chen, Z. A nonlinear approach for time-dependet settlement analysis of a single pile and pile groups. *Soil Mech. Found. Eng.* **2017**, *54*, 7–16. [CrossRef]
25. Li, X.-M.; Zhang, Q.-Q.; Liu, S.-W. Semianalytical solution for long-term settlement of a single pile embedded in fractional derivative viscoelastic soils. *Int. J. Geomech.* **2021**, *21*, 04020246. [CrossRef]
26. Zhang, Q.-Q.; Zhang, Z.-M.; He, J.-Y. A simplified approach for settlement analysis of single pile and pile groups considering interaction between identical piles in multilayered soils. *Comput. Geotech.* **2010**, *37*, 969–976. [CrossRef]
27. Liu, Q.-J.; Wang, J.-B.; Ma, J.-J.; Gao, W.-H. Vertically-loaded single floating pile in layered soils by thin annulus element method. *Chin. J. Geotech. Eng.* **2019**, *41*, 748–754. [CrossRef]
28. Abbas, M.I.; Chik, H.Z.; Taha, R.M.; Shafiqu, S.M.Q. Time-dependent lateral response of pile embedded in elasto-plastic soil. *J. Cent. South Univ. Technol. (Engl. Ed.)* **2010**, *17*, 372–380. [CrossRef]
29. Li, Y.; Li, W. Load-displacement behaviour of tapered piles: Theoretical modelling and analysis. *Geomech. Eng.* **2021**, *26*, 1–11. [CrossRef]
30. Xia, Z.-Q.; Zou, J.-F. Simplified approach for settlement analysis of vertically loaded pile. *J. Eng. Mech.* **2017**, *143*, 04017124. [CrossRef]
31. Li, L.; Deng, Y. Analysis of Settlement of Group Pile Foundation in Linear Viscoelastic Soil. *Adv. Civ. Eng.* **2023**, *2023*, 3207304. [CrossRef]
32. El Gendy, M. Deriving equations for analyzing friction piles in clay soil. *Ain Shams Eng. J.* **2007**, *42*, 1–27.
33. Chen, Q.; Li, Z.; Huang, X.; Zhang, C. Study of rheological-mechanical properties and vibration mechanics bandgap of row pile foundation. *Eur. J. Comput. Mech.* **2023**, *32*, 31–52. [CrossRef]
34. Ren, X.; Wang, L. Study on the time domain semi analytical method for horizontal vibration of pile in saturated clay. *Appl. Sci.* **2024**, *14*, 778. [CrossRef]
35. *SP 24.13330.2021 «Pile Foundations»*; Minstroy of Russia: Moscow, Russia, 2021; p. 113.
36. Voigt, W. *Ueber die Innere Reibung der Festen Körper, Insbesondere der Krystalle*; Abhandlungen der Königlichen Gesellschaft von Wissenschaften zu Göttingen: German, 1890; pp. 3–47.
37. Buliček, M.; Málek, J.; Rajagopal, K.R. On Kelvin-Voigt model and its generalizations. *Evol. Equ. Control Theory* **2012**, *1*, 17–42. [CrossRef]
38. Bronshtein, I.N.; Semendyaev, K.A. *Handbook of Mathematics*; State Publishing House of Physical and Mathematical Literature: Moscow, Russia, 1962; 608p.

buildings

MDPI

Article

Anchor Shear Strength Damage under Varying Sand Content, Freeze-Thaw Cycles, and Axial Pressure Conditions

Jie Dong [1], Yin-Chen Wang [1], Zhi-Hui Wu [1,*], Feng-Wu Gong [2], Ya-Dong Zhao [1] and Hong-Feng Zhang [1]

1 College of Civil Engineering, Hebei University of Architecture, Zhangjiakou 075000, China; dongjie1003@hotmail.com (J.D.); wangyinchen04@outlook.com (Y.-C.W.); zyd18336466820@outlook.com (Y.-D.Z.); zhf17861127976@outlook.com (H.-F.Z.)

2 No. 3 Geological Brigade, Hebei Provincial Geological Prospec Ting Bureau, Zhangjiakou 075000, China; gfw15830301682@126.com

* Correspondence: wuzhihui199245@hotmail.com; Tel.: +86-152-2278-3120

Abstract: Sandy soil in the north of Hebei region of China is widely distributed, the temperature difference between day and night is large, the phenomenon of freezing and thawing is obvious, and the soil body before and after the freezing and thawing cycle of sandy soil slopes is affected by the changes. This paper takes the stability of a sandy soil anchorage interface under a freeze-thaw cycle as the research background and, based on the self-developed anchor-soil interface shear device, analyses the influence of changing sand rate, confining pressure, and the number of freeze-thaw cycles on the shear characteristics of an anchor-soil interface in anchorage specimens. The research findings indicate that, at 50–60% sand contents, the shear strength increases with a higher sand content and is positively correlated with confining pressure within a higher range. A higher sand content stabilises the anchoring body, but an excessively high sand content can lead to failure. Increasing the sand content, confining pressure, and freeze-thaw cycle number all result in a reduction in the shear displacement at the peak strength. After 11 freeze-thaw cycles, the shear strength of the anchoring body stabilises, with a reduction in strength of approximately 32%, and a higher sand content effectively reduces the reduction in strength.

Keywords: sand anchors; shear characteristics; freeze-thaw cycles; complete stress-strain relationship curves; model tests; anchor-oil interface

1. Introduction

Anchor support is widely used in foundation pit engineering, slope engineering, and underground engineering, and the anchoring performance of anchors affects the stability and safety of engineering construction and the completed structures. Therefore, the shear deformation characteristics of an anchor-soil interface have remained a topic of study. Experts and scholars have carried out many experimental and theoretical studies. Zhuo Yang and Siu Chun Michael Ho et al. [1,2] directly monitored the internal force of anchors through intelligent monitoring to describe the stress distribution of anchors. Chase Barnard and Cherdsak Suksiripattanapong et al. [3,4] evaluated the mechanical properties of an anchor-soil interface in detail through on-site anchor pull-out testing. Aoxue Chen et al. [5–8] explored the anchor support system of a deep foundation pit and obtained the force and deformation characteristics of the anchors in the foundation pit, pointing out that strengthening the shallow support of the foundation pit can effectively improve the safety reserve. Weizhi Su [9] conducted anchor pull-out tests with two different model boxes and proposed an exponential anchor-soil interface model. Laura Blanco Martín [10] established a trilinear bond−slip prediction model from measured curves. M. Ghadimi [11] proposed a new analytical solution for predicting the displacement of fully grouted anchors, and the analytical method results agreed well with the numerical method results. Xiao Hua Xi [12] derived a shear stress model and an axial force model of full-length bonded anchors based on the Boussinesq formula, starting from the surrounding rock of the tunnel. Zhu Zheng-de et al. [13] analysed the distribution of shear stresses in full-length grouted paving rods by considering the shear characteristics of the interface between the slurry and the contact interface. Chen et al. [14] proposed a full-process shear displacement curve by independently developing an anchor-soil interface friction tester.

In addition to theoretical studies of the anchor-soil interface, many researchers have explored this interface using indoor modelling tests. Since the anchor-soil contact surface properties closely depend on the soil state, it is possible to extend the study of the freeze-thaw cycling of soil to the anchor-soil interface. Numerous studies have pointed out that freezing and thawing change the original structure of the soil, as well as the composition of the particles, thus changing the mechanical properties of the soil, leading to changes in the engineering properties of the soil [15–17]. Edwin J et al. [18–21] studied the changes in pore size and pore water in the soil in the process of freezing and thawing from a microscopic point of view. Leuther Frederic Eskisar [22] found that the compressive strength of both natural and saturated soils decreased after freeze-thaw cycles. In freeze-thaw-related studies, many experts and scholars have used cohesion and the internal friction angle as indicators of soil changes and investigated the freeze-thaw damage mechanism [23–25]. Jilin Qi et al. [26,27] carried out direct shear tests on soils and found that changes in the dry density affected the cohesion, and the cohesion decreased when the dry density exceeded the critical density. Jiankun Liu et al. [28] suggested that the cohesion and internal friction angle of fly ash decreased most in the first cycle and reached their minimum values after seven–nine cycles. Naji N. Khoury et al. [29] suggested that 12 freeze-thaw cycles can generally reflect the complete effect of freezing and thawing on fly ash parameters. Zhoufei Yao suggested that the cohesion of Shanghai clay gradually decreased and its internal friction angle increased after experiencing very-low-temperature freezing and thawing [30]. Jianqiao Mu et al. [31] noted that cohesion is more sensitive to freezing and thawing than the internal friction angle, establishing an exponential decay model to predict the trend of rock strength degradation in cold regions due to freeze-thaw cycling.

Moreover, some scholars have also carried out research on the shear characteristics of soil–structure interfaces through direct shear tests. Pengfei He et al. [32] studied the change in the shear strength of frozen soil and structure surfaces subjected to freeze-thaw cycles and found that the freeze-thaw cycles had little influence on the shear displacement curve. Rongkai Pan [33] carried out straight shear tests on sand structures after experiencing different freeze-thaw cycles and combined the results with fine-scale analysis to explain the reasons for the change in the shear strength of the sand body. Jie Dong et al. [34] pointed out that the shear strength of the interface gradually stabilised after a significant decrease in the shear strength of the interface in the first three freeze-thaw cycles, whereas the friction angle increased slightly and the soil particle structure changed. Jingjing Pan et al. [35] found that, in a clay–concrete interface, the changes in the clay particles and pore content were most obvious after the first three freeze-thaw cycles and that the directionality of particles in the shear zone was the main reason for the changes in the shear zone.

In summary, previous research has mostly focused on the influence of freeze-thaw cycles on the shear performance of soil–structure interfaces, mainly using direct shear testing methods. However, direct shear testing faces issues such as an unstable shear area, an unrealistic shear interface geometry, and difficulty in accurately controlling the anchorage force state. Although previous research has mainly focused on various types of soil, insufficient attention has been given to the influence of sand content within the soil, and the application in the relevant engineering practice is also low. For sandy soil, the particle size distribution curves vary widely, leading to different soil stabilities. The stability of the soil can be simplified by evaluating the ratio of sand content to clay content. This study optimised the testing method and developed a self-designed shear device for testing under confining pressure to more accurately reflect the actual stress state of anchor rods in engineering. This study analysed the influence of sand content and anchoring material on the soil−anchor interface under freeze-thaw cycles, controlling factors such as confining pressure, sand content, and freeze-thaw cycle frequency. This work examined the shear-stress–displacement relationship, including the peak shear strength and soil freeze-thaw damage patterns at the soil–anchor interface.

Research has revealed the impact of different sand contents in sandy soil layers on the anchoring performance, further enriching the methods for evaluating soil stability and providing an important parameter basis for the design of anchor rods in engineering projects, such as foundation pits and slopes. By determining the optimal sand content, the stability and bearing capacity of the anchoring system can be improved while meeting engineering safety requirements, optimising material usage, and enhancing economic benefits. In addition, this study conducted a detailed analysis on the influence of freeze-thaw cycles on the shear strength of the soil–anchor interface under different sand contents, which is helpful for the adaptive design of engineering projects in freeze-thaw environments and extending the service life of the projects.

2. Materials and Methods

2.1. Materials

This experiment analyses the content of sandy soil in the north Hebei region and uses powdery clay from the Zhangjiakou region for configuration. After air-drying and crushing, the fine-grained clay with less than 0.075 mm was taken and placed in a cool and dry place for storage. Standard sand was mixed with the fine clay to create three soil mixtures with sand contents of 50%, 55%, and 60%. After mixing the sand and clay uniformly, the resulting soil was dried and then subjected to compaction and direct shear tests. The experimental data pertaining to the fundamental physical characteristics of sandy soil are presented in Figure 1 and Table 1. For anchoring, M8 ribbed steel bars and cement mortar were used, with ordinary Portland cement of strength grade 42.5. The cement mortar mix ratio of standard sand–ordinary Portland cement–water was 1:1:0.45, and the uniaxial compressive strength of the cement mortar was measured to be 21 MPa.

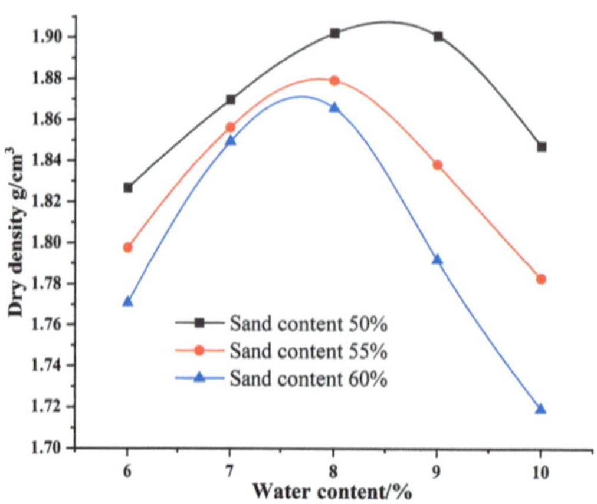

Figure 1. Compaction curve of the test soils.

Table 1. The cohesion and internal friction angle of the test soils.

Sand Content/%	Cohesion/kPa	Friction/°
50	53.03	25.93
55	55.04	27.25
60	50.03	26.15

2.2. Testing Instruments

This test apparatus is based on a permafrost triaxial apparatus, which contains a confining pressure system, a fixing system, a control system, and an acquisition system. The shear test system for anchors is shown in Figure 2. The whole instrumentation unit consists of a lifting platform, a steel base, a fixed support, a pumping cylinder wall, a confining pressure chamber, a fixed frame, an anchor centring device, an anchor fixture, a force transducer, and a counterforce frame beam. First, the solid anchor is maintained on the upper end of the fixed support through holes, and then a rubber film is installed on the wall of the pumping cylinder. The pumping machine is used to tightly attach the rubber film to the cylinder wall. The smooth cylinder wall is set on the solid anchor, the pumping machine is turned off so that the rubber film is tightly adhered to the solid anchor, and then rubber bands are placed around the rubber film to prevent water seepage. Afterwards, the fixed frame is installed and surrounded by the pressure chamber, and the different components are connected with a sealing ring to ensure that the device is sealed and airtight. During the pulling process, the bars are kept in the centre by means of an anchor centring device and clamps to ensure vertical pulling. After the specimen is installed, it is pressurised with liquid with a water-filled pressurising device. When the confining pressure reaches the target value, a pressure balancing device is activated to stabilise the confining pressure at the set value. Subsequently, the lifting device is activated, the steel base drives the solid anchor to move downwards at a rate of 1 mm/min, the force between the anchor and soil are transmitted to the acquisition system through the force sensor, and the force and displacement data are collected through the data acquisition instrument. Throughout the anchor pulling process, the shear of the anchor-soil contact surface is always kept constant to ensure the accuracy of the test.

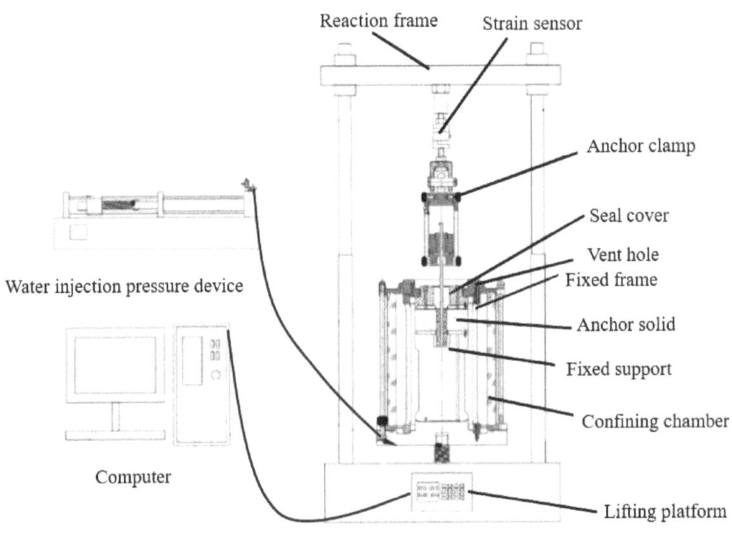

Figure 2. Bolt shear test system.

2.3. Test Method

Considering the research purpose of this paper, through the optimisation of the relevant testing method, a self-developed anchor-soil interface triaxial shear test device is used to collect more accurate shear stress and shear displacement data, and the influence of different confining pressures, freeze-thaw cycle numbers, and sand contents is investigated to elucidate the behaviour of sandy soil slope anchors in various slope conditions in a cold region. The analysis focuses on the shear stress-shear displacement at the anchor-soil interface, the peak shear strength and displacement, and the freeze-thaw damage to the soil body. The main test conditions are shown in Table 2.

Table 2. Experimental protocol.

Number	Sand Content/%	Pressure/kPa	Freeze-Thaw Cycles
MG1	50	50	0/1/3/8/11
MG2		100	
MG3		150	
MG4		200	
MG5	55	50	0/1/3/8/11
MG6		100	
MG7		150	
MG8		200	
MG9	60	50	0/1/3/8/11
MG10		100	
MG11		150	
MG12		200	

(1) Specimen production. The retrieved powdery clay is dried, crushed, and sieved again. A 0.075 mm sieve is used for sieving, and the sieved fine-grained soil is set aside for later use. The fine-grained soil is mixed uniformly with standard sand according to the different sand contents adopted for this work. Initial compaction experiments on soils with three different sand contents showed that the optimal moisture content varied, but all were concentrated at approximately 8%. The moisture content of the test specimens in this study is controlled at 8%. The soil mixture is prepared with an 8% moisture content to ensure even moisture distribution and minimal large clay particles. First, the compaction mould is assembled by fixing a hollow steel pipe to the base plate and connecting the cylinder wall to the bottom using bolts. Vaseline is applied to the surface of the hollow steel pipe to reduce friction with the compacted soil and facilitate smooth extraction, minimising interference with the soil. Once the compaction container is assembled, the soil is filled in five layers of a set amount of soil and compacted; compaction is stopped when the soil reaches the desired height. The compacted soil is scraped to make the layers more tightly bound. When the separation height is reached, filter paper and aluminium sheets are placed to separate the soil for later grouting and cutting, ensuring that the shear area of the anchor rod remains constant during the pull-out phase. Additional soil is poured in, and the compaction and scraping process is repeated until the soil sample is complete. The forces on the specimen are shown in Figure 3.

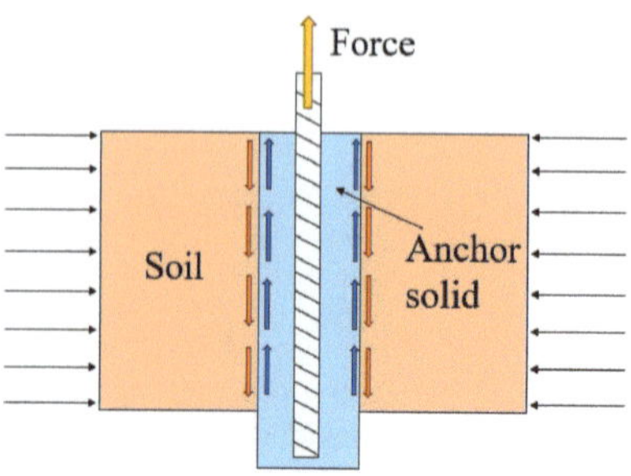

Figure 3. Specimen force diagram.

(2) Bolt shear test system production. The demoulding process is carried out using an electric hydraulic demoulding device. After demoulding, the sample is sealed with

plastic wrap and tightly wrapped with two hollow cylindrical acrylic plates around the anchor body. Rubber bands are then applied to the outer acrylic plate to provide circumferential restraint and prevent disturbance to the anchor body. The cement mortar mix ratio of standard sand–ordinary Portland cement–water is 1:1:0.45. The anchor positioning device is used to vertically place an M8 steel bar in the middle of the hole, and grouting material is made according to the mix ratio for soil anchor grouting. After thorough mixing, the grouting material is poured into the hollow soil body and compacted to prevent the formation of air bubbles in the mortar, which could affect the grouting strength. After the initial setting of the cement mortar, the specimens are cured at room temperature for 14 days. To minimise the boundary effect on the anchorage, the ratio of the soil body to the anchor body diameter and the steel bar diameter is greater than 5. The soil body has a diameter of 15 cm, the anchor body has a diameter of 3 cm, and the steel bar has a diameter of 8 mm. During the pull-out process, the soil body is subjected to circumferential confinement, which reduces the boundary effect on the anchor rod and ensures the accuracy of the test.

(3) Freeze-thaw cycling. First, a cured specimen is cut, and the soil is separated along the filter paper to ensure that the bottom of the soil is smooth. The surface of the specimen is wrapped with plastic wrap and fixed in a fixed container, which was placed in a freeze-thaw cycle box. The temperature is designed to be $-20°$ to $20°$. Previous freeze-thaw cycle tests have shown that soil reaches a new stable state after 10 freeze-thaw cycles [35]. Therefore, the number of freeze-thaw cycles is set to 0, 1, 3, 8, and 11 for a total of 5 different freeze-thaw cycles. The freeze-thaw test adopts a large freeze-thaw cycle box with a temperature control range of $-40°$ to $120°$ and an accuracy of $\pm 0.1°$, which can meet the test requirements. Temperature control adopts the 11 + 1 mode, which involves freezing for 11 h and stepped temperature adjustment for 1 h. The freezing and thawing stages are both 12 h to ensure that the sample can be fully frozen and thawed. Every cycle takes 24 h to complete, and after reaching the designed number of cycles is performed, the sample is removed from the box and rested for 3 h before shear testing.

(4) Confining pressure setting. At the end of the freezing and thawing treatment, the solid anchor is placed into the anchor support with a hollow groove in the centre and the fixing device and confining pressure chamber are installed. The confining pressure chamber is filled when ready. The pump is used to quickly inject water, and the upper part of the confining pressure chamber has holes to remove air during the water injection stage. The bolts are tightened to close the holes when the water is about to overflow the chamber so that the entire confining pressure chamber is kept in a sealed environment. Afterwards, a pressurising device is used to pressurise the injected water. After the confining pressure reaches the set value, the confining pressure is kept stable for a period. The control and acquisition software is opened in the data acquisition terminal, and the corresponding parameters are input to pull the bolt at a steady rate of displacement. The shear displacement and shear stress in the drawing process are transmitted to the computer terminal through the acquisition system in real time for subsequent data storage and processing.

3. Results and Analysis

3.1. Shear Characteristics Tests of Anchor Rods under Static Loading

This test is completed in the laboratory of the Hebei University of Architecture and Engineering with a frozen soil triaxial testing system manufactured by Nanjing Tekao, Co. (Nanjing, China). After the independent modification of this test device, the base is replaced to achieve different sizes of anchor-soil interface shear characteristics and the effect of static pull-out. The test is carried out in a sealed environment in the confining pressure chamber, where the solid anchor is pressurised by water injection, the base is controlled to lower at a rate of 1 mm/min, and the force transmission between the anchor-soil interface is collected and transmitted to the computer for display and reprocessing through the force transducer fixed on the counterforce frame.

3.1.1. Ultimate Pull-Out Force of an Anchor Rod under Static Loading

The pull-out test differs from the shear test in that it achieves pulling out for different specimens and installation bases. In the process of making the parts, the step of cutting the soil is eliminated to keep the soil sample intact. During installation, solid fixed supports are removed to gradually reduce the contact area between the anchor and the soil during the pulling process, consistent with the actual working conditions of pulled anchor rods. This work includes pull-out tests on nine samples with three different sand contents and three different confining pressures as variables, resulting in the pull-out curves shown in Figure 4.

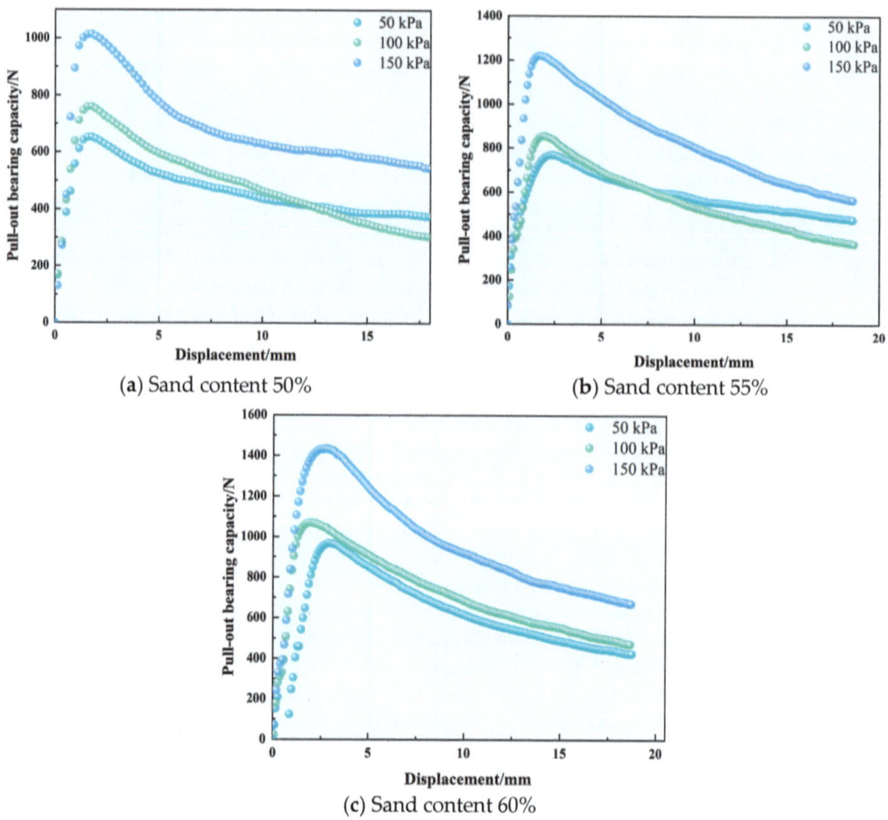

(**a**) Sand content 50%

(**b**) Sand content 55%

(**c**) Sand content 60%

Figure 4. Load-displacement curves at different sand content levels.

The pull-out curve initially rapidly increases to its peak value with a linear slope but then decreases at a faster rate of change. During the decreasing process, the slope of the curve gradually decreases while the pull-out load continues to decrease.

For the sample with a sand content of 50%, the peak load increases from 653.45 N to 1015.91 N; for the sample with a sand content of 55%, the peak load increases from 769.58 N to 1220.04 N; for the sample with a sand content of 60%, the peak load increases from 969.85 N to 1435.97 N. The confining pressure is positively correlated with the peak load, and appropriately increasing the radial stress of the anchoring body can increase the friction and particle interlocking force at the anchor-soil interface, thereby improving the bearing stability of the anchor rod. By controlling the constant confining pressure and observing the effect of increasing the sand content on the peak load, which is under 50 kPa, when the sand content of the specimen increases from 50% to 60%, the peak load increases by 316.39 N; under 100 kPa, the peak load increases by 308.71 N; and under 150 kPa, the peak load increases by 420.07 N. It can be seen that appropriately increasing the sand content also helps to increase the stability of the soil and improve the pull-out resistance of the anchor rod.

This paper concentrates on the static load-bearing characteristics of anchors, but we also acknowledge the significant impact of dynamic loads on the stability of anchor support systems in practical engineering. During construction, anchors are subjected to prestress as well as complex influences from the surrounding rock and soil. Throughout their service life, dynamic loads, such as earthquakes, traffic vibrations, and blasting, can cause damage or even failure to the rock, soil and support structures, posing a serious threat to project safety and leading to significant economic losses. Therefore, an in-depth study of the changes in the pull-out capacity of anchors under cyclic alternating loads and an accurate assessment of the actual pull-out capacity of in-service anchor bodies with prestress applied are crucial for ensuring the long-term stability of anchor support systems.

3.1.2. Constitutive Relationship of the Anchor-Soil Interface throughout the Test Process

This paper adopts a normalisation approach to fit the load-displacement curve of the anchor rod and the shear stress-shear displacement curve [14].

The load-displacement curve shown in the figure above is processed as follows: the load and displacement at different times are divided by the peak load p_m and peak displacement s_m, respectively, to obtain the load ratio y and displacement ratio x. That is,

$$y = p/p_m \tag{1}$$

$$x = s/s_m \tag{2}$$

By analysing the normalised curve, the shape of the curve can be fitted using the following equation:

$$y = \frac{(ax + (b-1)x^2)}{1 + (b-2)x + ax^2} \tag{3}$$

In this equation, there is an undetermined parameter. By substituting Equations (1) and (2) into Equation (3), the expression for the constitutive relationship of the entire process of anchor rod pull-out is obtained.

$$p = p_m[ass_m + (b-1)s^2]/[s^2 + (b-2)ss_m + as^2] \tag{4}$$

This gives the normalised fitting curve for the load-displacement relationship; the shear stress-shear displacement relationship described below is also processed with this approach. The experimental results and the normalised fitting results are shown in Tables 3 and 4. Some fitting curves and measured results are shown in Figures 5 and 6. The tests were conducted on unfrozen samples, where A, B, and C represent sand content rates of 50%, 55%, and 60% and 1, 2, and 3 represent the confining pressures of 50 kPa, 100 kPa, and 150 kPa, respectively. Clearly, Equation (4) can fit the pull-out curve and shear displacement curve well and has a high fitting accuracy for all sample curves.

Table 3. Load-displacement fitting results.

Number	A1	A2	A3	B1	B2	B3	C1	C2	C3
a	1.84	1.30	1.48	2.84	1.39	1.40	0.89	1.13	1.62
b	1.92	1.49	1.66	2.54	1.53	1.61	1.21	1.44	1.55
Correlation coefficient	0.98	0.93	0.98	0.95	0.96	0.94	0.94	0.96	0.99

Table 4. Shear stress-shear displacement fitting results.

Number	A1	A2	A3	B1	B2	B3	C1	C2	C3
a	9.17	2.52	1.32	8.82	2.82	2.94	6.49	1.19	0.94
b	7.98	2.77	1.58	8.36	2.91	2.29	5.17	1.75	1.32
Correlaion coefficient	0.99	0.96	0.99	0.99	0.98	0.99	0.98	0.97	0.98

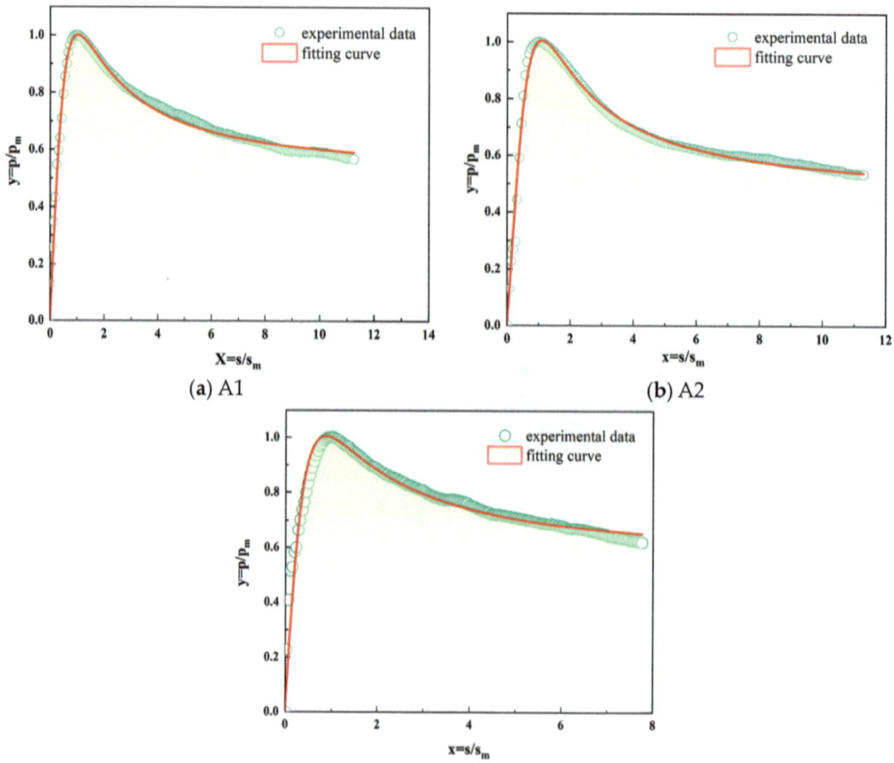

(**a**) A1 (**b**) A2 (**c**) B1

Figure 5. Comparison of the experimental pull-out data and fitted curve.

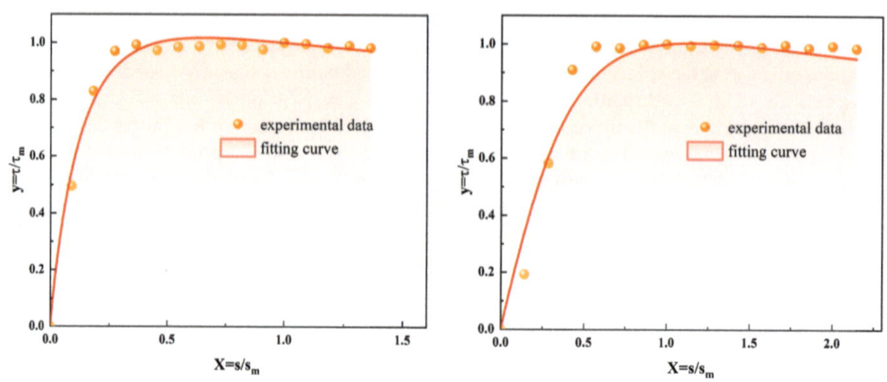

(**a**) A1 (**b**) A2

Figure 6. *Cont.*

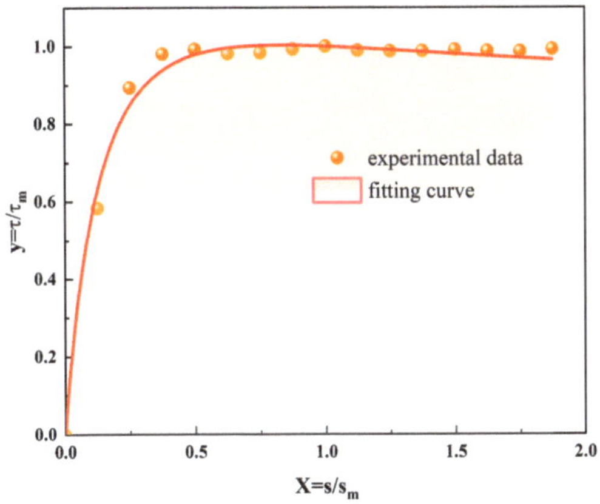

(c) B1

Figure 6. Comparison of the experimental shear displacement data and fitted curve.

3.2. Feature Analysis of the Shear Stress-Displacement Curve of the Anchor-Soil Contact Surface

3.2.1. Analysis of the Influence of Confining Pressure on the Shear Stress-Displacement Curve of an Anchor-Soil Contact Surface

To study the effect of confining pressure on the shear characteristics of the anchor-soil interface, Figure 7 shows the shear stress-displacement curves of soils with sand contents of 50%, 55% and 60%. In the corresponding legends, the value in the first column represents the number of freeze-thaw cycles, the value in the second column is the sand content, the value in the third column is the size of the applied confining pressure, and the curves in the figures are fitted curves.

For the specimens that did not undergo freezing and thawing, it can be seen from Figure 7 that even with different sand contents, the shear stress at the anchor-soil contact surface increases with increasing confining pressure, and the peak displacement also increases gradually, which shows that the anchor is subjected to the stress of the soil, which has a great influence on the shear stress at the interface. The maximum value of the curve into the residual strength stage is taken as the shear strength, which is 75.32 kPa, 99.67 kPa, 123.99 kPa, and 159.64 kPa for the specimens with 50 percent sand content with peripheral pressures of 50 kPa, 100 kPa, 150 kPa, and 200 kPa, respectively. The shear strengths at 100 kPa, 150 kPa, and 200 kPa are approximately 1.32, 1.64, and 2.12 times that of the shear strength at 50 kPa, respectively. For the 55% sand content specimen, the shear strength at 200 kPa is 2.14 times the shear strength at 50 kPa, and for the 60% sand content specimen, the shear strength at 200 kPa is 2.17 times the shear strength at 50 kPa. This shows that the shear strength is linearly correlated with the confining pressure, and the greater the confining pressure is, the greater the shear strength. The reason for this phenomenon is as follows: anchor-soil shear action will form a shear zone at the interface; in the pulling process, the solid anchor is not easy to deform due to its rigid body, so the soil particles within a certain distance around the anchor will rotate. With increasing confining pressure, the area and thickness of the shear zone increases, and the shear strength also increases.

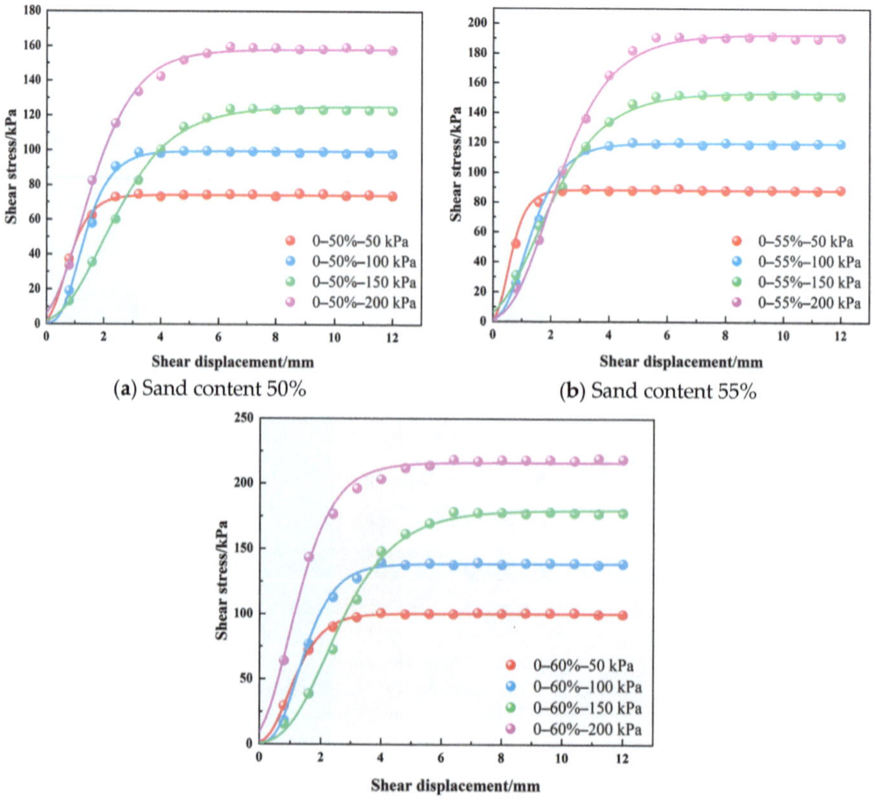

(**a**) Sand content 50%

(**b**) Sand content 55%

(**c**) Sand content 60%

Figure 7. Shear stress-shear displacement curves for different sand contents.

The variation in the displacements corresponding to the peak shear strength is shown in Figure 8. The shear displacements associated with different confining pressures (50 kPa, 100 kPa, 150 kPa, and 200 kPa) for varying sand content percentages (50%, 55%, and 60%) were measured. For a 50% sand content, the shear displacements were found to be 2.97 mm, 3.74 mm, 6.07 mm, and 6.48 mm, respectively. Similarly, for a 55% sand content, the corresponding shear displacements were 3.05 mm, 4.28 mm, 6.33 mm, and 6.49 mm. Finally, for a 60% sand content, the shear displacements at peak shear strengths were recorded as 3.57 mm, 3.98 mm, 6.39 mm, and 6.87 mm at the same confining pressures. It can be concluded that, the greater the applied confining pressure, the greater the displacement of the anchor-soil interface at the peak stress. The smaller circumferential constraints on the anchor and surrounding soil at low confining pressures result in smaller bonding and friction forces at the anchor-soil interface, making it easier for the anchor to reach the stage of plastic deformation and undergo a smaller displacement at the peak strength. At the same confining pressure, the displacement corresponding to the peak strength increases from 2.97 mm to 3.57 mm at 50 kPa when the sand content of the soil body increases from 50% to 60%; it increases from 3.74 mm to 3.98 mm at 100 kPa, from 6.07 mm to 6.39 mm at 150 kPa, and from 6.48 mm to 6.87 mm at 200 kPa. Under the same confining pressure, as the sand content increases, the peak value also slightly increases.

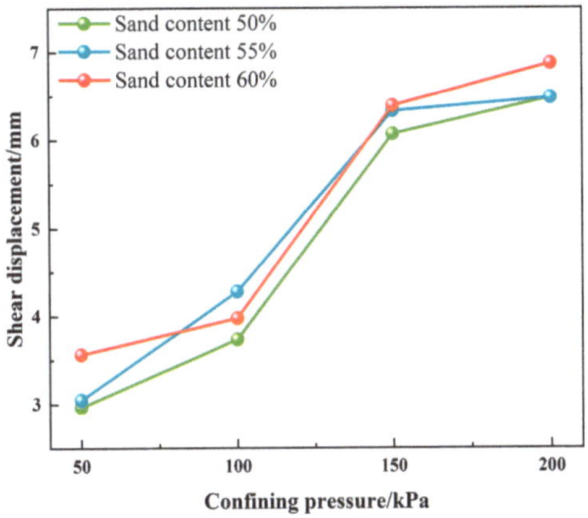

Figure 8. Variations in the displacements corresponding to the peak shear strengths for different sand contents.

3.2.2. Numerical Simulation

This paper employs the FLAC3D 6.0 finite difference software for numerical simulation. The numerical simulation matches the size of the indoor test specimens, and the ratio of the diameter of the test samples to the anchor body should be no less than five, which can effectively reduce the impact of boundary constraint effects. The model consists of two parts: the anchor body and the soil. The soil layer is silty sand, and the soil constitutive model uses an elastic–plastic model, with the elastic part described by a linear elastic model and the plastic part replaced by the Mohr–Coulomb model. The soil has a diameter of 15 cm and a height of 12 cm. The anchor body is simulated using an elastic model, and the effect of density on the gravitational field is neglected, as shown in Figure 9. The simulation is validated by shear tests on specimens with a 50% sand content, and the material parameter assignments are listed in Table 5. The soil is constrained vertically on all sides and the bottom. To prevent model oscillation due to the initial application of force, given the small size of the specimen model, the simulation process uses velocity loading. During the velocity loading process, the velocity does not remain constant; it starts from zero and increases with a constant slope to prevent excessive displacement at the beginning that could damage the model before then maintaining a predetermined value.

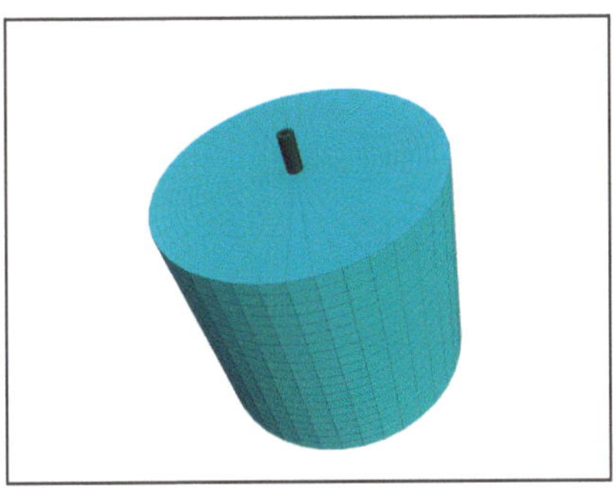

Figure 9. Numerical model.

Table 5. Physical and Mechanical Parameters.

Material Parameter	Young's Modulus (GPa)	Poisson's Ratio	Friction Angle (deg)	Cohesion (kPa)
Soil	0.02	0.35	25.93	53.03
Anchor	29	0.3		

The contact surface employs a statistical damage model, assuming that the micro-element failure follows a two-parameter Weibull distribution with the contact surface shear strain ($\gamma > 0$) as the random variable [36], and its probability density function is:

$$P(\gamma) = \frac{m}{n}\left(\frac{\gamma}{\eta}\right)^{m-1} \exp\left[-\left(\frac{\gamma}{\eta}\right)^{m}\right] \tag{5}$$

In the formula, m and η represent the shape parameter and the scale parameter, respectively. Integrating the probability density function yields the cumulative distribution function $P(\gamma)$. Since the total number of micro-elements on the contact surface is N, the number of damaged micro-elements on the contact surface at shear strain γ is:

$$N''(\gamma) = N\int_0^{\gamma} P(\gamma)d\gamma \tag{6}$$

Substituting Equation (5) into Equation (6) yields the evolution equation of the damage variable. That is,

$$D = 1 - \exp\left[-\left(\frac{\gamma}{\eta}\right)^{m}\right] \tag{7}$$

For the anchor-soil interface, the contact surface strength is jointly provided by the damaged and undamaged parts, and the shear stress relationship on the contact surface satisfies the following formula:

$$\tau = \tau'(1-D) + \tau'' D \tag{8}$$

In the formula, τ, τ', and τ'' represent the apparent shear stress on the contact surface, the shear stress of the undamaged part, and the shear stress of the damaged part, respectively. For the undamaged micro-elements, it is assumed that their constitutive relationship is a nonlinear elastic relationship, which is hyperbolic. For the damaged micro-elements, it is assumed that they can still provide a constant shear strength through the frictional resistance between the anchor and the soil under the action of normal stress.

Substituting Equation (7) into Equation (8) yields the statistical damage constitutive model for the anchor-soil contact surface.

$$\tau = \frac{\gamma}{a + b\gamma}\exp\left[-\left(\frac{\gamma}{\eta}\right)^{m}\right] + \tau''\left\{1 - \exp\left[-\left(\frac{\gamma}{\eta}\right)^{m}\right]\right\} \tag{9}$$

In numerical simulation, the contact surface employs the statistical damage model from the aforementioned formula. The FISH language is utilized for secondary development of the contact elements inherent in FLAC3D, and the developed model is applied to numerically simulate shear tests of the anchor-soil contact surface, analysing the relationship between shear stress and shear displacement.

Figure 10 shows the variation in shear stress for the anchor under four different confining pressures of 50, 100, 150, and 200 kPa, both in simulation and actual measurement. It can be observed from the figure that the simulated curves match well with the actual measured data points. Both simulation analysis and field measurement data consistently show that there is a positive correlation between the pull-out resistance of the anchor.

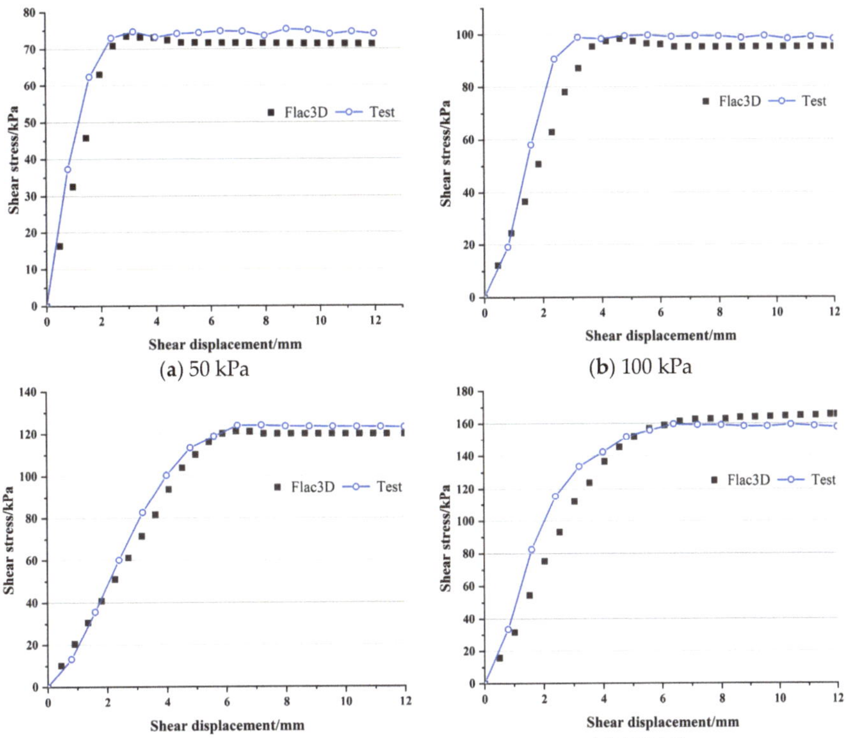

(a) 50 kPa (b) 100 kPa (c) 150 kPa (d) 200 kPa

Figure 10. Comparison of test data and model calculation results.

3.2.3. The Influence of Sand Content on the Shear Strength of an Anchor-Soil Contact Surface

As can be seen from the shear displacement curves in Figure 11, under the same confining pressure, an increase in sand content drives the increase in the shear strength of the anchor-soil interface and slightly increases the corresponding peak displacement, and the shear strength is positively correlated with the confining pressure. As shown in Figure 12, under the condition of 50 kPa, the shear strengths of the anchor-soil interfaces for specimens with a 50% sand content, 55% sand content, and 60% sand content are 75.32 kPa, 89.35 kPa, and 101.20 kPa, which are 1.19 times and 1.35 times higher than that for a specimen with a 50% sand content, respectively. The anchor-soil interface shear strength of a specimen with a 55% sand content increases by 14.02 kPa compared with that with a 50% sand content, and those with a 60% sand content increase by 11.84 kPa compared with those with a 55% sand content. The increase in sand content leads to a gradually diminishing increment in shear strength.

3.2.4. Influence of Freeze-Thaw Cycling on the Strength of a Contact Surface

Figure 13 displays three-dimensional scatter plots illustrating the variation in contact surface shear strength across different levels of confining pressures and varying numbers of freeze-thaw cycles.

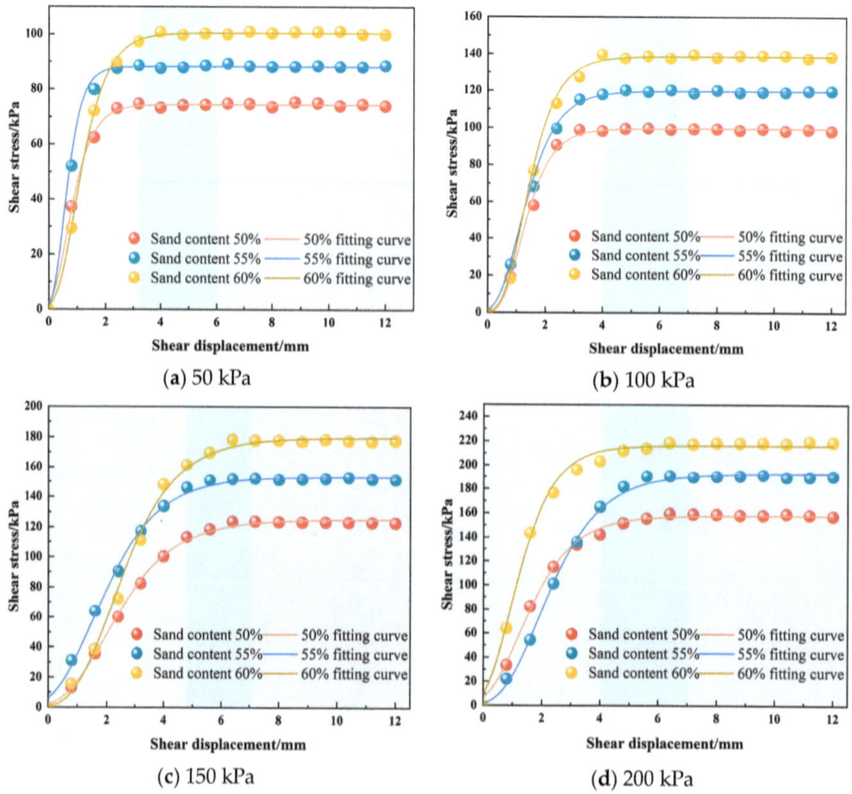

(**a**) 50 kPa (**b**) 100 kPa

(**c**) 150 kPa (**d**) 200 kPa

Figure 11. Shear displacement curves for different confining pressures.

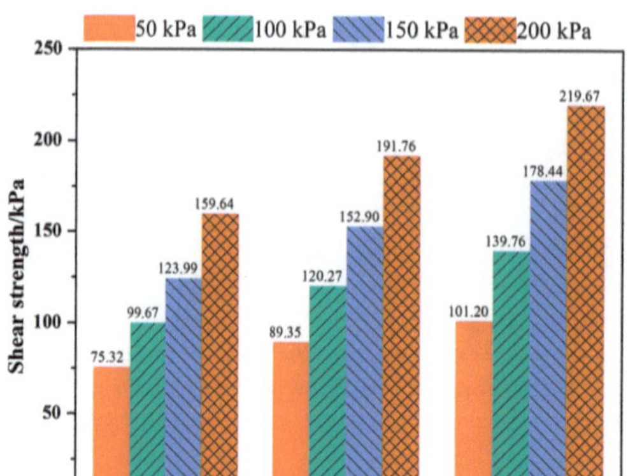

Figure 12. Shear strength results for different sand contents.

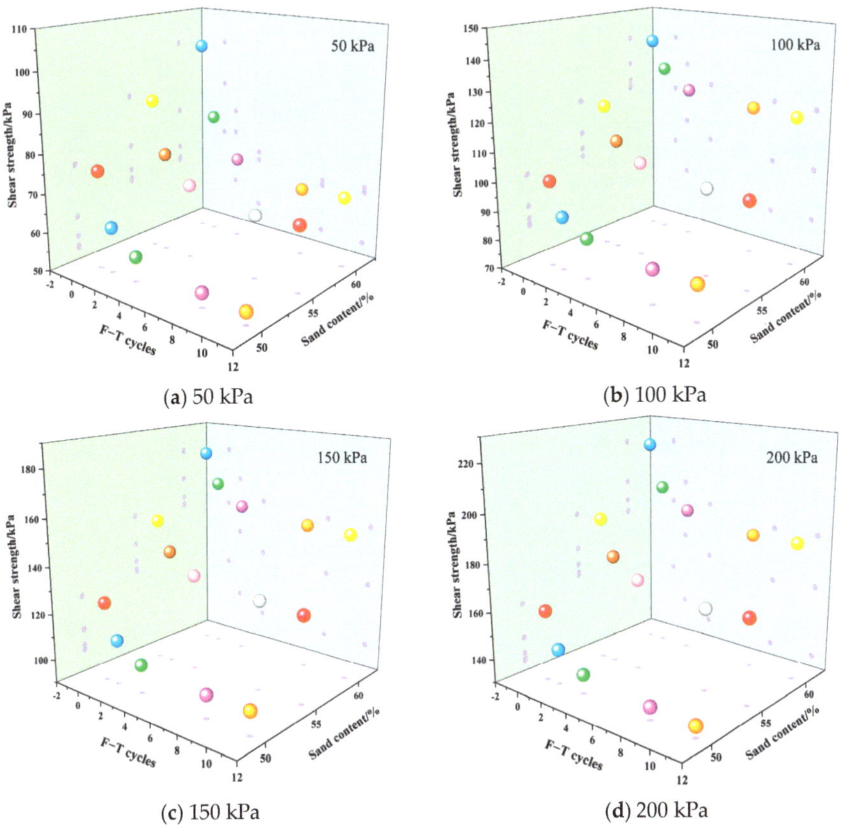

Figure 13. Shear strength of contact surface under different confining pressures and freeze-thaw cycles conditions.

Freeze-thaw cycles significantly reduce the shear strength of the anchor-soil interface. The reduction in shear strength due to freeze-thaw cycles increases with the number of cycles, reaching a stable level after eight cycles. As shown in Figure 13a, when the confining pressure is 50 kPa, the specimens with sand contents of 50%, 55%, and 60% are damaged by freeze-thaw cycles and the shear strength of the anchor-soil interface decreases by 29.3%, 25.1%, and 32.5%, respectively; when the confining pressure is 200 kPa, the shear strength of the anchor-soil interfaces for specimens with sand contents of 50%, 55%, and 60% decrease by 15.6%, 13.6%, and 14.6%, respectively. After stabilisation by 11 freeze-thaw cycles, the soil is subjected to freeze-thaw damage in the range of 15–30%. When the confining pressure increases, the freeze-thaw cycle damage decreases by nearly half, so it can be seen that the change in the confining pressure has a great influence on the damage during freeze-thaw cycles.

Comparing the results of freeze-thaw damage under different sand content, the freeze-thaw damage rates of 50–60% sand content under 100 kPa confining pressure were 17.7%, 16.8%, and 13.4%, respectively. The increase in sand content reduces the decrease in the shear strength due to freeze-thaw cycles and helps to improve the resistance of anchors to freeze-thaw effects. Because of the high sand content of the soil, the particle gradation is uniform, and the arrangement between particles is more compact. The sand particles have a high shear strength and freeze-thaw resistance, which can better withstand the stress and deformation caused by freeze-thaw cycles and, thus, are less prone to breakage or deformation than clay particles.

3.2.5. Influence of Freeze-Thaw Cycles on the Apparent Cohesion and Internal Friction Angle of the Contact Surface

Based on the shear strengths of the anchor-soil interface obtained in the previous section for different numbers of freeze-thaw cycles, the Coulomb damage criterion is used to fit the experimental data and determine the internal friction angle and cohesion at the anchor-soil interface.

Figure 14 shows the damage curve of cohesion at the anchor-soil interface under freeze-thaw cycles; with the increase in the number of freeze-thaw cycles, the cohesion decreases, and the cohesion decreases the most with the first freeze-thaw cycle. The cohesions of the specimens with sand contents of 50%, 55% and 60% decrease by 26.2%, 21.6%, and 23.9%, respectively. After the first three freeze-thaw cycles, the soil damage is basically completed, and the cohesion of the specimens with 50%, 55%, and 60% sand contents decreases by 35.3%, 32.4%, and 39.4%, respectively. Then, as the number of freeze-thaw cycles continues to increase, the cohesion decreases slightly, and the cohesion remains stable after eight cycles. When not subjected to freezing and thawing, the cohesive forces of the specimens with 50%, 55% and 60% sand content are 45.33 kPa, 53.60 kPa and 61.24 kPa, respectively. As the sand content of the specimen increases, the volume and surface area of sand also increase, leading to a larger contact area between the particles. This results in a greater distribution of the water film on the surface of the sand particles, which in turn generates a higher adhesive cohesion.

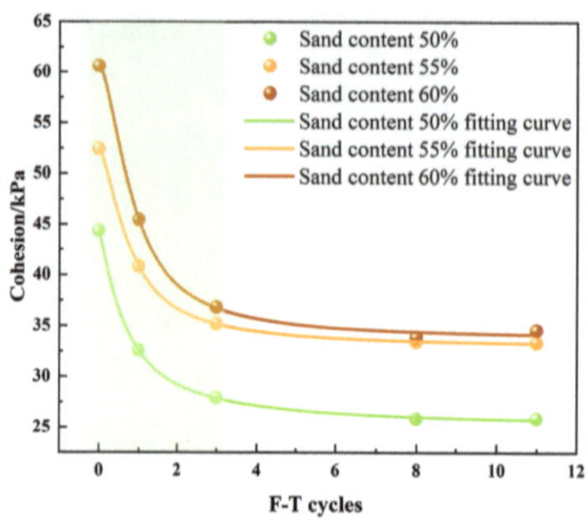

Figure 14. Change in cohesion with the number of freeze-thaw cycles.

Figure 15 shows the curve of the variation in the internal friction angle. The internal friction angle is affected by the first few freeze-thaw cycles, but only slightly. The internal friction angle does not change much during the freeze-thaw process, the internal friction angle of the soil with a 50% sand content decreases from 29.1° to 27.7°, the internal friction angle of the soil with a 55% sand content decreases from 34.2° to 32.9°, and the internal friction angle of the soil with a 60% sand content decreases from 38.2° to only 37.9°. In general, under the action of freeze-thaw cycles, the internal friction angle at the anchor-soil interface of samples with different sand content will slightly decrease, but the decrease is not significant.

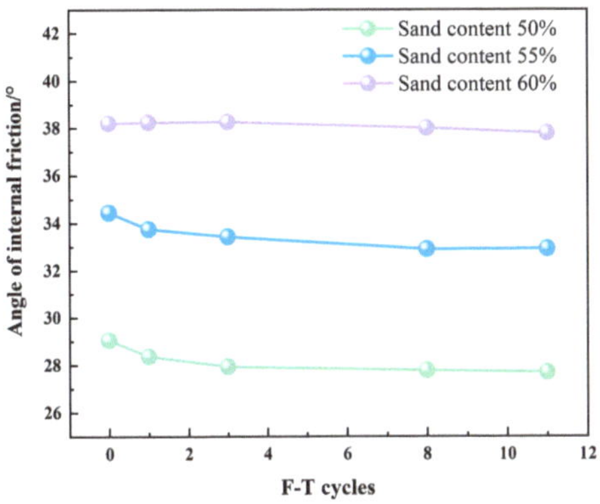

Figure 15. Variation in internal friction angle with freezing-thawing cycles.

4. Discussion

The current research on the shear performance between soil and structures often employs direct shear tests [33]. However, these tests have several limitations, including unstable shear areas, interface shapes that do not match actual conditions, and difficulties in precisely controlling the forces on the anchor body, leading to failure in accurately reflecting the stress state of anchors in engineering applications.

This paper conducted an in-depth analysis of the shear characteristics of anchors in sandy soil under freeze-thaw cycles in northern China and found that, as the sand content in the specimens increased, the interface shear strength between the anchor and soil significantly improved. Specifically, under a confining pressure of 50 kPa, the interface shear strength at a sand content of 60% was 1.34 times that at a sand content of 50%. This indicates that an increase in sand content helps to enhance soil density and strengthen the interlocking effect between sand particles, thereby increasing cohesion and the angle of internal friction, and thus enhancing shear strength.

The impact of freeze-thaw cycles on the strength of the anchor interface was also discussed. After the first freeze-thaw cycle, the interface strength suffered the greatest damage, but as the number of cycles increased, the cohesion and angle of internal friction gradually decreased, stabilising after eight cycles. This pattern is similar to the research findings of Jiankun Liu [28]. This is because the volume expansion and contraction caused by the phase change in ice and water during the freeze-thaw process damaged the connections between soil particles, leading to a reduction in cohesion and the angle of internal friction.

Despite providing valuable insights, this paper has some limitations. Firstly, the standardized sandy soil used in the experiments may not fully simulate the complex conditions on site. Secondly, although the mechanism of the shear strength of the anchor-soil interface was explained from a macro perspective, the processes of ice–water phase change, soil deformation, and their superimposed modes at the micro level are not yet clear.

To overcome these limitations, future research should delve into the following areas: further testing of the microstructure of specimens under different freeze-thaw cycles and at room temperature using scanning electron microscopy (SEM) to observe changes in soil particle morphology, pore structure, and aggregate structure, as well as to obtain a more accurate assessment of the shear strength of the anchor-soil interface through quantitative analysis of micro parameters. We should expand the scope of the research to consider the changes in the internal forces of anchors and the overall structural stability in the support of slopes with frame anchors in the seasonal freezing area, helping to build a connection between anchor research and engineering applications to provide more comprehensive guidance for engineering practice.

5. Conclusions

In the context of complex freeze-thaw conditions and severe soil sandification in the cold regions of northern China, the safety and stability of foundation-pit-anchoring engineering is of utmost importance. This paper conducts experiments using a self-developed anchor-soil shear-pull device to study the influence of factors such as sand content, confining pressure, and freeze-thaw cycles on the shear characteristics of the anchor-soil interface. Indoor model tests are used to analyse the pull-out and shear results and to con-struct a unified model of the anchor-soil interface behaviour. The main conclusions are as follows:

(1) The pull-out model of anchor rods is different than the shear model, and both curves can be fitted using a comprehensive model and normalized, with the fitting results showing a high degree of agreement with the measured curves.
(2) The increase in the confining pressure has a significant effect on the shear strength of the anchor-soil interface. With a 50% sand content, the shear strengths under 100 kPa, 150 kPa, and 200 kPa confining pressures are approximately 1.32, 1.64, and 2.12 times higher than that under 50 kPa pressure, respectively, and the corresponding peak displacement is also increased by more than two times. The pattern of peak displacements observed in soils with different sand content is consistent, with sand content having a small effect on peak displacements.
(3) Under the same confining pressure, the shear strength of the anchor-soil interface is greatly affected by the sand content of the soil. Under the condition of 50 kPa, the anchor-soil interface shear strength with a 55% sand content and 60% sand content is 1.18 times and 1.34 times that of with a 50% sand content, respectively.
(4) Freeze-thaw cycles significantly reduce the shear strength of an anchor-soil interface. Comparing the results of the decreases in shear strength for different sand content rates, the results with 50–60% sand contents are 19.3%, 16.7%, and 14.5%, respectively, under 50 kPa. The appropriate increase in sand content will improve the freeze-thaw resistance of the anchored soil body. The anchor soil with a sand content of 60% experiences an increase in confining pressure from 50kPa to 200kPa, resulting in a decrease in shear strength damage from 32.5% to 14.6%. After a freeze-thaw cycle, the cohesion of the soil specimens with different sand contents is reduced by more than 30%, and the decrease in the internal friction angle is very small.

Author Contributions: Conceptualisation, Y.-C.W.; methodology, Y.-C.W. and Z.-H.W.; validation, Y.-D.Z. and H.-F.Z.; formal analysis, F.-W.G., Y.-C.W. and Z.-H.W.; investigation, J.D. and Y.-D.Z.; resources, Y.-C.W. and H.-F.Z.; data curation, Y.-C.W., F.-W.G. and Z.-H.W.; writing—original draft preparation, Y.-C.W. and F.-W.G.; writing—review and editing, H.-F.Z. and Z.-H.W.; funding acquisition, J.D. All authors have read and agreed to the published version of the manuscript.

Funding: The research described in this paper was financially supported by the Natural Science Foundation of China (NO. 51878242), The Natural Science Foundation of the Hebei Province of China (NO. E2020404007), The Research Project of Young Top Talent in Hebei Province (BJK2024116), The Project of Research Start-up Fund (B-202307).

Institutional Review Board Statement: Not applicable.

Informed Consent Statement: Not applicable.

Data Availability Statement: The data presented in this study are available upon request from the corresponding author.

Conflicts of Interest: The authors declare no conflicts of interest.

References

1. Yang, Z.; Peng, W.; Cao, J.; Xie, D. Application of Wireless Sensor Network in Rock Bolt Pulling Force Distribution Pattern Monitoring. *Iop Conf. Ser. Earth Environ. Sci.* **2019**, *267*, 042005. [CrossRef]
2. Ho, S.C.M.; Li, W.; Wang, B.; Song, G. A load measuring anchor plate for rock bolt using fiber optic sensor. *Smart Mater. Struct* **2017**, *26*, 057003. [CrossRef]

3. Barnard, C.; Kallu, R.R.; Warren, S.; Thareja, R. Inflatable rock bolt bond strength versus rock mass rating (RMR): A comparative analysis of pull-out testing data from underground mines in Nevada. *Int. J. Min. Sci. Technol.* **2016**, *26*, 19–22. [CrossRef]
4. Suksiripattanapong, C.; Horpibulsuk, S.; Udomchai, A.; Arulrajah, A.; Tangsutthinon, T. Pullout resistance mechanism of bearing reinforcement embedded in coarse-grained soils: Laboratory and field investigations. *Transp. Geotech.* **2020**, *22*, 100297. [CrossRef]
5. Chen, A.; Wang, Q.; Chen, Z.; Chen, J.; Chen, Z.; Yang, J. Investigating pile anchor support system for deep foundation pit in a congested area of Changchun. *Bull. Eng. Geol. Environ.* **2020**, *80*, 1125–1136. [CrossRef]
6. Sun, Y.; Li, Z. Analysis of Deep Foundation Pit Pile-Anchor Supporting System Based on FLAC3D. *Geofluids* **2022**, *2022*, 1699292. [CrossRef]
7. Su, T.; Zhou, Y.; Wang, Z.; Ye, S. Large Scale Model Test Study of Foundation Pit Supported by Pile Anchors. *Appl. Sci.* **2022**, *12*, 9792. [CrossRef]
8. Cui, G.; Ma, S.; Liu, Z.; Liu, S.; Xi, C.; Cheng, Z. Effect of freeze-thaw cycles on deformation properties of deep foundation pit supported by pile-anchor in Harbin. *Rev. Adv. Mater. Sci.* **2022**, *61*, 756–768. [CrossRef]
9. Su, W.; Fragaszy, R.J. Uplift Testing of Model Anchors. *J. Geotech. Eng.* **1988**, *114*, 961–983. [CrossRef]
10. Martín, L.B.; Tijani, M.; Hadj-hassen, F. A new analytical solution to the mechanical behaviour of fully grouted rockbolts subjected to pull-out tests. *Constr. Build. Mater.* **2011**, *25*, 749–755. [CrossRef]
11. Ghadimi, M.; Shahriar, K.; Jalalifar, H. A new analytical solution for the displacement of fully grouted rock bolt in rock joints and experimental and numerical verifications. *Tunn. Undergr. Space Technol.* **2015**, *50*, 143–151. [CrossRef]
12. Xi, X.H.; Gu, S.C. Shearing Stress Model of Damage Bolt in Tunnel. *Appl. Mech. Mater.* **2011**, *90–93*, 1761–1767. [CrossRef]
13. Zhu, Z.; Shu, X.; Li, Z.; Tian, H.; Tian, Y. Stress analysis of full-length grouted bolt under shear deformation of anchor interface. *J. Mt. Sci.* **2022**, *19*, 3286–3301. [CrossRef]
14. Chen, C.; Liang, G.; Tang, Y.; Xu, Y. A new method for indoor testing of interface characteristics between anchor solid and soil for anchor rods. *J. Geotech. Eng.* **2015**, *37*, 1115–1122. (In Chinese)
15. Chang, D.; Liu, J. Review of the influence of freeze-thaw cycles on the physical and mechanical properties of soil. *Sci. Cold Arid. Reg.* **2013**, *5*, 457–460.
16. Qi, J.; Vermeer, P.A.; Cheng, G. A review of the influence of freeze-thaw cycles on soil geotechnical properties. *Permafr. Periglac. Process.* **2006**, *17*, 245–252. [CrossRef]
17. Konrad, J. Physical processes during freeze-thaw cycles in clayey silts. *Cold Reg. Sci. Technol.* **1989**, *16*, 291–303. [CrossRef]
18. Chamberlain, E.J.; Gow, A.J. Effect of freezing and thawing on the permeability and structure of soils. *Eng. Geol.* **1979**, *13*, 73–92. [CrossRef]
19. Leuther, F.; Schlüter, S. Impact of freeze-thaw cycles on soil structure and soil hydraulic properties. *Soil* **2021**, *7*, 179–191. [CrossRef]
20. Nartowska, E.; Kozłowski, T. The Effect of Freeze-Thaw Cycling and the Initial Mass of Water on the Unfrozen Water Content of Calcium Bentonites Modified by Copper Ions. *Minerals* **2022**, *12*, 66. [CrossRef]
21. Zhao, Y.; Hu, X. How Do Freeze-Thaw Cycles Affect the Soil Pore Structure in Alpine Meadows Considering Soil Aggregate and Soil Column Scales? *J. Soil Sci. Plant Nutr.* **2022**, *22*, 4207–4216. [CrossRef]
22. Eskişar, T.; Altun, S.; Kalıpcılar, İ. Assessment of strength development and freeze-thaw performance of cement treated clays at different water contents. *Cold Reg. Sci. Technol.* **2015**, *111*, 50–59. [CrossRef]
23. Zhao, L.; Peng, J.; Ma, P.; Leng, Y.; Ma, Z. Microstructure response to shear strength deterioration in loess after freeze-thaw cycles. *Eng. Geol.* **2023**, *323*, 107229. [CrossRef]
24. Xu, J.; Li, Y.; Lan, W.; Wang, S. Shear strength and damage mechanism of saline intact loess after freeze-thaw cycling. *Cold Reg. Sci. Technol.* **2019**, *164*, 102779. [CrossRef]
25. Xu, J.; Ren, J.; Wang, Z.; Wang, S.; Yuan, J. Strength behaviors and meso-structural characters of loess after freeze-thaw. *Cold Reg. Sci. Technol.* **2018**, *148*, 104–120. [CrossRef]
26. Qi, J.; Ma, W.; Song, C. Influence of freeze-thaw on engineering properties of a silty soil. *Cold Reg. Sci. Technol.* **2008**, *53*, 397–404. [CrossRef]
27. Qu, Y.; Ni, W.; Niu, F.; Mu, Y.; Luo, J.; He, H. Shear Properties and Mechanism of Freeze-Thaw Interface in Unsaturated Coarse-Grained Soil from Qinghai-Tibet Plateau. *Adv. Civ. Eng.* **2021**, *2021*, 6688724. [CrossRef]
28. Liu, J.; Chang, D.; Yu, Q. Influence of freeze-thaw cycles on mechanical properties of a silty sand. *Eng. Geol.* **2016**, *210*, 23–32. [CrossRef]
29. Khoury, N.N.; Zaman, M.M. Effect of Wet-Dry Cycles on Resilient Modulus of Class C Coal Fly Ash-Stabilized Aggregate Base. *Transp. Res. Rec. J. Transp. Res. Board* **2002**, *1787*, 13–21. [CrossRef]
30. Yao, Z.; Du, X.; Zhang, Z. Effects of ultra-freezing/thawing cycles on the mechanical properties and microstructure of Shanghai clay. In Proceedings of the Third International Conference on Optoelectronic Science and Materials (icosm 2021), Spie, Hefei, China, 10–12 September 2021.
31. Mu, J.; Pei, X.; Huang, R.; Rengers, N.; Zou, X. Degradation characteristics of shear strength of joints in three rock types due to cyclic freezing and thawing. *Cold Reg. Sci. Technol.* **2017**, *138*, 91–97. [CrossRef]
32. He, P.; Mu, Y.; Yang, Z.; Ma, W.; Dong, J.; Huang, Y. Freeze-thaw cycling impact on the shear behavior of frozen soil-concrete interface. *Cold Reg. Sci. Technol.* **2020**, *173*, 103024. [CrossRef]

33. Pan, R.; Yang, P.; Shi, X.; Zhang, T. Effects of freeze-thaw cycles on the shear stress induced on the cemented sand–structur interface. *Constr. Build. Mater.* **2023**, *371*, 130671. [CrossRef]
34. Dong, J.; Tao, C.; Zhong, S.; Wu, Z.; Yan, X. Shear behaviour of anchor solid-soil interface under confining pressure an freeze-thaw cycles. *Arab. J. Geosci.* **2022**, *15*, 879. [CrossRef]
35. Pan, J.; Wang, B.; Wang, Q.; Ling, X.; Liu, J.; Fang, R.; Wang, Z. Deformation characteristics of the shear band of silty clay–concret interface under the influence of freeze-thaw. *Cold Reg. Sci. Technol.* **2023**, *206*, 103750. [CrossRef]
36. Yang, L.; Liu, Q. Statistical damage constitutive model for soil-structure contact surface. *J. Undergr. Space Eng.* **2006**, 79–82+8 (In Chinese)

buildings

MDPI

Article

Nonlinear Dynamic Response of Galfenol Cantilever Energy Harvester Considering Geometric Nonlinear with a Nonlinear Energy Sink

Lingzhi Wang, Chao Liu, Weidong Liu, Zhitao Yan * and Xiaochun Nie *

School of Civil Engineering and Architecture, Chongqing University of Science and Technology, Chongqing 401331, China; lzwang@cqust.edu.cn (L.W.); 2022206027@cqust.edu.cn (C.L.); 2021206040@cqust.edu.cn (W.L.)

* Correspondence: yanzhitao@cqu.edu.cn (Z.Y.); xiaochunn@cqust.edu.cn (X.N.)

Abstract: The nonlinear energy sink (NES) and Galfenol material can achieve vibration suppression and energy harvesting of the structure, respectively. Compared with a linear structure, the geometric nonlinearity can affect the output performances of the cantilever beam structure. This investigation aims to present a coupled system consisting of a nonlinear energy sink (NES) and a cantilever Galfenol energy harvesting beam with geometric nonlinearity. Based on Hamilton's principle, linear constitutive equations of magnetostrictive material, and Faraday's law of electromagnetic induction, the theoretical dynamic model of the coupled system is proposed. Utilizing the Galliakin decomposition method and Runge–Kutta method, the harvested power of the external load resistance, and tip vibration displacements of the Galfenol energy harvesting model are analyzed. The influences of the external excitation, external resistance, and NES parameters on the output characteristic of the proposed coupling system have been investigated. Results reveal that introducing NES can reduce the cantilever beam's vibration while considering the geometric nonlinearity of the cantilever beam can induce a nonlinear softening phenomenon for the output behaviors. Compared to the linear system without NES, the coupling model proposed in this work can achieve dual efficacy goals over a wide range of excitation frequencies when selecting appropriate parameters. In general, large excitation amplitude and NES stiffness, small external resistance, and small or large NES damping values can achieve the effect of broadband energy harvesting.

Keywords: Galfenol; nonlinear energy sink (NES); geometric nonlinearity; energy harvesting; vibration suppression

1. Introduction

With the development of micro-electro-mechanical devices towards miniaturization, intelligence, wireless, and low power consumption, traditional chemical battery power supply methods are no longer applicable. Therefore, thanks to the priorities of high energy conversion efficiency, easy miniaturization, and strong applicability utilizing smart materials to transform ambient mechanical energy into usable electrical energy has become the current research focus for scholars. The harvested electric energy can power low-power online monitoring and sensing devices in practical engineering, such as vehicle tracking, structural health monitoring, environmental monitoring, intelligent building, and aerospace fields [1–3]. In the literature, several energy harvesting mechanisms have been introduced to transform such vibration energy into electricity, for instance, electrostatic [4,5], electromagnetic [6,7], piezoelectric [8–12], and magnetostrictive [13,14].

The magnetostrictive vibration energy harvester has attracted growing attention thanks to its peculiar properties. For instance, high magnetostriction, high permeability, small brittleness, high tensile strength, and a wide working temperature range are reported in the previous works [2,15]. Through the inverse effect (Virari effect) of magnetostrictive materials, including Terfenol-D and Galfenol material, the vibration energy can be effectively transferred into electric energy by the magnetostrictive vibration energy harvester [16–18]. Compared to Terfenol-D, Galfenol has better magnetic and mechanical properties, namely, high magnetostriction and magnetic permeability, ideal tensile and compressive mechanical properties, and a wide operating temperature range [19–22]. These superior features of Galfenol make it more suitable for energy harvesting. Yoo and Flatau [18] established the mechanical equation of the lumped parameter model for the Galfenol cantilever beam harvester and the constitutive relationship of the magnetostrictive material. The output results of the proposed harvester had been tested by the indoor experiment. Cao et al. [23,24] developed a cantilever-based energy harvester using the Galfenol material and derived the nonlinear coupled dynamic equations for the designed energy harvester. The output performances of the proposed device were numerically calculated by the Newmark method and verified by the experimental data. Clemente et al. [25] presented a force-activated kinetic energy conversion device and analyzed the energy harvester's output performances through experimental tests as well as FEM COMSOL Multiphysics simulation. Jin et al. [26] proposed a cylindrical wave energy harvester, of which the Galfenol sheet is chosen as the core material. The motion characteristics and the induction voltage of the Galfenol device were analyzed by theoretical derivation analysis, the ANSYS Workbench, as well as experimental tests. Wang et al. [15] established the electromechanically coupled model of the Galfenol-based cantilever energy harvester. In their work, the explicit analytical formulations of the output performances for the proposed energy harvesting model were obtained by the Galerkin decomposition method and electro-mechanical decoupling method.

Excessive structural vibration response will damage the structure; therefore, vibration suppression of structures has always been a hot research topic for global scholars and engineers. Due to the benefits of simple structure, broad operating frequency bandwidth, and great robustness, the nonlinear energy sink (NES) has attracted extensive focus from vibration control researchers [27,28]. The NES is a new kind of nonlinear dynamic absorber proposed by Vakakis [29] and Gendelman et al. [30] in 2001. Numerous investigations have proved that an NES with third-order nonlinear stiffness can remarkably reduce the vibration amplitude of the main vibration structure [31–34]. Recently, scholars have combined the NES and energy harvesters to achieve the dual efficacy goal, namely, vibration suppression and energy harvesting [35,36]. Kremer and Liu [37] studied the steady performances of the coupled system for an NES attached to an energy harvester by analytical and experimental methods. The study showed that the developed system can achieve dual efficacy goals in a broadband range. Xiong et al. [38] designed an NES-piezoelectric energy harvesting system. The mixed multi-scale and harmonic balance methods, as well as the Newton-Raphson harmonic balance method, were employed to analyze the nonlinear dynamic behavior and harvested power of the constructed coupling system. The results revealed that the coupled system could decline the vibration from the main structure and harvest energy in a wide frequency band. Tian et al. [39] coupled an energy-harvester-enhanced NES system into the cantilever trapezoidal plate to control the aeroelastic behaviors of the plate and harvest the vibration energy. Compared to the system with only the NES, the suppression effect of the energy-harvester-enhanced NES has been demonstrated better. Zhang's team conducted extensive research by integrating the NES and giant magnetostrictive material (GMM) to achieve vibration suppression and energy capture. The traditional NES [40] and lever-type NES [41] coupled by rod-type GMM [42] and cantilever-type GMM [43] had been investigated for application in the one degree of freedom primary system [44], two degrees of freedom primary system [45], and aerospace industry [46].

In the current theories, the high-order terms in the first derivative of vibration displacement were generally ignored for the derivation of the energy harvesting model. Therefore, the influence of the geometric nonlinear characteristics of the cantilever beam on the free end motion was omitted [47]. A study was performed to discuss the relationship between structural size parameters and the features of cantilever-based piezoelectric energy harvesters by Tan et al. [48]. The results showed that once the wind speed surpasses a certain value, the geometric nonlinearity will significantly impact the output responses of the energy harvester. Li et al. [47] investigated the influence of geometric nonlinear characteristics on the dynamic response of the beam-mass-foundation system and concluded that larger errors were produced when the effect of the geometrical nonlinearity was neglected. According to the Euler-Bernoulli beam theory and the extended Hamilton principle, Shooshtari et al. [49] discussed the impacts of the geometric nonlinearity of the cantilever structure on the free vibration behaviors of the coupled energy harvesting system. The piezoelectric layer's length was found to have a greater effect on the response of the analyzed system. Li et al. [50] discussed the effects of the dual nonlinear properties on the performances of the L-shaped harvester. It was concluded that with the increase in the excitation amplitude, the geometric nonlinearity will significantly impact the output features. Nie et al. [51–54] introduced the structure's and piezoelectric materials' geometric nonlinearities into the L-shaped piezoelectric energy harvester. The analysis indicated that the energy harvester with the geometric nonlinearities considered could harvest power more efficiently with a reduced vibration displacement over a wider frequency bandwidth.

From the above previous literature, we can find that considering geometric nonlinear effects can enhance the prediction exactitude of the output properties of the energy harvester, especially for large-amplitude vibrations and high electric field conditions. Combining NES and energy harvesting modules can simultaneously achieve both vibration reduction and energy harvesting. As a new type of magnetostrictive material, Galfenol has become a particularly promising transducer material due to its excellent characteristics and is gradually being applied in vibration energy harvesting. In our previous work, the research focus was more on introducing geometric nonlinearity into piezoelectric energy harvesters [51–54]. In 2023, based on Galfenol materials, we proposed a magneto-electric energy harvester without taking into account the cantilever structure's geometric nonlinearity [15]. To the best of the authors' knowledge, there have been no reports on the simultaneous introduction of NES and geometric nonlinearity into the Galfenol energy harvester. Therefore, an electromechanically coupled system integrated by the NES and Galfenol energy harvester with geometric nonlinearity considered is introduced in this work. For various structures in engineering, the prerequisite for analyzing the performance impact mechanisms is to establish more accurate static and dynamic models. For example, Han et al. [55–57] improved the efficiency and accuracy of analyzing bearing systems by establishing and modifying nonlinear dynamic models of bearing systems. Wang et al. [58] constructed the kinematic model of a bistable origami flexible gripper based on the relationship between crease angles and unfolding angles in auxiliary spherical triangles to analyze and improve the grabbing capacity. Then the drawbacks of traditional gripper models, such as slow response speed and low grasping efficiency, were settled. Therefore, to analyze the output characteristic of the proposed coupling system, the electromechanically coupled dynamic equations of the proposed coupling system are first established according to the Hamiltons principle, the linear constitutive relationships of Galfenol materials, and Faraday's law of electromagnetic induction. The Galerkin approximation is adopted for separating the time and displacement variables, and the Runge-Kutta method is used to analyze the output performances of the linear and nonlinear coupled systems. In addition, the influences of various parameters of the designed system on the output responses are explored. Through parameter analysis, it is revealed that the proposed coupled model can realize the dual effect goals of vibration suppression and energy capture when the model parameters are set in a certain range.

2. Mathematical Model

As illustrated in Figure 1, the proposed cantilever vibration suppression and energy harvesting system, which is referred to as the NES-Galfenol coupled system, consists of two main components: the cantilever energy harvesting system and the NES system. The cantilever energy harvesting system is composed of an aluminum and Galfenol composite cantilever beam wrapped with coils, a lumped mass, and an external circuit. Note that the structural aluminum layer of the composite cantilever is bounded by a Galfenol sheet and tightly wound with a coil. Besides, the lumped mass M is fixed at the end of the composite cantilever beam, and an external circuit is connected to each end of the coil. It should be pointed out that, as plotted in Figure 1, the fixed end of the cantilever structure can be considered as the elastic foundation. Then the proposed NES-Galfenol-coupled system can be treated as a cantilevered beam-like structure laying on an elastic foundation and can be modeled by the Winkler spring model. The NES system is comprised of a lightweight lumped mass, a viscous damping, and a spring with cubic nonlinear stiffness. Under the external harmonic base excitation, as the composite cantilever beam vibrates, the magnetic induction intensity of the coil will alter, benefiting from the inverse magnetostriction effect. Then an induced current is produced in the coil based on Faraday's law of electromagnetic induction. As a consequence, the vibration energy will finally be converted to electrical energy by the Villari effect of the Galfenol layer and dissipated by the NES. It was noted that the introduction of the NES system can dissipate the vibration energy transmitted from the base and protect the structure of the cantilever energy harvesting system. This means that the model proposed in this work can achieve the dual-effect goal of reducing vibration transmitted from the base (main structure) and capturing the abandoned vibration energy. Additionally, the continuous green and environmentally friendly electrical energy harvested by the cantilever energy harvesting system can provide power to low-power electronic devices. These electronic devices have broad application prospects in structural health detection of transmission towers, building structures, and bridge structures, vehicle tracking on roads, and indoor and outdoor environmental monitoring.

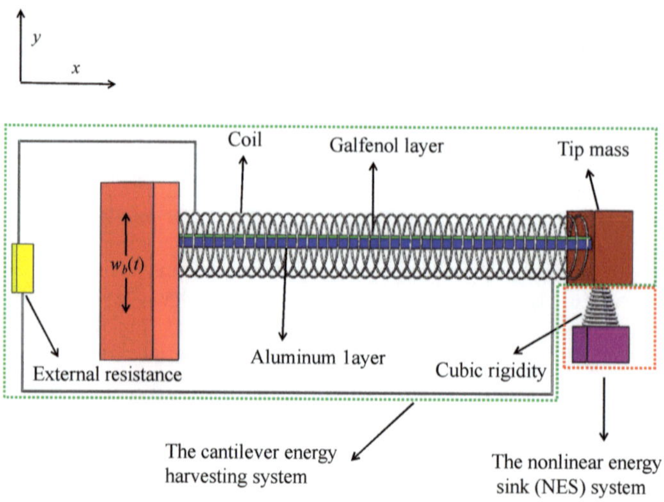

Figure 1. Schematic representation of the NES-Galfenol coupled system.

2.1. Coupled Mechanical Equations of the Cantilever Energy Harvesting System

To obtain the nonlinear electromechanical-coupled governing equations of the cantilever energy harvesting system for the coupled model, the extended Hamilton's principle [51] is utilized as

$$\int_{t_1}^{t_2} (\delta T - \delta V + \delta W_{nc})dt = 0 \tag{1}$$

where T, V, and W_{nc} are respectively the kinetic energy, potential energy, and virtual work done by the nonconservative forces of the energy harvester model. It is worth noting that the cantilever structure is regarded as a uniform composite beam, satisfying the Euler-

Bernoulli beam assumption since the cantilever structure is designed as a considerably thin beam. Therefore, the shear effect and rotational inertia of the cantilever composite structure are ignored, and the cantilever structure always satisfies the plane section assumption during the vibration deformation process. Additionally, the Galfenol layer is assumed to be tightly tied to the structural layer, so the deformation of the two material layers remains coordinated. Based on the above assumptions, the kinetic energy T of the coupling system is written as

$$\begin{aligned} T &= \frac{1}{2}\int_0^L m\left[\left(\frac{\partial u}{\partial t}\right)^2 + \left(\frac{\partial w}{\partial t} + \frac{\partial w_b}{\partial t}\right)^2\right]dx \\ &+ \frac{1}{2}M\left[\left(\frac{\partial u(L)}{\partial t}\right)^2 + \left(\frac{\partial w(L)}{\partial t} + \frac{\partial w_b}{\partial t}\right)^2\right] \\ &+ \frac{1}{2}J\left[\left(\frac{\partial^2 w(L)}{\partial x \partial t}\right)^2 + \left(\frac{\partial^2 w(L)}{\partial x \partial t}\right)^2\left(\frac{\partial w(L)}{\partial x}\right)^2\right] \end{aligned} \tag{2}$$

where m is the mass per unit length of the composite cantilever structure and given by $m = b_s h_s \rho_s + b_m h_m \rho_m$, of which ρ_s, ρ_m, b_s, b_m, h_s, and h_m are respectively the density, width, and thickness of the aluminum and the Galfenol layer. L is the total length of the composite cantilever beam. M and J are the mass and rotational inertia of the lumped mass. $w(x,t)$ and $u(x,t)$ are the transverse and axial vibration displacements of the cantilever beam at section x with respect to the fixed end of the cantilever structure, respectively. $w_b(t)$ is the harmonic vibration displacement of the base excitation. Note that $w(x,t)$, $u(x,t)$, and $w_b(t)$ are abbreviated as w, u, and w_b in the subsequent content for convenience, respectively. Additionally, the effect of gravitational potential energy is neglected since it is relatively small in comparison with the elastic strain potential energy of the beam. To account for the geometric nonlinearities of the coupled system's cantilever beam, the horizontal and vertical vibration velocities of both the beam and the lumped mass have been taken into consideration.

Assuming that the stress and strain of the structural layer follow Hooke's law and the stress-strain of the Galfenol layer follows a linear constitutive equation [15], the potential energy V of the coupling system can be given by

$$V = \int_0^L (EI)\left(\left(\frac{\partial^2 w}{\partial x^2}\right)^2 + \left(\frac{\partial^2 w}{\partial x^2}\right)^2\left(\frac{\partial w}{\partial x}\right)^2\right)dx \tag{3}$$

where EI is the elasticity stiffness of the composite cantilever beam and expressed as $EI = \frac{1}{3}E_s b_s\left({h_b}^3 - {h_a}^3\right) + \frac{1}{3}E_m b_m\left({h_c}^3 - {h_b}^3\right)$, of which E_s and E_m are the elasticity modulus of the aluminum layer and the Galfenol layer, and I is the cross-section inertia moment of the composite cantilever structure. $h_a = -[(E_m h_m h_m + E_m h_m h_s)/(2E_m h_m + 2E_s h_s) + h_s/2]$ and $h_b = h_s + h_a$ are coordinates of the lower and upper boundaries for the aluminum layer, and $h_c = h_s + h_m + h_a$ and $h_{pc} = \frac{h_b + h_c}{2}$ are respectively the coordinates of the top and the neutral axis for the Galfenol layer.

According to Nie et al. [52], the geometrically nonlinear deformation relation of the transverse and axial vibration displacement for the cantilever beams can be denoted by $w(x,t) = \int_0^x \sin(\varphi(\eta,t))d\eta, u(x,t) = \int_0^x \cos(\varphi(\eta,t))d\eta$, of which, $\varphi(\eta,t)$ represents the rotation angle of section η. Eliminating higher-order terms by Taylor expansion yields

$$u(x,t) = \int_0^x \left(1 - \frac{1}{2}w'(\eta,t)^2\right)d\eta \tag{4}$$

$$\frac{d\varphi(x,t)}{dt} = \frac{\partial^2 w(x,t)}{\partial x \partial t}\left(1 + \frac{1}{2}\left(\frac{\partial w(x,t)}{\partial x}\right)^2\right) \tag{5}$$

$$\frac{\partial \varphi(x,t)}{\partial x} == \frac{\partial^2 w(x,t)}{\partial x^2}\left(1 + \frac{1}{2}\left(\frac{\partial w(x,t)}{\partial x}\right)^2\right) \tag{6}$$

It points out that the higher-order terms in Equations (4)–(6) are regarded as far less than one and are thus ignored.

The virtual work performed by the nonconservative force is given by

$$W_{nc} = W_{ele} + W_{damp} \tag{7}$$

where W_{ele} and W_{damp} are respectively the virtual works done by the electric and damping forces, of which the moment caused by the electric effect W_{ele} is formulated as

$$W_{ele} = -\int_0^L M_{ele}\frac{\partial^2 w}{\partial x^2}dx - \int_0^L M_{ele}\frac{\partial^2 w}{\partial x^2}\left(\frac{1}{2}\left(\frac{\partial w}{\partial x}\right)^2\right)dx \tag{8}$$

where, $M_{ele} = \vartheta_m(H_b + NI(t)/L)[H(x) - H(x-L)]$ is the moment due to the electrical effect [15]. It should be pointed out that in the derivation of M_{ele}, the magnetic field in the Galfenol layer is presumed to be uniformly distributed along the axial and lateral directions of the cantilever structure. Simultaneously, the hysteresis phenomena in the Galfenol layer are neglected [59], and the bias field H_b is assumed to have a constant value [18]. Besides, N is the coil turn number, and $I(t)$ is the induced current in the coil. $H(x)$ is the Heaviside step function, ϑ_m is the magnetoelectric coupling coefficient and expressed as $\vartheta_m = E_m d b_m \frac{1}{2}\left(h_c{}^2 - h_b{}^2\right)$, of which d is the coupling coefficient of the Galfenol layer.

Moreover, by adopting the assumption of proportional damping for the internal damping mechanism of the cantilever composite beam structure [23], the virtual work according to the damping force W_{damp} can be written as

$$W_{damp} = \int_0^L -\left(c_s I\frac{\partial^3 w}{\partial x^2\partial t}\right)\frac{\partial^2 w}{\partial x^2}dx + \int_0^L -c_a\frac{\partial w}{\partial t}wdx \tag{9}$$

where c_s and c_a are respectively the strain-rate damping and viscous-air damping coefficients of the composite cantilever structure.

By integrating Equations (2), (3), and (7) into Equation (1), the governing equations of the cantilever energy harvesting system are derived as

$$\begin{aligned}
&\left[m\frac{\partial^2}{\partial t^2}\left(\int_0^x \frac{w'(\eta,t)^2}{2}d\eta\right)+M\delta(x-L)\frac{\partial^2}{\partial t^2}\left(\int_0^L \frac{w'(\eta,t)^2}{2}d\eta\right)\right]\frac{\partial w}{\partial x}\\
&-\left[\int_x^L m\frac{\partial^2}{\partial t^2}\left(\int_0^\xi \frac{w'(\eta,t)^2}{2}d\eta\right)d\xi + \mathrm{MH}(x-L)\frac{\partial^2}{\partial t^2}\left(\int_0^L \frac{w'(\eta,t)^2}{2}d\eta\right)\right]\frac{\partial^2 w}{\partial x^2}\\
&+m\frac{\partial^2 w}{\partial t^2}+EI\left[\frac{\partial^4 w}{\partial x^4}+\left(\frac{\partial^2 w}{\partial x^2}\right)^3+4\frac{\partial^3 w}{\partial x^3}\frac{\partial^2 w}{\partial x^2}\frac{\partial w}{\partial x}+\frac{\partial^4 w}{\partial x^4}\left(\frac{\partial w}{\partial x}\right)^2\right]\\
&+\vartheta_m(H_b+NI(t)/L)\left[\frac{\partial\delta(x)}{\partial x}-\frac{\partial\delta(x-L)}{\partial x}\right]\\
&+\frac{1}{2}\vartheta_m(H_b+NI(t)/L)\left[\frac{\partial\delta(x)}{\partial x}-\frac{\partial\delta(x-L)}{\partial x}\right]\left(\frac{\partial w}{\partial x}\right)^2\\
&+\vartheta_m(H_b+NI(t)/L)[\delta(x)-\delta(x-L)]\frac{\partial w}{\partial x}\frac{\partial^2 w}{\partial x^2}\\
&+\frac{\partial^2}{\partial x^2}\left(c_s I\cdot\frac{\partial^3 w}{\partial x^2\partial t}\right)+c_a\cdot\frac{\partial w}{\partial t}=-\left[m\frac{\partial^2 w_b}{\partial t^2}+M\delta(x-L)\frac{\partial^2 w_b}{\partial t^2}\right]
\end{aligned} \tag{10}$$

Moreover, the linear boundary conditions are defined by

$$
\begin{aligned}
&\frac{\partial w(0,t)}{\partial x}=0, \qquad w(0,t)=0,\\
&M\frac{\partial^2 w(L,t)}{\partial t^2}-EI\frac{\partial^3 w(L,t)}{\partial x^3}-EI\frac{\partial w(L,t)}{\partial x}\left(\frac{\partial^2 w(L,t)}{\partial x^2}\right)^2-EI\frac{\partial^3 w(L,t)}{\partial x^3}\left(\frac{\partial w(L,t)}{\partial x}\right)^2\\
&-\vartheta_m(H_b+NI(t)/L)[\delta(x)-\delta(x-L)]\\
&-\frac{1}{2}\vartheta_m(H_b+NI(t)/L)[\delta(x)-\delta(x-L)]\left(\frac{\partial w(L,t)}{\partial x}\right)^2\\
&-\frac{\partial}{\partial x}\left(c_s I\cdot\frac{\partial^3 w(L,t)}{\partial x^2\partial t}\right)=0,\\
&J\frac{\partial^3 w(L,t)}{\partial x\partial t^2}+J\frac{\partial^3 w(L,t)}{\partial x\partial t^2}\left(\frac{\partial w(L,t)}{\partial x}\right)^2+J\left(\frac{\partial^2 w(L,t)}{\partial x\partial t}\right)^2\frac{\partial w(L,t)}{\partial x}+EI\frac{\partial^2 w(L,t)}{\partial x^2}\\
&+EI\left(\frac{\partial w(L,t)}{\partial x}\right)^2\frac{\partial^2 w(L,t)}{\partial x^2}+\vartheta_m(H_b+NI(t)/L)[H(x)-H(x-L)]\\
&+\frac{1}{2}\vartheta_m(H_b+NI(t)/L)[H(x)-H(x-L)]\left(\frac{\partial w(L,t)}{\partial x}\right)^2+c_s I\cdot\frac{\partial^3 w(L,t)}{\partial x^2\partial t}=0
\end{aligned}
\tag{11}
$$

In the above derivation process, the edge effect is neglected for the reason that the coil is assumed to be long enough. According to Ampere's law [43], the magnetic field strength H applied longitudinally to the Galfenol layer can be given by $H=H_b+NI(t)/L$. In addition, the linear constitutive relationships of the Galfenol layer $\varepsilon=\sigma_m/E_m+dH$ and $B=d\sigma_m+\mu H$ are adopted, of which σ_m and μ are the axial stress and material permeability of the Galfenol layer. B is the magnetic induction strength of the Galfenol layer. Note that in Equations (10) and (11), $w(L,t)$ is the displacement of transverse vibration for the section $x=L$ at time t. $\delta(x)$ is the Dirac function and has a relationship with the Heaviside function $H(x)$ as $\frac{\partial H(x)}{\partial x}=\delta(x)$.

2.2. Coupled Electrical Circuit Equations of the Cantilever Energy Harvesting System

According to Faraday's law of electromagnetic induction [18], the induced voltage generated in the Galfenol layer with length Δl for the cantilever energy harvesting system can be expressed as $\Delta V(t)=-\Delta l\frac{NA}{L}\frac{dB}{dt}$. Integrating over the length with respect to x, the total induced voltage can be obtained as

$$
V(t)=-\frac{NdE_mA}{L}\int_0^L\frac{d\varepsilon_{pc}}{dt}dx-L_0\frac{dI(t)}{dt}
\tag{12}
$$

where $L_0=(\mu-d^2E_m)N^2\frac{A}{L}$ is the equivalent inductance of the coil, ε_{pc} is the neutral layer strain of the Galfenol layer and expressed as $\varepsilon_{pc}=-h_{pc}\frac{\partial^2 w}{\partial x^2}\left(1+\frac{1}{2}\left(\frac{\partial w}{\partial x}\right)^2\right)$ to take into account the geometric nonlinearity of the composite cantilever structure. Substituting the expressions of the B and the neutral layer strain of the Galfenol layer into Equation (12), the electrical circuit equation is derived as

$$
G\int_0^L\frac{d^3w}{dx^2dt}dx+\frac{G}{2}\int_0^L\frac{d^3w}{dx^2dt}\left(\frac{dw}{dx}\right)^2dx+G\int_0^L\frac{d^2w}{dx^2}\frac{d^2w}{dxdt}\frac{dw}{dx}dx-L_0\frac{dI(t)}{dt}=\mathrm{V(t)}
\tag{13}
$$

where $G=NdE_mA\frac{h_{pc}}{L}$ is the force-electric coupling coefficient of the cantilever structure, and A is the coil's section area.

2.3. Coupled Mechanical Equations of the NES System

For the NES system, the mass is M_2, the damping is c_2, and the nonlinear elastic stiffness is k_2. The transverse vibration displacement of the NES along the y direction with respect to the fixed end of the composite cantilever beam is $w_2(t)$ and abbreviated as w_2. Therefore, the kinetic energy T_2, potential energy V_2, and virtual work W_{nc2} due to the non-conservative forces of the NES system are expressed as

$$
T_2=\frac{1}{2}M_2\left(\frac{\partial w_2}{\partial t}+\frac{\partial w_b}{\partial t}\right)^2
\tag{14}
$$

$$V_2 = \frac{k_2(w_2 - w(L,t))^4}{4} \tag{15}$$

$$W_{nc2} = -c_2\left(\frac{\partial w_2}{\partial t} - \frac{\partial w(\mathrm{L},\mathrm{t})}{\partial t}\right)(w_2 - w(\mathrm{L},\mathrm{t})) \tag{16}$$

Substituting Equations (14)–(16) into the extended Hamilton's expression, i.e., Equation (1), the additional governing equation for NES is obtained as

$$\begin{aligned}
&-\int_{t_1}^{t_2} M_2 \frac{\partial^2 w_2}{\partial t^2}\delta w_2 dt - \int_{t_1}^{t_2} M_2 \frac{\partial^2 w_b}{\partial t^2}\delta w_2 dt \\
&+\int_{t_1}^{t_2} k_2\left[\left(w_2{}^3 + 3w_2 w(L,t)^2 - 3w_2{}^2 w(L,t) - w(L,t)^3\right)\delta w(L,t)\right]dt \\
&+\int_{t_1}^{t_2} k_2\left\{\left(w(L,t)^3 - 3w_2 w(L,t)^2 + 3w_2{}^2 w(L,t) - w_2{}^3\right)\delta w_2\right\}dt \\
&-\int_{t_1}^{t_2} c_2\left(\frac{\partial w_2}{\partial t} - \frac{\partial w(L,t)}{\partial t}\right)(\delta w_2)dt + \int_{t_1}^{t_2} c_2\left(\frac{\partial w_2}{\partial t} - \frac{\partial w(L,t)}{\partial t}\right)(\delta w(\mathrm{L},\mathrm{t}))dt = 0
\end{aligned} \tag{17}$$

3. Representative Model of Output Responses for the Coupled System

To conduct the nonlinear analyses, we adopt the Galerkin decomposition method to decompose into the transverse displacement of the composite cantilever structure. This means that $w(x,t)$ can be discretized into the product of the spatial and time variables as

$$w(x,t) = \phi_r(x) q_r(t) \tag{18}$$

where $\phi_r(x)$ and $q_r(t)$ are the r mode shape and modal coordinates of the composite cantilever structure. Note that the modal shape function $\phi_r(x)$ is the same as that of the undamped free vibration system [60] and can be expressed as

$$\phi_r(x) = A_r \sin \beta_r x + B_r \sin \beta_r x + C_r \sin \beta_r x + D_r \sin \beta_r x \tag{19}$$

Adopting the proportional damping assumption, the boundary conditions are simplified and written as follows

$$\begin{aligned}
&\phi_r(0) = 0, \quad \frac{\partial \phi_r(0)}{\partial x} = 0, \\
&EI\frac{\partial^3 \phi_r(L)}{\partial x^3} + \omega^2 M \phi_r(L) = 0, \quad EI\frac{\partial^2 \phi_r(L)}{\partial x^2} - \omega^2 J \frac{\partial \phi_r(L)}{\partial x} = 0.
\end{aligned} \tag{20}$$

and the orthogonality conditions given by

$$\begin{aligned}
&\int_0^L \phi_r(x) m \phi_s(x) dx + \phi_r(x)(M\phi_s(x))|_{x=L} + \frac{d\phi_r(x)}{dx}\left(J\frac{\partial \phi_s(x)}{\partial x}\right)|_{x=L} = \delta_{rs}, \\
&\int_0^L \phi_s(x) \frac{d^2}{dx^2}\left(EI\frac{d^2\phi_r(x)}{dx^2}\right)dx - \phi_s(x)\frac{d}{dx}\left(EI\frac{d^2\phi_r(x)}{dx^2}\right)|_{x=L} + \frac{d\phi_s(x)}{dx} EI \frac{d^2\phi_r(x)}{dx^2}|_{x=L} = \omega_r^2 \delta_{rs}, \\
&\int_0^L \frac{d^2\phi_s(x)}{dx^2} EI \frac{d^2\phi_r(x)}{dx^2} dx = \omega_r^2 \delta_{rs}.
\end{aligned} \tag{21}$$

where $\omega_r = \beta_r{}^2\sqrt{\frac{EI}{m}}$ is the r th order intrinsic frequency of the composite cantilever beam. δ_{rs} is the Kronecker delta and equal to 1 for $r = s$ and 0 for $r \neq s$. Substituting Equation (18) into Equations (10), (13), and (17).

For cantilever beam structures with a concentrated mass fixed at the free end, it can be proved that the contribution of the distributed mass of the beam and the attached mass in the energy of the first mode, particularly the potential and elastic strain energies, is drastically higher than that of the other modes energy. Based on this, the first mode has been widely adopted to analyze the response of the cantilever energy harvester with a concentrated mass attached to the free end in many studies [61–64]. However, it should be emphasized that if the position of the concentrated mass changes, such as being placed at the mid-span of the cantilever beam, the contribution of the second-order mode in the dynamical response of the system will significantly increase. In this work, the concentrated mass is placed at the free end of the cantilever beam for the proposed system as a consequence, only the first vibration mode is considered. Then, by using orthogonal conditions Equation (20) and boundary conditions Equation (21), the governing equations of the NES-Galfenol coupled system are easily reduced to

$$\begin{aligned}&\ddot{q}(t)+\omega^2 q(t)+C_m\dot{q}(t)+\eta_1 q^3(t)+\eta_2 q(t)\left[\dot{q}(t)^2+q(t)\ddot{q}(t)\right]\\&+\chi_{r1}+\chi_{r2}q^2(t)+\theta_{p1}I(t)+\theta_{p2}I(t)q^2(t)+M_2\ddot{w}_2(t)\eta_3=P*a_0(t)\end{aligned} \tag{22}$$

$$M_2\ddot{w}_2(t)+M_2a_0(t)+k_2(w_2(t)-w(L))^3+c_2[\dot{w}_2(t)-\dot{w}(L)]=0 \tag{23}$$

$$L_0\dot{I}(t)+I(t)R_L+I(t)R_c-\theta_{p1}\dot{q}(t)-\theta_{p3}\dot{q}(t)q(t)^2=0 \tag{24}$$

It should be pointed out that Equations (22)–(24) represent the governing equations of a dual nonlinear system, which is a vibration reduction and energy harvesting coupling system with both NES and geometric nonlinear characteristics considered. In the above equations, $P=-\left[m\int_0^L\phi(x)dx+M\phi(L)+M_2\phi(L)\right]$, $C_m=2\xi\omega$, ξ and ω are respectively the mechanical damping ratio and first natural frequency of the composite cantilever beam. $a(t)$ is the acceleration of the external excitation. R_L and R_c are, respectively, the external load resistance and internal resistance of the coil. The geometric nonlinearity coefficients η_1, η_2, and η_3 are respectively expressed as

$$\begin{aligned}&\eta_1=\int_0^L EI\phi(x)\left(\frac{\partial^4\phi(x)}{\partial x^4}\left(\frac{\partial\phi(x)}{\partial x}\right)^2+4\frac{\partial^3\phi(x)}{\partial x^3}\frac{\partial^2\phi(x)}{\partial x^2}\frac{\partial\phi(x)}{\partial x}+\left(\frac{\partial^2\phi(x)}{\partial x^2}\right)^3\right)dx,\\&\eta_2=\int_0^L m\left(\int_0^x\phi'^2(x)d\eta\right)^2dx+M\left(\int_0^L\phi'^2(x)d\eta\right)^2,\\&\eta_3=\phi(L).\end{aligned} \tag{25}$$

while the magnetostrictive coupling coefficients χ_{r1}, χ_{r2}, θ_{p1}, θ_{p2}, and θ_{p3} have the following form

$$\begin{aligned}\chi_{r1}&=E_m db_m h_m h_{pc} H_b\phi'(L),\\\chi_{r2}&=\frac{E_m db_m h_m h_{pc} H_b}{2}\phi'(L)^3,\\\theta_{p1}&=\frac{NE_m db_m h_m h_{pc}\phi'(L)}{L},\\\theta_{p2}&=\frac{NE_m db_m h_m h_{pc}\phi'(L)^3}{2L},\\\theta_{p3}&=\frac{3NE_m db_m h_m h_{pc}}{2L}\int_0^L\phi''(x)\phi'(x)^2dx.\end{aligned} \tag{26}$$

To determine the output performances of the coupled system, the following state variables are introduced:

$$\mathbf{X}=\begin{bmatrix}X_1\\X_2\\X_3\\X_4\\X_5\end{bmatrix}=\begin{bmatrix}q\\\dot{q}\\w_2\\\dot{w}_2\\I\end{bmatrix} \tag{27}$$

According to Equations (22)–(24), the governing equations can be rewritten as

$$\begin{aligned}&\dot{X}_2+\omega_r^2X_1+C_{mr}X_2+\eta_1X_1{}^3+\eta_2X_1\left[X_2{}^2+X_1\dot{X}_2\right]+\chi_{r1}+\chi_{r2}X_1{}^2\\&+\theta_{p1}X_5+\theta_{p2}X_5X_1{}^2+M_2\dot{X}_4\eta_3=Pa_0(t)\end{aligned} \tag{28}$$

$$M_2\dot{X}_4+M_2a_0(t)+k_2(X_3-\eta_3X_1)^3+c_2(X_4-\eta_3X_2)=0 \tag{29}$$

$$L_0\dot{X}_5+X_5R_L+X_5R_c-\theta_{p1}X_2-\theta_{p3}X_2X_1^2=0 \tag{30}$$

The output responses of the NES-Galfenol-coupled system, **X**, can be rewritten in the form of

$$\dot{\mathbf{X}} = \mathbf{B}\mathbf{X} + \mathbf{C}(\mathbf{X}^2, \mathbf{X}^3) \tag{31}$$

in which $\mathbf{C}(\mathbf{X}^2, \mathbf{X}^3)$ is the nonlinear vector of the state variables; **B** is the linear coefficient matrix of the state variables and can be expressed as

$$\mathbf{B} = \left\{ \begin{array}{ccccc} 0 & 1 & 0 & 0 & 0 \\ -\omega^2 & -2\zeta\omega & 0 & 0 & -\theta 1 \\ 0 & 0 & 0 & 1 & 0 \\ 0 & c_2\eta_3/M_2 & 0 & -c_2/M_2 & 0 \\ 0 & \theta_{p1}/L_0 & 0 & 0 & -R/L_0 \end{array} \right\} \tag{32}$$

Obviously, all parameters that affect the linear part of the coupled system have been included in matrix **B**. The third and fourth rows, as well as the third and fourth columns in matrix **B**, are the relevant parameters of the NES system. Ignoring the relevant parameters of the NES system, a new matrix **B** about the composite cantilever energy harvesting system is obtained. The global natural frequency and damping of the electromechanically coupled energy harvesting system can be determined by the new matrix **B**. Note that the output responses of the NES-Galfenol-coupled system are calculated by the Runge-Kutta method for all case studies in the following section.

4. Results and Discussion

4.1. Model Validation

To assess the reliability of the proposed model, a degenerated cantilever Galfenol harvesting system is adopted by omitting the geometric nonlinearity characteristics of the cantilever structure and the NES coupling system for the proposed NES-Galfenol coupling system. The harvested power of the degenerated cantilever Galfenol harvesting system proposed in this paper is compared with those of the previous research [15,65], as shown as plotted in Figure 2. Note that the same physical and geometric parameters are used to obtain the numerical results of the degenerated cantilever Galfenol harvesting system by the Runge-Kutta method. As plotted in Figure 2, the harvested power of the discussed cantilever Galfenol harvesting system obtained by the proposed method is in good agreement with the theoretical analysis results and numerical solutions in other works [15,65]. This means that the proposed model and the derived results are reliable and applicable.

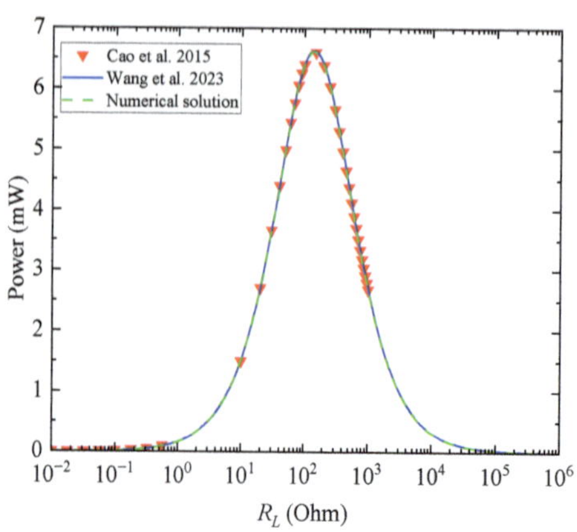

Figure 2. Comparison of the harvested power calculated by this work, Cao et al. [65], and Wang et al. [15].

4.2. Comparison between the Linear and Nonlinear Systems

In order to analyze the difference between linear and nonlinear systems, four cases are adopted and discussed in this section. The four cases are pure linear systems, nonlinear systems that only consider NES, nonlinear systems that only consider geometric large deformations, and dual nonlinear systems that consider both NES and geometric large deformations. Among them, the dual nonlinear system is the proposed NES-Galfenol coupling system in this work. It is obvious that the governing equations of the other three systems (apart from the proposed NES-Galfenol coupling system) can be easily obtained by eliminating the corresponding nonlinear terms from the proposed coupling model, as shown in Equations (22)–(24). Note that the physical and geometric properties of the four mentioned cases, as well as all the cases discussed in the following sections, are shown in Table 1.

Table 1. Physical and geometric properties of the NES-Galfenol coupled system.

Parameter	Description	Value
L	Length of the aluminum layer and Galfenol layer(mm)	130
b_s, b_m	Width of the aluminum layer and Galfenol layer (mm)	20
h_s	Thickness of the aluminum layer and (mm)	0.7
h_m	Thickness of the Galfenol layer (mm)	0.3
E_s	Young's Modulus of the aluminum layer (GN/m^2)	68
E_g	Young's Modulus of the Galfenol layer (GN/m^2)	70
ρ_s	Density of the aluminum layer (kg/m^3)	2700
ρ_g	Density of the Galfenol layer (kg/m^3)	7496
ξ	Mechanical damping ratio of the first modal	0.014
M	Mass of the tip lumped mass (g)	70
μ	magnetic permeability(H/m)	$920\pi \times 10^{-7}$
d_0	Piezomagnetic coefficient (T/Gpa)	34
H_b	bias magnetic field strength (kA/m)	3.58
N	number of the coil	1000
R_C	Internal resistance of the coil (Ohm)	36.4
M_2	Mass of the NES (g)	4
c_2	damping of the NES (N·s/m)	0.4
k_2	elastic stiffness of the NES (N/m^3)	10^5

Figure 3 illustrates the variations of the output performances for the four mentioned coupled systems with the excitation frequency ratio (the ratio of the excitation frequency to the natural frequency, β), at $R_L = 10$ Ohm and $a_0 = 2.1$ m/s^2. As plotted in Figure 3, for the pure linear system, the tip displacement and the harvested power reach the maximum value, i.e., 33.63 mm and 10.15 mW, at an excitation frequency ratio of 1.002. At this time, the excitation frequency is exactly equal to the natural frequency of the coupled structure. For nonlinear systems that only consider large geometric deformations, the maximum tip displacement of 31.45 mm and harvested power of 8.71 mW are obtained when the external excitation frequency ratio is 0.961. In particular, a softening phenomenon occurs, i.e., the output performances show a jumping phenomenon when the excitation frequency ratio reaches 0.954. For the nonlinear systems that only consider NES, the maximum tip displacement and harvested power are respectively 20.17 mm and 3.49 mW, with the excitation frequency ratio equal to 0.974. For the dual nonlinear systems that consider both NES and geometric large deformations, the maximum value of output properties are achieved, respectively, at 20.84 mm and 3.62 mW, with the excitation frequency ratio equal to 0.954.

The conclusion can be drawn from Figure 3 that the NES can effectively reduce the tip displacement as well as the harvested power. Besides, the excitation frequency ratio corresponding to the maximum output response shifts significantly to the left for the four discussed cases. This means that, after considering geometric nonlinearity, the optimal external excitation frequency ratio for the maximum output responses decreases and will result in a jump phenomenon and a nonlinear softening phenomenon. Moreover, when considering both the NES and the large deformation, the output performances are effectively reduced, as is the optimal external excitation frequency ratio, which corresponds to the maximum output response. However, the jumping phenomenon disappeared since the relatively small vibration amplitude cannot excite the geometric deformation effect. In other words, for the coupled model that introduces the NES, a larger external excitation amplitude is required to excite the geometric nonlinearity properties of the cantilever structure for the energy harvesting system.

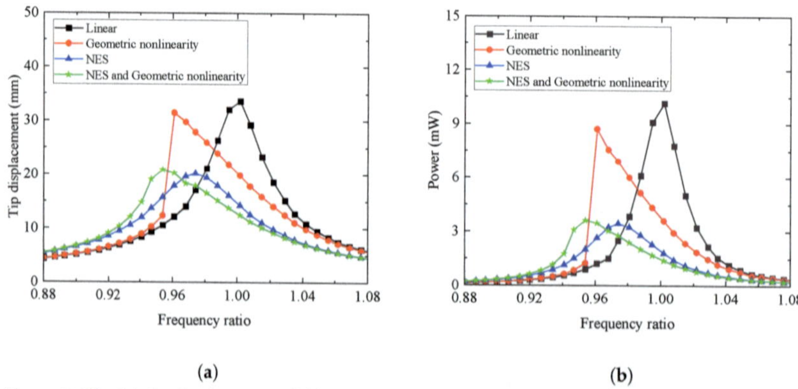

Figure 3. The (**a**) tip displacement (**b**) harvesting power with the external excitation frequency ratio for different coupled systems.

4.3. *Effect of External Load Resistance on Global Damping and Frequency*

The variation trends of the modified frequency and damping for the NES-Galfenol-coupled system considering dual nonlinearity with the external load resistance are presented in Figure 4. The modified frequency and global damping of the coupled system are gained by calculating the eigenvalue of the simplified matrix **B** as mentioned below Equation (32). As depicted in Figure 4, with the presence of electromechanical coupling for the Galfenol layer, the external load resistance will affect the global frequency and damping. Specifically, when the external resistance increases, the modified frequency first decreases, then increases, and finally stabilizes. The minimum value of the global frequency is reached at 30 Ohm. In contrast, the global damping of the proposed coupling system decreases with the increase in external load resistance. In general, the influence on the global frequency and damping of the coupled system is not obvious, which is inconsistent with the effect of resistance on piezoelectric energy harvesters considering geometric nonlinearity as illustrated in the previous work [51].

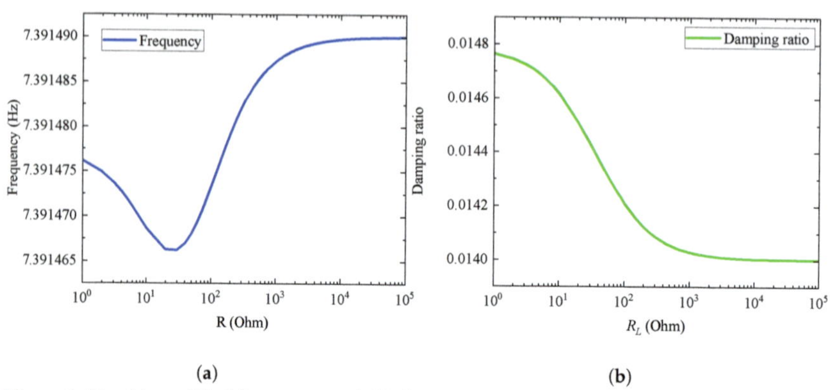

Figure 4. The (**a**) modified frequency and (**b**) damping with the external load resistance for the coupled system.

4.4. Parametric Analysis

4.4.1. Effects of External Resistance and Excitation Frequency Ratio

Figure 5 presents the relationships of tip displacement and harvested power with the excitation frequency ratio for different external resistances at $a_0 = 2\ \mathrm{m/s^2}$, $k_2 = 10^5\ \mathrm{N/m^3}$, and $c_2 = 0.4\ \mathrm{N{\cdot}s/m}$. Inspecting Figure 5, we note that with the increase of the external excitation frequency ratio, the tip displacement and output power of the coupled system increase with the excitation frequency ratio less than 0.961. Then the output performances decrease when the excitation frequency ratio exceeds 0.961. Moreover, as the external resistance rises, the harvested power gradually decreases while the tip displacement at the end of the cantilever structure is slightly increased. This can be explained by the fact that with the increase in resistance, the electrical damping of the coupled system faintly reduced, as illustrated in Figure 4. Additionally, inspecting Figure 4, the output results of the proposed NES-Galfenol coupling system show a certain left-bending trend, which is known as the nonlinear softening characteristics. As a comparison, the responding output results for the linear coupled system are symmetrical at an excitation frequency ratio of around 1.0, as presented in our previous work [15]. In general, external resistance has a relatively small impact on vibration displacement and a greater influence on harvested power. There always exists an optimal excitation frequency ratio corresponding to the maximum output performance. The nonlinear softening characteristics appeared with the introduction of geometric nonlinearity in cantilever structures.

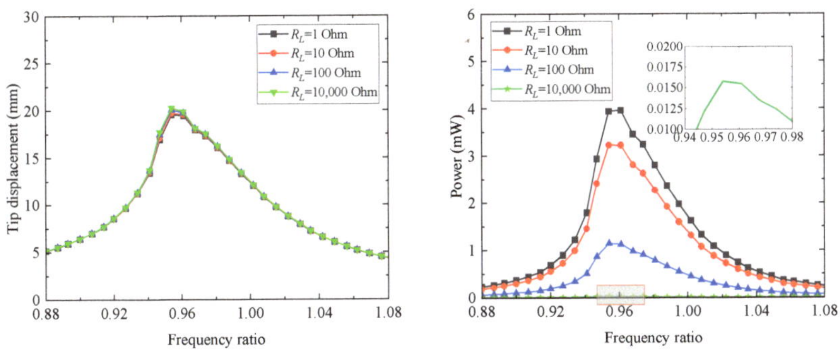

(a) (b)

Figure 5. The (**a**) tip displacement and (**b**) harvested power with the excitation frequency ratio for different external load resistance.

4.4.2. Effects of Excitation Acceleration Amplitude and Excitation Frequency Ratio

The influences of the excitation frequency ratio and the excitation acceleration on the output performances are plotted in Figure 6 for the case of $R_L = 10$ Ohm, $k_2 = 10^5\ \mathrm{N/m^3}$, and $c_2 = 0.4\ \mathrm{N{\cdot}s/m}$. Inspecting Figure 6, as the excitation frequency ratio increases, the tip displacement and harvested power both first increase and then reduce to zero. Besides, with the increase in excitation acceleration, the vibration placement as well as the output power increase sequentially. It is noted that when the amplitude of the excitation acceleration reaches $2.8\ \mathrm{m/s^2}$, the output performances present a drastic change, as indicated by the black dashed arrow in the figure. This means that the jumping phenomenon occurs, i.e., the vibration characteristics of the coupled system jump from a stable motion with a smaller amplitude to another stable motion with a larger amplitude. This jumping phenomenon does not occur in the linear Galfenol energy harvesting coupled systems, as reported in the work [15]. In general, the plots show that when the amplitude of external excitation exceeds a certain value, a softening jump phenomenon of the output response will occur. The reason for this is that the geometric nonlinearity of the coupled system will be excited by increasing the excitation acceleration amplitude. As a result, when the softening phenomenon occurs, the output results show a clear left-bending trend. At this point, the working frequency range of the coupled system dramatically shifts to the left and widens, which is the output effect that this model aims to achieve.

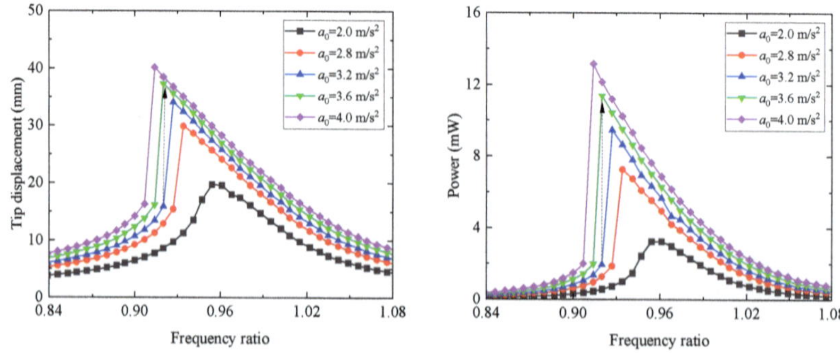

(a) (b)

Figure 6. The (**a**) tip displacement and (**b**) harvested power with the excitation frequency ratio for different amplitudes of the excitation acceleration.

4.4.3. Effects of External Resistance and Excitation Acceleration Amplitude

Figure 7 illustrates the variations of the output performances with the amplitudes of the excitation acceleration for different external load resistances when the excitation frequency ratio is fixed at 0.920, $k_2 = 10^5$ N/m^3, and $c_2 = 0.4$ N·s/m. As shown in Figure 7, the output performances of the coupled system first increase with the increase of the excitation acceleration amplitude. When the amplitude of the excitation acceleration reaches about 3.4 m/s, the vibration of the coupled system will jump to the upper branch. This branch corresponds to the output performance being stable in a periodic motion with a larger amplitude. It means that the jumping phenomenon occurs when the excitation acceleration exceeds a certain value. In addition, as depicted in Figure 7, the harvested power of the coupled system declines with the external load resistance increasing from 1 Ohm to 10 Ohm, 10^2 Ohm, and 10^4 Ohm. At this time, as the resistance of the external circuit changes, the vibration displacement remains unchanged. This indicates that the value of the external resistance has little effect on the vibration displacement and has a reverse effect on the harvested power. As a consequence, to obtain a larger output power, the resistance of the external circuit and the excitation acceleration should be taken as smaller and larger values, respectively.

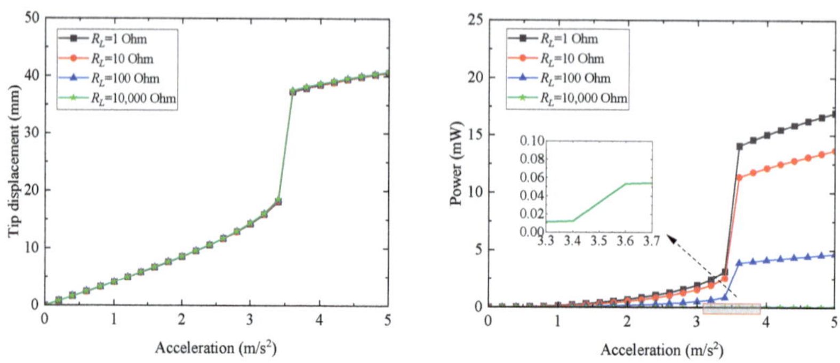

(a) (b)

Figure 7. The (**a**) tip displacement and (**b**) harvested power with the amplitudes of the excitation acceleration for different external load resistance.

Figure 8 shows the changes of the tip displacement and harvested power for the coupled system with the excitation acceleration amplitude when the external load resistance is respectively set as 1, 10, 10^2, and 10^4 Ohm and the excitation frequency ratio is equal to 0.988, $k_2 = 10^5$ N/m^3, and $c_2 = 0.4$ N·s/m. As presented in Figure 8, the output performances of the coupled system gradually increase with the increase of the excitation acceleration. As the external load resistance increases, the harvested power decreases significantly. The effect of the external resistance on the tip displacement of the coupled system can be almost negligible. Compared to the responses in Figure 7, the output responses of the coupled system still increase with the increase in excitation acceleration, but no nonlinear jumping phenomenon occurs. This can be explained by the reason that with the excitation frequency ratio closer to 1, the output result is farther away from the left-bending region (the softening region) plotted in Figures 5 and 6. Therefore, it can be concluded that the nonlinear jumping phenomenon will not be induced with the excitation frequency ratio far away from the softening region.

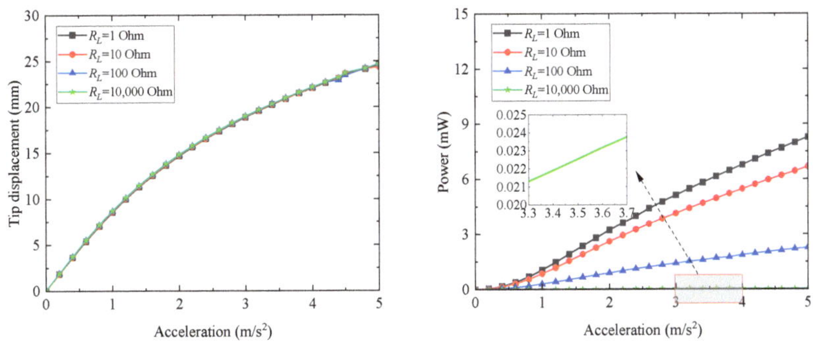

Figure 8. The (**a**) tip displacement and (**b**) harvested power with the amplitudes of the excitation acceleration for different external load resistance.

4.4.4. Effects of the Damping and Nonlinear Stiffness for the NES

Figure 9 presents the variations of the output performances with the damping of the NES for the cases of $R_L = 10$ Ohm, $a_0 = 2$ m/s^2, and $k_2 = 10^5$ N/m^3. As illustrated in Figure 9, the output responses of the coupled system first increase as the excitation frequency ratio approaches the optimal frequency ratio, then decrease as the excitation frequency ratio gradually shifts away from the optimal frequency ratio. In addition, as the damping increases from 0.1 N·s/m to 0.6 N·s/m, the vibration displacement and harvested power first decrease and then increase. With the increase in the NES damping, the optimal excitation frequency ratio corresponding to the maximum output response increases first and then declines. Moreover, the softening phenomenon of the output responses shows a trend of first weakening and then strengthening as the NES damping increases. This means that the damping of NES should be taken as a smaller or larger value to achieve better output performance for a wider excitation frequency range.

The varied relationships of the tip displacement and harvested power with the excitation frequency ratio for different NES stiffnesses are depicted in Figure 10 for the case of $R_L = 10$ Ohm, $a_0 = 2$ m/s^2, and $c_2 = 0.4$ N·s/m. Inspecting Figure 10, it is obvious that the changing trend of the output properties to the excitation frequency ratio is consistent with those in Figure 9. That is, the output responses increase first and then decrease with the growth of the excitation frequency ratio. As the NES stiffness rises from 10^4 N/m^3 to 10^7 N/m^3, the tip displacement and harvested power first decline and then increase sequentially. Additionally, the softening phenomenon of output responses is gradually emerging and becomes apparent with the growth of the NES stiffness. The optimal excitation frequency ratio corresponding to the maximum output performance shows a trend of left shift as the NES stiffness increases. Therefore, in order to harvest more energy within a wider range of excitation frequencies, the stiffness of NES should be taken as high as possible.

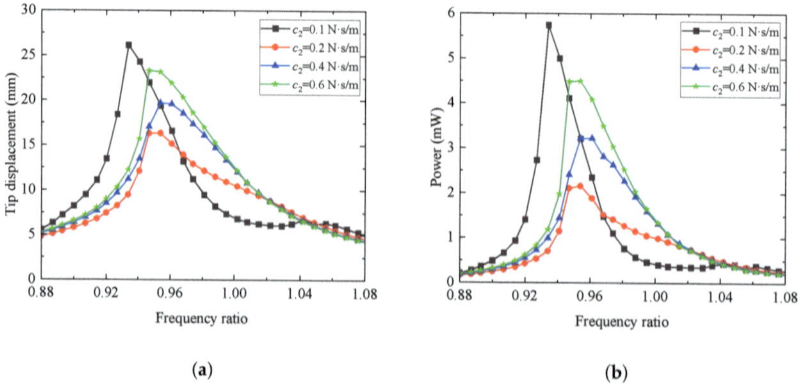

Figure 9. The (**a**) tip displacement and (**b**) harvested power with the excitation frequency ratio for different NES dampings.

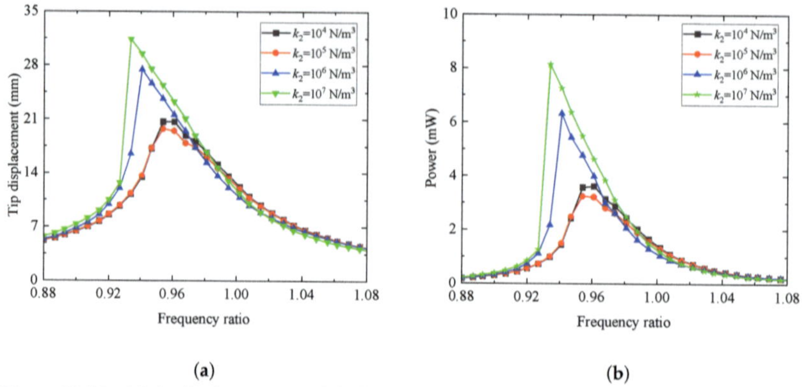

Figure 10. The (**a**) tip displacement and (**b**) harvested power with the excitation frequency ratio for different NES stiffnesses.

4.4.5. Effects of the Initial Excitation Condition

To clarify the influences of the initial excitation condition on the output responses of the proposed NES-Galfenol coupling system, the excitation acceleration amplitude, the external resistance, the NES damping, and the NES stiffness are respectively set as $a_0 = 3.5\ \text{m/s}^2$, $R_L = 10$ Ohm, $c_2 = 0.4\ \text{N·s/m}$, and $k_2 = 10^5\ \text{N/m}^3$. Two initial excitation conditions, i.e., the small and large initial excitation displacements, are selected for analysis. The tip displacement and harvested power of the coupled system with the mentioned two initial excitation conditions are presented in Figure 11. It is noted that the symbol ID in the figure represents the initial displacement of the discussed coupling system. The green and blue arrows represent the direction of the forward or reverse sweep frequency, respectively.

As observed in Figure 11, the changing curves of the vibration displacement and harvested power both exhibit jumping phenomena with the change of the excitation frequency ratio. In the case of forward sweep frequency analysis with a small initial displacement, the response of the coupled system increases gradually when the excitation frequency ratio is less than 0.920. Then it jumps from the lower branch to the upper branch while the excitation frequency ratio reaches 0.920, and finally gradually decreases when the excitation frequency ratio is larger than 0.920. In the case of reverse sweep frequency analysis with the large initial displacement, as the excitation frequency ratio decreases, the response of the coupled system grows gradually until the excitation frequency ratio is greater than 0.873. Then it decreases gradually for the excitation frequency ratio less than 0.873. When the excitation frequency ratio is equal to 0.873, the output performance jumps from the upper branch to the lower branch. This indicates that 0.920 and 0.873 are, respectively, the jumping thresholds of the output responses for the small initial condition and the large initial condition. Comparing the output performances of the two discussed cases, the output responses for the large ID are dramatically greater than those for the small ID when the excitation frequency ratio is in the jumping range. The jumping range can be called the softening region for the output responses. In this region, two steady-state solutions of the output performances for the coupled system can be obtained. Consequently, the wider the range of jump intervals, the better the output performance of the proposed NES-Galfenol coupling system.

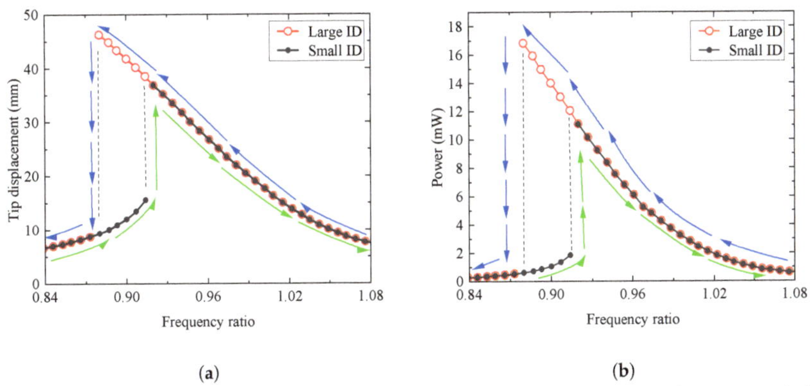

Figure 11. The (**a**) tip displacement and (**b**) harvested power with the excitation frequency ratio for different initial excitation conditions. The green and blue arrows represent the forward and reverse sweep frequency direction, respectively.

To further clarify the effect of initial conditions on the output responses, the time history curve and phase portrait curve for the tip displacement are illustrated in Figure 12. The results plotted in Figure 12 are obtained by the external excitation frequency ratio, which is respectively set at 0.920 and 0.988. As can be seen from Figure 11, the two adopted excitation frequency ratios are respectively located in and outside the jumping region. Inspecting Figure 12a,c, we found that with the excitation frequency ratio inside the jumping region, there are indeed two steady-state solutions for the vibration responses. For the case of the excitation frequency ratio equal to 0.988, Figure 12b,d show that the steady-state vibration responses for the different initial conditions are in good agreement. It means that the initial conditions no longer affect the vibration response when the excitation frequency ratio is outside the jumping range. In general, when the excitation frequency ratio is located in the jumping range, two stable vibration responses will be found for various initial conditions. When the excitation frequency is beyond the jumping region, only one stable vibration response can be observed for different initial conditions.

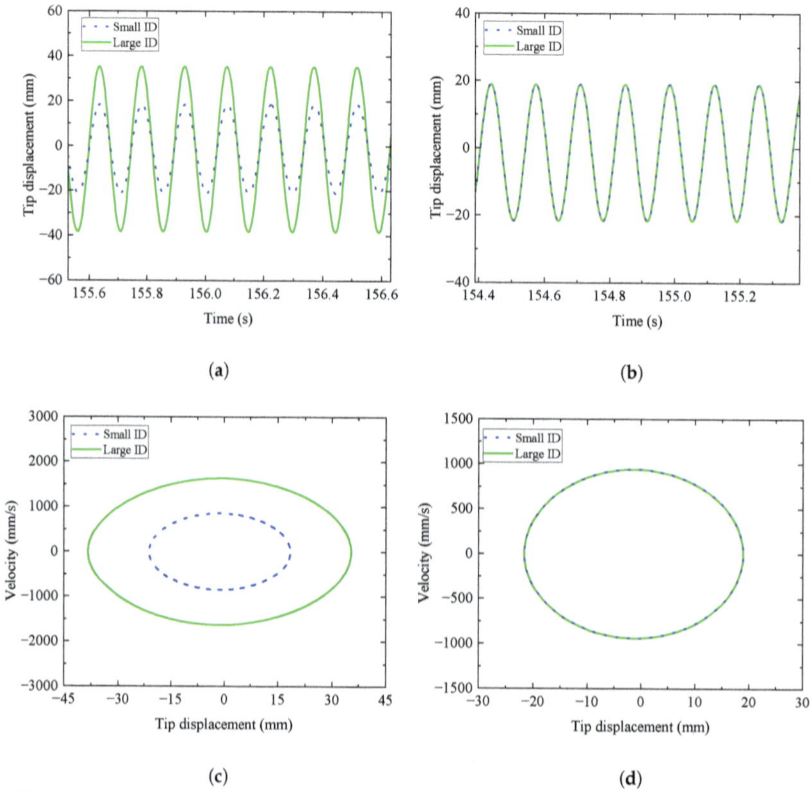

Figure 12. Time history curve and phase portrait for the tip displacement when external excitation frequency ratios are respectively set at 0.920 and 0.988. (**a**) $\beta = 0.920$. (**b**) $\beta = 0.988$. (**c**) $\beta = 0.920$. (**d**) $\beta = 0.988$.

5. Conclusions

In recent years, Galfenol, a novel magnetostrictive material, has increasingly been utilized in the field of vibration energy harvesting. NES, commonly used for vibration reduction, has garnered interest from researchers in energy harvesting because of its simple structure, wide working frequency band, and high robustness. Inspecting the previous works, researchers mainly devote themselves to introducing the NES or geometric nonlinearity into piezoelectric energy harvesting systems to obtain better output performances. This work proposes the NES-Galfenol coupling system by combining the advantages of NES and Galfenol materials. The proposed model can achieve the multifunctional goal of suppressing the vibration of the main structure and protecting the safety of the cantilever energy harvesting system while capturing energy. Based on the inverse magnetostrictive effect, Faraday electromagnetic induction effect of Galfenol material, and Hamilton variational principle, the electromechanically coupled distributed parameter governing equations of the proposed coupling system are formulated. It is noted that the geometric nonlinear characteristics of cantilever beams and stiffness nonlinear characteristics of NES are considered in the deriving process. According to the Galerkin discretization method, the reduced-order equations are deduced. The numerical solutions of the proposed NES-Galfenol coupling system are calculated by the Runge-Kutta method. A comparative analysis was conducted on the effects of geometric nonlinearity and NES on the output performances of the cantilever Galfenol energy harvesting system. Moreover, the influences of the excitation acceleration amplitude, excitation frequency ratio, external load resistance,

NES damping and stiffness, and the initial excitation condition on the output performances of the proposed coupling system are conducted. The main conclusions drawn from this work are as follows:

(1) The numerical calculation solutions of the proposed coupling system reveal that the addition of an NES can effectively reduce the vibration behaviors of the cantilever Galfenol energy harvesting system. After considering the geometric nonlinearity of the cantilever beam, the jumping phenomenon will occur relative to the linear system.
(2) For any fixed external load resistance, excitation acceleration amplitude, NES damping, or NES nonlinear stiffness, the corresponding optimal range of the excitation frequency ratio, around 0.96, could be identified to maximize the harvested power.
(3) When the external excitation acceleration, NES damping, and nonlinear stiffness exceed a certain threshold, the system output responses will exhibit a jumping phenomenon, namely the nonlinear softening phenomenon. In the nonlinear softening range, two branching curves exist, corresponding to the responses of small and large initial conditions, respectively. In the non-softening range, there is only one branching curve. This means that the output response of the coupled system is independent of the initial conditions when the excitation is outside the jumping region.
(4) In summary, to achieve better energy harvesting characteristics, the external resistance should be taken as small (such as $R_L < 10$ Oom) and the external excitation amplitude should be larger (such as $a_0 > 2.8\ \mathrm{m/s^2}$). The NES damping should be taken as a smaller or larger value (such as $c_2 = 0.1$ N· s/m or $c_2 > 0.4$ N· s/m), and the stiffness of the NES should be as large as possible (such as $k_2 > 10^6\ \mathrm{N/m^3}$). Then the appearance of the nonlinear softening phenomenon will occur and effectively expand the vibration energy-capturing frequency bandwidth of the proposed coupling system.

In addition, it is particularly pointed out that the correct selection of calculation methods will significantly affect the computational efficiency and accuracy of the coupled model. Zhang et al. [66] proposed a novel stochastic homotopy method to discuss the structure-static features with random variables distributed arbitrarily. It is concluded that this proposed new method has excellent accuracy and stability. In this work, through the numerical parameter analysis, the intrinsical characteristics of the influence of some parameters on the output response of the coupled system can be preliminarily obtained. However, the specific optimal parameter values cannot be directly determined to achieve the best harvesting performance. Excitedly, with the development of computer technology and big data technology, the optimization analysis can be effectively trained through machine learning-based and artificial neural network-based algorithms (DL and ANN) [67,68]. This means that in view of the DL and ANN method, the specified optimal parameters of coupled system can be confirmed by effectively training after the dynamic governing equation of the NES-Galfenol coupled system is determined. This will greatly improve the optimization analysis efficiency of the dual nonlinear coupling model proposed in this work, which will be the research direction of our team in the future. Similarly, the DL and ANN methods can provide technical support for parameter optimization analysis of nonlinear dynamic systems in other fields. Therefore, the DL and ANN methods deserve more attention and discussion for parameter optimization analysis in the future.

Author Contributions: Author Contributions: Conceptualization, L.W.; Methodology, C.L.; Software, C.L. and L.W.; Validation, C.L.; Investigation, C.L.; Data curation, C.L. and W.L.; Writing—original draft, C.L.; Writing—review & editing, C.L. and L.W.; Visualization, Z.Y. and X.N.; Supervision, Z.Y. and X.N. All authors have read and agreed to the published version of the manuscript.

Funding: This study is supported by the National Natural Science Foundation of China (Grant No.: 52178458), Natural Science Foundation of Chongqing Science and Technology Commission (Grant No.: CSTB2022NSCQ-MSX0469), Science and Technology Research Project of Chongqing Education Commission (Grant No.: KJQN202101521), Research Project of Chongqing Doctoral "Through Train" of China (Grant No.: CSTB2022BSXM-JCX0164), Research support project of Chongqing University of Science and Technology (Grant No.:ckrc2020011), China Postdoctoral Science Foundation (Grant No.:

2022M720021), Chongqing Natural Science Foundation General Project (CSTB2022NSCQ-MSX0518), Chongqing Technological Innovation and Application Development Project (CSTB2022TIAD-KPX0144), Chongqing Construction Science and Technology Plan Project: Chengkezi 2023 No.1-1, and Natural Science Foundation of Chongqing of China (No.: CSTB2023NSCQ-LZX0051).

Data Availability Statement: The original contributions presented in the study are included in the article, further inquiries can be directed to the corresponding author.

Conflicts of Interest: The authors declare that they have no known competing financial interests or personal relationships that could have appeared to influence the work.

References

1. Alavi, A.H.; Hasni, H.; Lajnef, N.; Chatti, K.; Faridazar, F. An intelligent structural damage detection approach based on sel powered wireless sensor data. *Automat. Constr.* **2016**, *62*, 24–44. [CrossRef]
2. Clemente, C.S.; Mahgoub, A.; Davino, D.; Visone, C. Multiphysics circuit of a magnetostrictive energy harvesting device. *J. Inte Mat. Syst. Str.* **2017**, *28*, 2317–2330. [CrossRef]
3. Wang, L.Z.; Tan, T.; Yan, Z.M.; Yan, Z.T. Tapered galloping energy harvester for power enhancement and vibration reductior *J. Intel. Mat. Syst. Str.* **2019**, *30*, 2853–2869. [CrossRef]
4. Dragunov, V.P.; Ostertak, D.I.; Sinitskiy, R.E. New modifications of a Bennet doubler circuit-based electrostatic vibrational energ harvester. *Sens. Actuat. A-Phys.* **2020**, *302*, 111812. [CrossRef]
5. Tan, Y.; Dong, Y.; Wang, X. Review of MEMS Electromagnetic Vibration Energy Harvester. *J. Microelectromech. Syst.* **2017**, *26*, 1–1 [CrossRef]
6. Budhathoki, S.; Sapkota, A.; Law, K.M.; Nepal, B.; Ranjit, S.; Kc, S.; Mewes, T.; Hauser, A.J. Low Gilbert damping and linewidth i magnetostrictive FeGa thin films. *J. Magn. Magn. Mater.* **2020**, *496*, 165906. [CrossRef]
7. Moradian, K.; Sheikholeslami, T.F.; Raghebi, M. Investigation of a spherical pendulum electromagnetic generator for harvestin energy from environmental vibrations and optimization using response surface methodology. *Energ. Convers. Manag.* **2022**, *26* 1158–1124. [CrossRef]
8. Jin, Y.; Xiao, S.; Zhang, Y. Enhancement of tristable energy harvesting using stochastic resonance. *J. Stat. Mech. Theory Exp.* **201** *23*, 123–211. [CrossRef]
9. Yang, T.; Cao, Q. Novel multi-stable energy harvester by exploring the benefits of geometric nonlinearity. *J. Stat. Mech. Theor Exp.* **2019**, *7*, 033405. [CrossRef]
10. Wang, W.; Cao, J.; Wei, Z.H.; Litak, G. Approximate Fokker-Planck-Kolmo-gorovequation analysis for asymmetric multistabl energy harvesters excited by white noise. *J. Stat. Mech. Theory Exp.* **2021**, *3*, 023407. [CrossRef]
11. Chand, R.R.; Tyagi1, A. Investigation of the Effects of the Piezoelectric Patch Thickness and Tapering on the Nonlinearity of Parabolic Converging Width Vibration Energy Harvester. *J. Vib. Eng. Technol.* **2021**, *10*, 1–18. [CrossRef]
12. Ghouli, Z.; Litak, Z. Effect of High-Frequency Excitation on a Bistable Energy Harvesting System. *J. Vib. Eng. Technol.* **2023**, *11* 99–106. [CrossRef]
13. Lin, C.H.; Lin, Y.Z. Analysis of nonlinear piezomagnetism for magnetostrictive terfenol-D composites. *J. Magn. Magn. Mate* **2021**, *540*, 168490. [CrossRef]
14. Nakajima, K.; Tanaka, S.; Mori, K.; Kurita, H.; Narita, F. Effects of Heat Treatment and Cr Content on the Microstructures Magnetostriction, and Energy Harvesting Performance of Cr-Doped-Fe-Co Alloys. *Adv. Eng. Mater.* **2021**, *24*, 2101036. [CrossRef]
15. Wang, L.Z.; Lian, C.L.; Shu, D.L.; Yan, Z.T.; Nie, X.C. Analytical solution and optimal design for the output performance o Galfenol cantilever energy harvester considering electromechanical coupling effect. *Sci. Rep.* **2023**, *13*, 12857. [CrossRef] [PubMed]
16. Yang, Z.; Tan, Y.; Zu, J. A multi-impact frequency up-converted magnetostrictive transducer for harvesting energy from finge tapping. *Int. J. Mech. Sci.* **2017**, *126*, 235–241. [CrossRef]
17. Wu, J.; Hu, Z.; Gao, X.; Cheng, M.; Zhao, X.; Su, W.; Li, X.; Wang, Z.; Zhou, Z.; Dong, S.; et al. Electrode shape dependence of the barbell-shaped magneto-mechano-electric energy harvester for low-frequency applications. *Sens. Actuat. A-Phys.* **2019**, *297* 111535. [CrossRef]
18. Yoo, J.H.; Flatau, A.B. A bending-mode galfenol electric power harvester. *J. Intell. Mater. Syst. Struct.* **2012**, *23*, 647–654. [CrossRef]
19. Atulasimha, J.; Flatau, A.B. A review of magnetostrictive iron-gallium alloys. *Smart. Mater. Struct.* **2011**, *20*, 043001. [CrossRef]
20. Davino, D.; Giustiniani, A.; Visone, C.; Adly, A.A. Energy Harvesting Tests With Galfenol at Variable Magneto-Mechanical Conditions. *IEEE Trans. Magn.* **2012**, *48*, 3096–3309. [CrossRef]
21. Clark, A.E.; Hathaway, K.B.; Wun-Fogle, M.; Restorff, J.B.; Lograsso, T.A.; Keppens, V.M.; Petculescu, G.; Taylor, R.A. Extraordinary magnetoelasticity and lattice softening in bcc Fe-Ga alloys. *J. Appl. Phys.* **2003**, *93*, 8621–8623. [CrossRef]
22. Javed, A.; Morley, N.A.; Gibbs, M.R.J. Structure, magnetic and magnetostrictive properties of as-deposited Fe-Ga thin films *J. Magn. Magn. Mater.* **2009**, *321*, 2877–2882. [CrossRef]
23. Cao, S.; Zheng, J.; Guo, Y.; Li, Q.; Sang, J.; Wang, B.; Yan, R. Dynamic Characteristics of Galfenol Cantilever Energy Harvester *IEEE Trans. Magn.* **2015**, *51*, 1–4. [CrossRef]
24. Cao, S.Y.; Sun, S.S.; Zheng, J.J.; Wang, B.; Wan, L.L.; Pan, R.Z.; Zhao, R.; Zhang, C.G. Modeling and analysis of Galfenol cantilever vibration energy harvester with nonlinear magnetic force. *AIP Adv.* **2018**, *8*, 056718. [CrossRef]

25. Clemente, C.S.; Davino, D.; Loschiavo, V.P. Analysis of a Magnetostrictive Harvester with a Fully Coupled Nonlinear FEM Modeling. *IEEE Trans. Magn.* **2021**, *57*, 4001201. [CrossRef]
26. Jin, S.; Meng, A.; Li, M.; Xu, Z.; Wu, S.; Chen, Y. Modeling and Analysis of Wave Energy Harvester with Symmetrically Distributed Galfenol Cantilever Beams. *Materials* **2023**, *16*, 5585. [CrossRef] [PubMed]
27. Vakakis, A.F.; Manevitch, L.I.; Gendelman, O.; Bergman, L. Dynamics of linear discrete systems connected to local essentially nonlinear attachments. *J. Sound. Vib.* **2003**, *264*, 559–577. [CrossRef]
28. Panagopoulos, P.N.; Vakakis, A.F.; Tsakirtzis, S. Transient resonant interactions of linear chains with essentially nonlinear end attachments leading to passive energy pumping. *Int. J. Solids. Struct.* **2004**, *41*, 6505–6528. [CrossRef]
29. Vakakis, A.F. Inducing passive nonlinear energy sinks in vibrating systems. *J. Vib. Acoust.* **2001**, *123*, 324–332. [CrossRef]
30. Gendelman, O.V.; Manevitch, L.I.; Vakakis, A.F.; Closkey, R.M. Energy pumping in nonlinear mechanical oscillators: Part I-Dynamics of the underlying Hamiltonian systems. *J. Appl. Mech.* **2001**, *68*, 34–41. [CrossRef]
31. Georgiades, F.; Vakakis, A.F. Dynamics of a linear beam with an attached local nonlinear energy sink. *Commun. Nonlinear. Sci.* **2007**, *12*, 643–651. [CrossRef]
32. Gendelman, O.V.; Sapsis, T.; Vakakis, A.F.; Bergman, L.A. Enhanced passive targeted energy transfer in strongly nonlinear mechanical oscillators. *J. Sound. Vib.* **2011**, *330*, 1–8. [CrossRef]
33. Kani, M.; Khadem, S.E.; Pashaei, M.H.; Dardel, M. Vibration control of a nonlinear beam with a nonlinear energy sink. *Nonlinear Dynam.* **2016**, *83*, 1–22. [CrossRef]
34. Parseh, M.; Dardel, M.; Ghasemi, M.H.; Pashaei, M.H. Steady state dynamics of a non-linear beam coupled to a non-linear energy sink. *Int. J. Nonlin. Mech.* **2016**, *79*, 48–65. [CrossRef]
35. Ahmadabadi, Z.N.; Khadem, S.E. Nonlinear vibration control and energy harvesting of a beam using a nonlinear energy sink and a piezoelectric device. *J. Sound. Vib.* **2014**, *333*, 4444–4457. [CrossRef]
36. Kremer, D.; Liu, K. A nonlinear energy sink with an energy harvester: Transient responses. *J. Sound. Vib.* **2014**, *333*, 4859–4880. [CrossRef]
37. Kremer, D.; Liu, K. A nonlinear energy sink with an energy harvester: Harmonically forced responses. *J. Sound. Vib.* **2017**, *410*, 287–302. [CrossRef]
38. Xiong, L.; Tang, L.; Liu, K.; Mace, B.R. Broadband piezoelectric vibration energy harvesting using a nonlinear energy sink. *J. Phys. D Appl. Phys.* **2018**, *51*, 185502. [CrossRef]
39. Tian, W.; Li, Y.M.; Yang, Z.C.; Li, P.; Zhao, T. Suppression of nonlinear aeroelastic responses for a cantilevered trapezoidal plate in hypersonic airflow using an energy harvester enhanced nonlinear energy sink. *Int. J. Mech. Sci.* **2020**, *172*, 105417. [CrossRef]
40. Fang, Z.W.; Zhang, Y.W.; Li, X.; Ding, H.; Chen, L.Q. Integration of a nonlinear energy sink and a giant magnetostrictive energy harvester. *J. Sound. Vib.* **2017**, *391*, 35–49. [CrossRef]
41. Wang, Z.J.; Zang, J.; Zhang, Y.W. Method for Controlling Vibration and Harvesting Energy by Spacecraft: Theory and Experiment. *AIAA J.* **2022**, *60*, 6097–6115. [CrossRef]
42. Xu, K.F.; Zhang, Y.W.; Niu, M.Q.; Zang, J.; Xue, J.; Chen, L.Q. An improved nonlinear energy sink with electromagnetic damping and energy harvesting. *Int. J. Appl. Mech.* **2022**, *14*, 2250055. [CrossRef]
43. Zhang, Y.W.; Gao, C.Q.; Zhang, Z.; Zang, J. Dynamic Analysis of Vibration Reduction and Energy Harvesting Using a Composite Cantilever Beam with Galfenol and a Nonlinear Energy Sink. *Int. J. Mech. Sci.* **2021**, *13*, 2150089. [CrossRef]
44. Xue, J.R.; Zhang, Y.W.; Niu, M.Q.; Chen, L.Q. Harvesting electricity from random vibrations via a nonlinear energy sink. *J. Vib. Control* **2023**, *29*, 5398–5412. [CrossRef]
45. Xu, K.F.; Zhang, Y.W.; Zang, J.; Niu, M.Q.; Chen, L.Q. Integration of vibration control and energy harvesting for whole-spacecraft: Experiments and theory. *Mech. Syst. Signal. Process.* **2021**, *161*, 107956. [CrossRef]
46. Zhang, Y.W.; Su, C.; Ni, Z.Y.; Zang, J.; Chen, L.Q. A multifunctional lattice sandwich structure with energy harvesting and nonlinear vibration control. *Compos. Struct.* **2019**, *221*, 110875. [CrossRef]
47. Li, J.; He, X.; Yang, X.; Liu, Y. A consistent geometrically nonlinear model of cantilevered piezoelectric vibration energy harvesters. *J. Sound. Vib.* **2020**, *486*, 115614. [CrossRef]
48. Tan, T.; Yan, Z.M.; Lei, H.; Sun, W.P. Geometric nonlinear distributed parameter model for cantilever beam piezoelectric energy harvesters and structural dimension analysis for galloping mode. *J. Intell. Mater. Syst. Struct.* **2017**, *28*, 3066–3078. [CrossRef]
49. Shooshtari, A.; Hoseini, S.M.; Mahmoodi, S.N.; Kalhori, H. Analytical solution for nonlinear free vibrations of viscoelastic microcantilevers covered with a piezoelectric layer. *Smart. Mater. Struct.* **2012**, *21*, 075015. [CrossRef]
50. Li, H.S.; Sun, H.X.; Song, B.Y.; Zhang, D.; Shang, X.C.; Liu, D.H. Nonlinear dynamic response of an L-shaped beam-mass piezoelectric energy harvester. *J. Sound. Vib.* **2021**, *499*, 116004. [CrossRef]
51. Nie, X.C.; Tan, T.; Yan, Z.M.; Yan, Z.T.; Hajj, M.R. Broadband and high-efficient L-shaped piezoelectric energy harvester based on internal resonance. *Int. J. Mech. Sci.* **2019**, *159*, 287–305. [CrossRef]
52. Nie, X.C.; Tan, T.; Yan, Z.M.; Yan, Z.T.; Zhang, W.M. Ultra-wideband piezoelectric energy harvester based on Stockbridge damper and its application in smart grid. *Appl. Energ.* **2020**, *267*, 114898. [CrossRef]
53. Nie, X.C.; Gao, X.; Wang, L.Z.; Tan, T.; Yan, Z.T.; Yan, Z.M.; Liu, X. Nonlinear analysis of the internal resonance response of an L-shaped beam structure considering quadratic and cubic nonlinearity. *J. Stat. Mech. Theory Exp.* **2022**, *2*, 023204. [CrossRef]
54. Nie, X.C.; Tan, T.; Yan, Z.M.; Yan, Z.T.; Wang, L.Z. Revised method of multiple scales for 1:2 internal resonance piezoelectric vibration energy harvester considering the coupled frequency. *Commun. Nonlinear Sci.* **2023**, *118*, 107018. [CrossRef]

55. Han, Q.K.; Li, X.L.; Chu, F.L. Skidding behavior of cylindrical roller bearings under time-variable load conditions. *Int. J. Mech Sci.* **2018**, *135*, 203–214. [CrossRef]
56. Han, Q.K.; Chu, F.L. Nonlinear dynamic model for skidding behavior of angular contact ball bearings. *J. Sound. Vib.* **2015**, *354*, 219–235. [CrossRef]
57. Wang, H.; Han, Q.K.; Zhou, D. Nonlinear dynamic modeling of rotor system supported by angular contact ball bearings. *Mech Syst. Signal. Process.* **2017**, *85*, 16–40. [CrossRef]
58. Liu, W.R.; Bai, X.; Yang, H.; Bao, R.X.; Liu, J.Q. Tendon driven bistable origami flexible gripper for high-speed adaptive grasping *IEEE Robot. Autom. Lett.* **2024**, *9*, 5417–5424. [CrossRef]
59. Clemente, C.; Davino, D. Modeling and characterization of a kinetic energy harvesting device based on galfenol. *Materials* **2019**, *12*, 3199. [CrossRef] [PubMed]
60. Erturk, A.; Inman, D.J. *Piezoelectric Energy Harvesting-Base Excitation Problem for Cantilevered Structures and Correction of the Lumped-Parameter Electromechanical Model*; John Wiley and Sons: Hoboken, NJ, USA, 2011; pp. 19–48. [CrossRef]
61. Erturk, A.; Inman, D. An experimentally validated bimorph cantilever model for piezoelectric energy harvesting from base excitations. *Smart Mater. Struct.* **2009**, *18*, 025009. [CrossRef]
62. Erturk, A.; Hoffmann, J.; Inman, D.J. A piezomagnetoelastic structure for broadband vibration energy harvesting. *Appl. Phys. Lett.* **2009**, *94*, 254102. [CrossRef]
63. Bibo, A.; Abdelkefi, A.; Daqaq, M.F. Modeling and characterization of a piezoelectric energy harvester under combined aerodynamic and base excitations. *J. Vin. Acoust.* **2015**, *137*, 031017. [CrossRef]
64. Khazaee, M.; Rezaniakolaei, A.; Rosendahl, L. An experimental study to determine damping of piezoelectric harvesters using transient analysis of unified electromechanical voltage equation. *Energ. Convers. Manag.* **2021**, *227*, 113567. [CrossRef]
65. Cao, S.; Sang, J.; Zheng, J.; Wang, B. Electromechanical coupling dynamic model of galfenol cantilever energy harvester. *Proc CSEE* **2015**, *35*, 5623–5631. (In Chinese)
66. Zhang, H.; Xiang, X.; Huang, B.; Wu, Z.F.; Chen, H. Static homotopy response analysis of structure with random variables of arbitrary distributions by minimizing stochastic residual error. *Comput. Struct.* **2023**, *288*, 107153. [CrossRef]
67. Du, H.; Du, S.; Li, W. Probabilistic time series forecasting with deep non-linear state space models. *CAAI Trans. Intell. Technol.* **2023**, *8*, 3–13. [CrossRef]
68. Wang, G.; Hao, Z.H.; Li, H.S.; Zhang, B. An activated variable parameter gradient-based neural network for time-variant constrained quadratic programming and its applications. *CAAI Trans. Intell. Technol.* **2023**, *8*, 670–679. [CrossRef]

buildings

MDPI

Article

Experimental Study on the Mechanical Properties of Rock–Concrete Composite Specimens under Cyclic Loading

Hongjun Li [1], Baoyun Zhao [2,*], Zhengjun Hou [3] and Hongyao Min [2]

1 China Anergy Group Third Engineering Bureau Co., Ltd., Chengdu 610036, China; lihj09@sina.com
2 School of Civil Engineering and Architecture, Chongqing University of Science and Technology, Chongqing 401331, China; 2023206005@cqust.edu.cn
3 Chengdu Sunvison Intelligent Technology Co., Ltd., Chengdu 610065, China
* Correspondence: baoyun666@163.com

Abstract: The foundations of bridges and other tall buildings are often subjected to cyclic loads. Therefore, it is essential to investigate the mechanical properties of rock–concrete composite foundations under cyclic loads. In this paper, uniaxial cyclic loading and unloading tests were conducted on rock–concrete composite specimens using the TFD-2000 microcomputer servo-controlled rock triaxial testing machine. The stress–strain curves, elastic modulus variation, and energy dissipation were analyzed. The results showed that the stress–strain curves of composite specimens under uniaxial cyclic loading and unloading conditions formed hysteresis loops. The hysteresis loop exhibited a sparse–dense–sparse pattern under the upper stress of 27.44 MPa, which was 90% of the uniaxial strength. The elastic modulus, as well as the dissipated energy, decreased rapidly in the first few cycles and then gradually decreased at a constant rate, with the upper stress increasing to 27.44 MPa. Both the elastic modulus and the dissipated energy exhibited an accelerated stage before specimen failure. The primary failure mode of the composite specimen was split failure from concrete to sandstone. A damage variable was derived to better reflect the laws governing the damage evolution of the composite under cyclic loads.

Keywords: rock–concrete complex; cyclic loading; damage evolution

1. Introduction

Building foundations are often composed of both concrete and rock, especially in areas with complex geological conditions. The interaction between the concrete foundation and the underlying rock foundation plays a critical role in the overall stability of the structure. However, the mechanical properties of the rock–concrete composite system under cyclic loads, such as those experienced during earthquakes or other dynamic events, are not well understood. Therefore, it is important to investigate the behavior of rock and concrete composite specimens under cyclic loads to enhance the design and construction of building foundations.

Numerous scholars have studied the effects of cyclic loading on the mechanical properties of rock masses in recent decades [1]. Based on the type of loading applied to rocks, the cyclic loading tests can be classified as cyclic uniaxial or triaxial compressive tests, cyclic tensile tests, cyclic shear tests, and cyclic flexural tests [2]. In the research field of rock fatigue characteristics, it has been found that factors such as loading frequency, temperature, stress amplitude, and cycle number in cyclic loading and unloading tests mainly affect the strength of rock mass [3–5]. With more in-depth research, many scholars have become concerned about fatigue damage [6–8] and the energy dissipation mechanism [9–11] of rock during cyclic loading. Sun et al. [12] deduced a damage variable according to the strength characteristics and energy evolution law in the cyclic loading process. Meng et al. [13] found that the energy density and hysteresis energy variation in different loading conditions show a decrease in the first few cycles and then start to increase up to failure. In addition, different types of experimental techniques, such as optics and acoustics, have been employed to observe the failure process of rock materials during cyclic loading.

As early as 1934, O. Graf and E. Brenner [14] began to study the effect of loading frequency on the fatigue life of concrete. Since then, numerous scholars have conducted experimental, theoretical, and numerical modeling studies on the cyclic loading behavior of concrete. Most of these studies have focused on common concrete [15–17], reinforced concrete [18,19], recycled concrete [20], and foamed concrete [21]. Cook et al. [15] conducted a pioneering study on the impact of sustained constant load and fatigue loading history on the compressive strength of common concrete and revealed that the application of a constant load leads to an increase in both the compressive strength and elastic modulus of concrete. Bian et al. [18] conducted uniaxial compression tests on basalt-fiber-reinforced concrete (BFRC) specimens with five different strength grades to investigate their stress–strain relationship. Feng et al. [21] investigated the dynamic mechanical properties and damage characteristics of lightweight foamed concrete under impact loading. They found that the material's dynamic mechanical properties showed a significant strain rate enhancement effect and density dependence.

In recent years, there has been growing interest in the study of the cyclic loading mechanical properties of composite specimens, which are composed of concrete and other materials. Researchers have conducted experiments to investigate the behavior of such composite materials under dynamic loading conditions, aiming to understand their response to cyclic loading and its implications for engineering practice. He et al. [22] examined the cyclic loading test properties of composite laminated beams and found that the stiffness of the specimen, which failed in the cyclic loading test, decreased by about 30%, while the bearing capacity decreased by about 10%. Zhang et al. [23] investigated the cyclic behavior of FRP–concrete–steel double-skin tubular columns under a combined loading condition of axial compression and cyclic lateral loading.

Despite efforts in the field of rocks, concrete, and composite materials, there is still a lack of comprehensive understanding of the behavior of rock and concrete composite specimens under cyclic loading, particularly in the context of building foundation systems. Therefore, this study aims to contribute to the existing body of knowledge by investigating the mechanical properties of composite specimens under cyclic loading. By doing so, the study aims to offer valuable insights into the behavior of rock and concrete composite foundations under dynamic loading conditions and to enhance the design and construction of building foundations in practice.

2. Materials and Method

2.1. Specimen Preparation

Sandstone blocks were extracted from a foundation engineering site in southern Sichuan Province. The sandstone was compact and uniform in structure, displaying a white-gray appearance. Its mineral composition mainly consisted of quartz, feldspar, calcite, and a small amount of iron, among others. Its chemical composition primarily included SiO_2, Al_2O_3, CaO, and Fe_2O_3, classifying it as a low-strength, brittle rock. First, the surface of one sandstone block was cleaned, and it was treated to achieve a certain degree of roughness. Then, concrete was poured on the surface. After 28 days of curing, the composite block was cored in the laboratory, and the composite specimens with a diameter of 50 mm and a height of 50 + 50 mm were prepared. The specimen preparation diagram and prepared specimens are shown in Figure 1.

2.2. Testing Equipment

The TFD-2000 microcomputer servo-controlled rock triaxial testing machine which was manufactured by Changchun Keyi Testing Instrument Co., Changchun, China was used in the test, as shown in Figure 2. The maximum axial compressive force of the test machine is 2000 kN. The measurement resolution is 0.001 MPa, and the pressure measurement error range is ±0.5%. During the test, the test data can be automatically collected by a microcomputer.

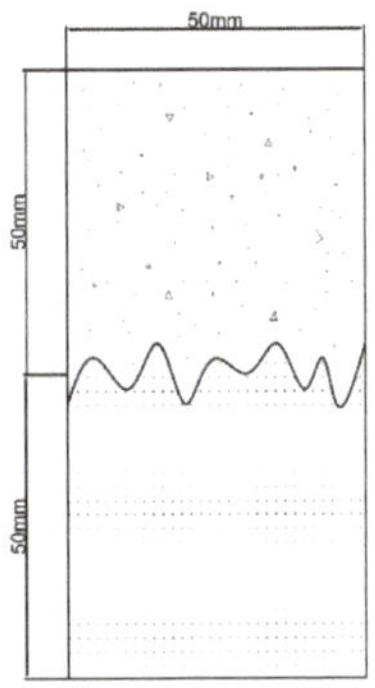

(a) (b)

Figure 1. Pictures and schematic diagram of rock–concrete composite specimens. (**a**) Schematic diagram of rock–concrete composite specimens; (**b**) pictures of rock–concrete composite specimens.

Figure 2. TFD-2000 microcomputer servo-controlled triaxial testing machine.

2.3. Preliminary Test

Before conducting the uniaxial cyclic loading and unloading test, the uniaxial compression tests of a sandstone and concrete composite specimen were performed at a displacement control rate of 0.01 mm/min. The test results are shown in Figure 3. It can be observed that the uniaxial compression test curve under different cycles can be divided into four stages. The first stage was the compaction stage of the micro-cracks (Stage I). The original cracks gradually closed under axial pressure in this stage, and the slope of the stress–strain curve gradually increased, indicating that the original cracks closed gradually with the increase in load. The second stage was the elastic deformation stage (Stage II), where the slope of the curve in the elastic stage remained approximately constant. The third stage was the plastic yield deformation stage (Stage III). The decrease in the slope of the curve at the end of the elastic deformation stage indicated that the specimen continued to form new cracks and progress towards an unstable state. The fourth stage was the failure stage following the peak strength (Stage IV). The stress decreased rapidly and then gradually decreased to a constant value. The uniaxial compressive strength of the composite specimen was 30.52 MPa.

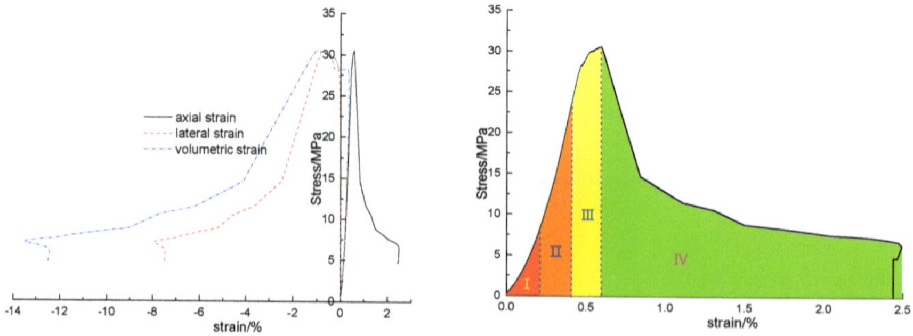

(a) (b)

Figure 3. Uniaxial stress−strain curves. (**a**) The uniaxial compression stress−strain curve; (**b**) the four-stage diagram of uniaxial compression.

2.4. The Cyclic Loading and Unloading Test Method

According to uniaxial compression test results, 18.29, 21.34, 34.39, and 27.44 MPa were selected as the cyclic upper limit stress amplitudes. These values corresponded to 60%, 70%, 80%, and 90% of the uniaxial compressive strength, respectively. Additionally, 6.11 MPa (20% of the uniaxial compressive strength) was the cyclic lower limit stress. Each specimen underwent 200 loading and unloading cycles. A force-controlled rate of 0.2 kN/s was adopted in the loading and unloading cycles. Figure 4 shows the stress path diagram of cyclic loading and unloading tests, while Table 1 presents the detailed test parameters.

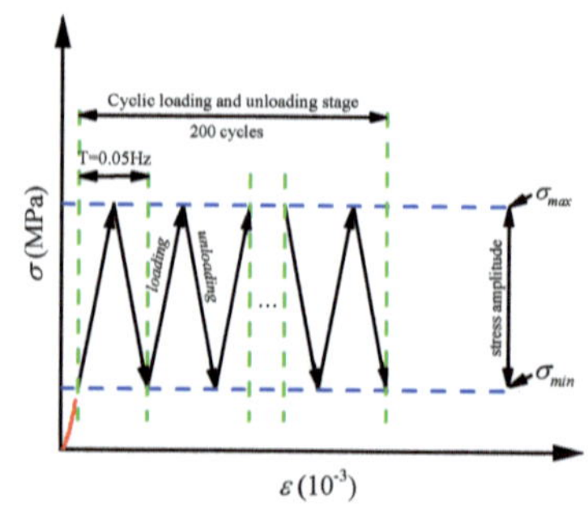

Figure 4. Stress path diagram of the cyclic loading and unloading test.

Table 1. Uniaxial cyclic loading and unloading test plan.

Specimen/No	Upper Stress/MPa	Lower Stress/MPa	Stress Amplitude/MPa	Cycles/N	Loading Speed/(kN/s)
RC-1	18.29	6.11	12.18	200	0.2
RC-2	21.34	6.11	15.03	200	0.2
RC-3	24.39	6.11	18.28	200	0.2
RC-4	27.44	6.11	21.33	200	0.2

3. Results and Discussion

3.1. Mechanical Properties of Composite Specimens under Cyclic Loading and Unloading

3.1.1. Deformation Characteristics

Figure 5a–d shows the stress–strain curves of rock–concrete composite specimens under uniaxial cyclic loading and unloading for tests conducted at stress amplitudes of 12.18, 15.03, 18.28, and 21.33 MPa.

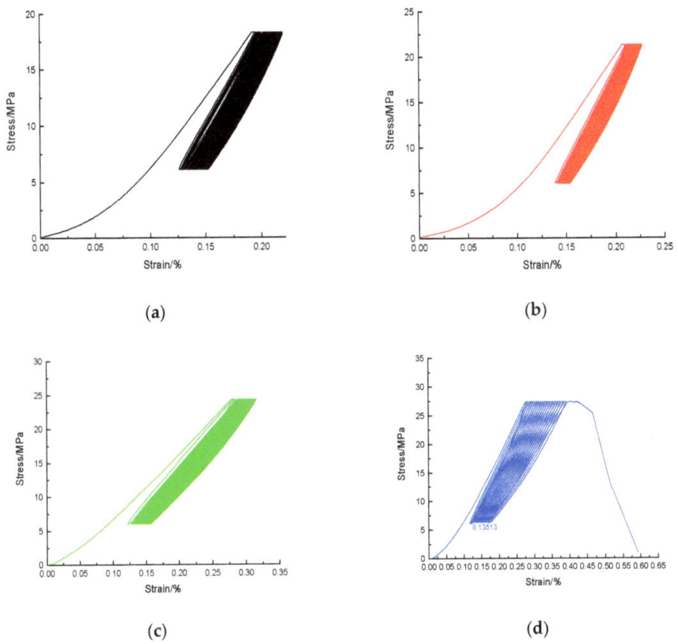

Figure 5. Stress–strain curve of a rock–concrete composite specimen under different upper cycle loads. (**a**) Stress amplitude of 12.18 MPa; (**b**) stress amplitude of 15.03 MPa; (**c**) stress amplitude of 18.28 MPa; (**d**) stress amplitude of 21.33 MPa.

It can be seen from Figure 5 that the stress and axial strain curves of the rock–concrete composite specimens formed hysteresis loops under cyclic loads. From the perspective of the entire experiment process, the axial strain hysteresis loop formed by the stress–strain curve demonstrated a trend from sparse to dense when the stress amplitude was less than 21.33 MPa. However, the axial strain hysteresis loop exhibited an evolution pattern from sparse to dense to sparse when the stress amplitude reached 21.33 MPa. The cyclic loading only endured for 19 cycles before the specimen was destroyed. It can be concluded that the axial deformation of the composite specimen develops rapidly, and the distance between the hysteresis loops is relatively larger, resulting in a larger area of hysteresis loop at the beginning of the cycle. This indicates that more energy is consumed under each loading and unloading cycle, and the cumulative fatigue damage inside the composite specimen is larger at this stage. After the first few cycles, the development rate of composite specimen deformation gradually decreases and enters the stable growth stage. The distance between each hysteresis loop becomes very dense, the area of the hysteresis loop is correspondingly reduced, and the fatigue damage caused by each loading and unloading cycle is significantly reduced. When the upper stress reaches a certain value, the axial deformation rate accelerates, leading to significant deformation and a more spread-out hysteresis loop.

Figure 6 shows the curve of residual strain with cycles for the rock–concrete composite specimen. It can be seen from Figure 6 that each cycle produced residual strain. The residual strain had a nonlinear relationship with the number of cycles, which was very similar to the typical three-stage rock creep curve. Referring to the division method of the rock creep curve, the curve of residual strain and cycles can be divided into two stages for stress amplitudes less than 21.33 MPa: the deceleration accumulation stage and the stable accumulation stage, as shown in Figure 7. The curve of residual strain and cycles can be divided into three stages for a stress amplitude of 21.33 MPa: the deceleration accumulation stage (I), the stable accumulation stage (II), and the acceleration accumulation stage, as shown in Figure 8. It can be concluded that during the early cycles (III), the residual strain increases relatively rapidly as the specimen adjusts to the cyclic stresses. However, as the number of cycles continues to increase, the rate of residual strain accumulation may slow down, reaching a more stable or saturated level. After the residual strain accumulates to a certain value, micro-cracks in the rock gradually connect. Consequently, the residual strain increases rapidly before failure.

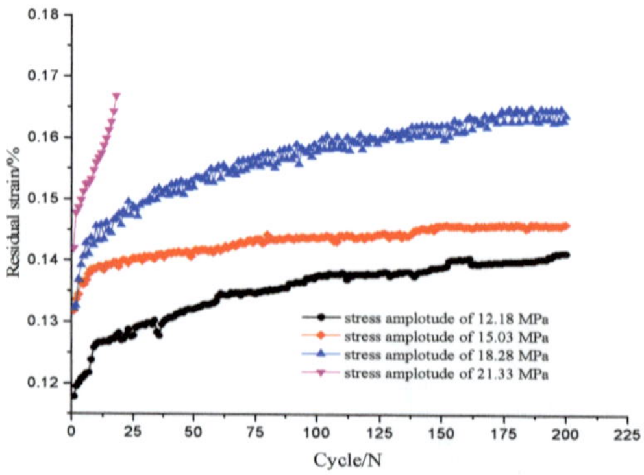

Figure 6. The curve of residual strain vs. cycles under cyclic loads.

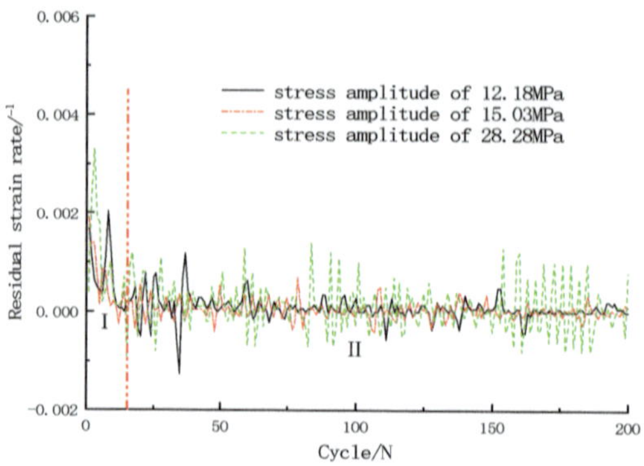

Figure 7. The curve of residual strain rate vs. cycles under cyclic loads (stress amplitudes less than 21.33 MPa).

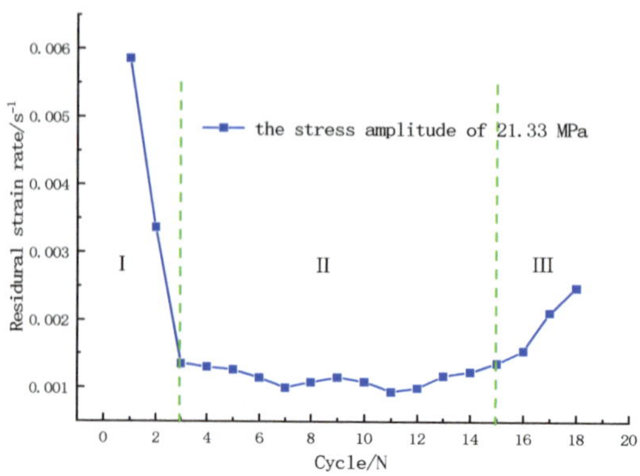

Figure 8. The curve of residual strain rate vs. cycles under cyclic loads (stress amplitude of 21.33 MPa).

3.1.2. The Elastic Modulus: Changing Characteristics

The Young's modulus (E) is one of the most important mechanical properties of materials and is often used to describe their elastic behavior. Elastic modulus is one of the most important parameters for evaluating the mechanical properties of rocks. According to the Regulation for testing the physical and mechanical properties of rock—Part 19: Test for determining the deformability of rock in uniaxial compression (DZ/T 0276.19-2015 19) [24], the elastic modulus in each load cycle was calculated by Equation (1):

$$E = \frac{\sigma_{max} - \sigma_{min}}{\varepsilon_{max} - \varepsilon_{min}} \tag{1}$$

where E is the elastic modulus, MPa; σ_{max} is the maximum axial stress in each cycle, MPa; σ_{min} is the minimum axial stress in each cycle; ε_{max} is the maximum axial strain; and ε_{min} is the minimum axial strain.

The elastic modulus of the composite specimen under different stress amplitudes can be calculated using Equation (1). The curves depicting the evolution of the elastic modulus are shown in Figure 9. It can be seen in the figure that the evolution trends of the elastic modulus under different stress amplitudes decreased rapidly in the first three cycles, after which the elastic modulus gradually decreased at a constant rate. With the increase in stress amplitude, the elastic modulus showed a rapid decreasing trend, indicating that higher stress amplitudes caused irreversible fatigue damage. Under the same stress amplitude, the primary pores inside the composite specimen are gradually compacted, the cracks are gradually closed in the first few cycles, and the composite specimen enters the elastic phase. This is the main reason that the elastic modulus of the composite specimen gradually tends to a constant value after decreasing in the first few cycles. The elastic strain of the specimen played a major role, resulting in a stable elastic modulus. Before failure, new cracks formed inside the composite specimens, leading to the destruction of the specimen's integrity. This resulted in a rapid decrease in the loading elastic modulus under significantly higher stress amplitudes.

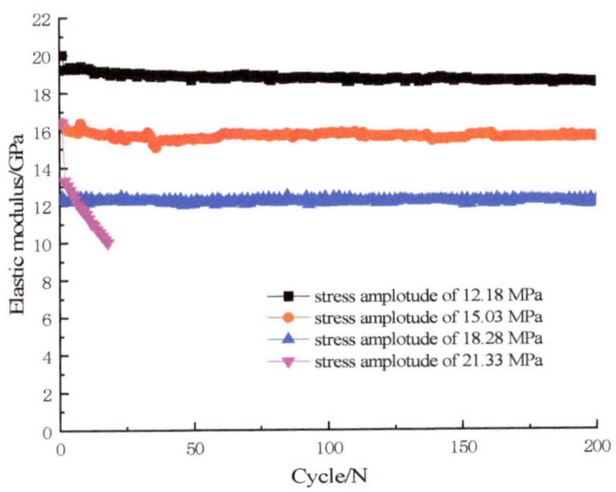

Figure 9. The evolution curves of the elastic modulus of the composite specimen under different stress amplitudes.

3.2. Failure Mode

Figure 10 displays the apparent damage pattern of the rock–concrete composite specimen after various levels of cyclic loading. It can be seen from Figure 10 that the main forms of damage were split failures from concrete to sandstone, and a partial blocking phenomenon appeared in both the concrete and sandstone sections. During the cyclic loading test, the micro-cracks inside the rock–concrete composite specimen were consistently opening and closing with each load–unload cycle. This phenomenon created room for the development of additional micro-cracks. During a long test period, the micro-cracks in rock–concrete composite specimens can grow significantly. Moreover, because the maximum stress from cyclic loading was 27.44 MPa, which was lower than the conventional compression strength of the rock–concrete composite specimen, the propagation of large cracks in the specimen was limited during the cyclic loading test. Instead, the development of micro-cracks prevailed. With the propagation and coalescence of the micro-cracks, the rock–concrete composite specimen eventually failed.

Figure 10. The damage evolution of the rock–concrete composite specimen under cyclic loads.

The rock and concrete portions of the failed samples were microscopically scanned with the aid of a scanning electron microscope (SEM), yielding the microstructures of the rock and concrete at the same magnification, as illustrated in Figure 11. The microstructures of the two blocks were evidently very different. The rock blocks were composed of a dense array of nested silicon dioxide crystals with very few additional large ratios of impurities. The compact and regular crystal arrangement formed the microstructure. The concrete's coarse and fine aggregates were interspersed with numerous micropores and micro-cracks; these meso-fracture characteristics elucidate the primary mechanism by which the failure characteristics of the rock–concrete composite specimen transitioned from the concrete to the rock portion.

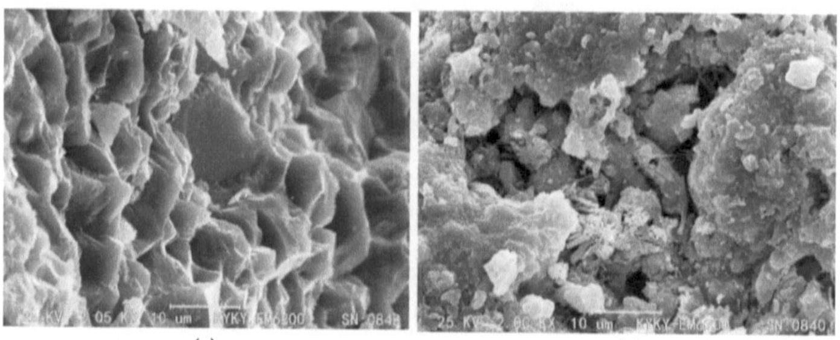

(a) (b)

Figure 11. The microstructures of the rock–concrete composite specimen after failure. (**a**) Rock; (**b**) concrete.

4. The Dissipated Energy Evolution and Analysis of Damage Characteristics

4.1. The Dissipated Energy

The specimen underwent cyclic loading and unloading, which is essentially a process of energy input and dissipation. The area of OABE which below the loading section OAB and the *X*-axis represents the input energy U, the area of BCDE which below the unloading section BC and the *X*-axis represents the elastic strain energy U_{m}, and the area OAB represents the dissipation energy density U_{W}, as illustrated in Figure 12. The energy of each part of the specimen can be defined as follows [24]:

$$U = \int_{\varepsilon_0}^{\varepsilon_B} \sigma_{\mathrm{L}} \mathrm{d}\varepsilon \tag{2}$$

$$U_{\mathrm{m}} = \int_{\varepsilon_{\mathrm{C}}}^{\varepsilon_{\mathrm{B}}} \sigma_{\mathrm{U}} \mathrm{d}\varepsilon \tag{3}$$

$$U_w = U - U_m \tag{4}$$

where σ_L is the stress of each cycle loading segment; σ_U is the stress of the unloading section for each cycle; and ε_0, ε_{B}, and ε_{C} are the axial strains at points of O, B, and C, respectively.

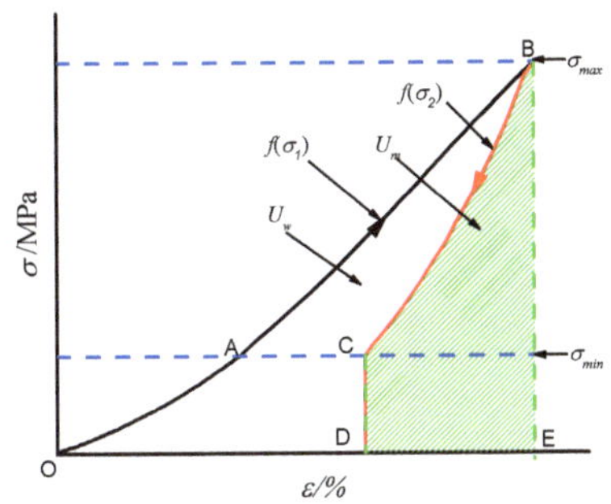

Figure 12. The schematic diagram of sandstone energy calculation under cyclic loading and unloading conditions.

According to Equations (2)–(4), the energy of each cycle of the composite specimen under different stress amplitudes was calculated. The energy evolution curve of the composite specimen under different stress amplitudes can be seen in Figures 13 and 14. It can be seen from Figure 13 that the energy evolution curves showed a sudden-drop stage (I) and a steady-speed stage (II) when the stress amplitude was less than 21.33 MPa. The curve of energy evolution can be divided into three stages for a stress amplitude of 21.33 MPa: the sudden-drop stage (I), the steady-speed stage (II), and the accelerated development stage (III). During the sudden-drop stage, the energy dissipation decreased rapidly in the first few cycles. Subsequently, the energy dissipation rate gradually stabilized at a certain value. This phenomenon primarily occurred due to energy consumption during the closure of micro-defects inside the specimen in the initial cycles. As the cycles progressed, the specimen became more compact, leading to a decrease in energy dissipation. In the steady-speed stage, the energy was mainly consumed in the form of acoustic energy, thermal energy, and plastic deformation. The dissipation energy rate gradually approached a constant speed with the increase in cycles. During the accelerated development stage, the upper stress exceeded the yield strength. The energy was primarily consumed in initiating micro-cracks, facilitating effective expansion, and creating new plastic areas. Energy dissipation escalated rapidly at this stage.

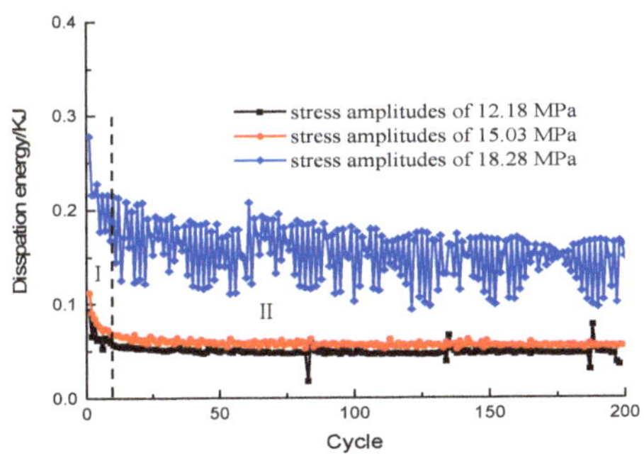

Figure 13. The energy evolution curve of the composite specimen under different stress amplitudes less than 21.33 MPa.

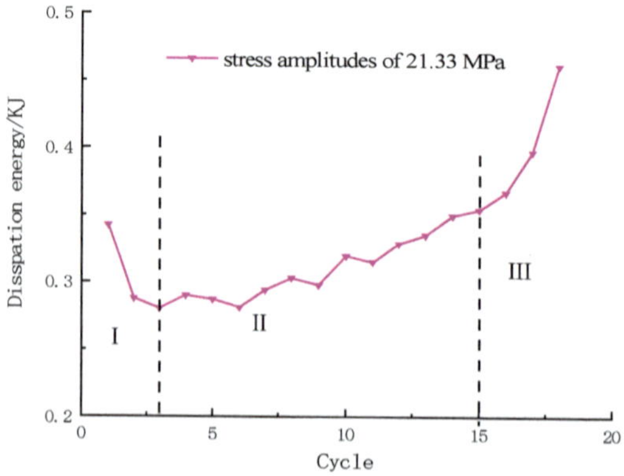

Figure 14. The energy evolution curve of composite specimens under a stress amplitude of 21.33 MPa.

4.2. Damage Evolution Law

With the increase in fatigue cycles, the internal pores, micro-cracks, and irreversible deformation gradually accumulated, leading to continuous damage and deterioration of the specimen. Based on the equivalent strain hypothesis of a continuously damaged medium, the damage constitutive equation of a rock–concrete composite specimen under uniaxial compression can be expressed as follows [25]:

$$1 - D = \frac{\sigma}{E\varepsilon} \tag{5}$$

where D is the damage variable, σ is the stress of a non-destructive material, E is the elastic modulus of the non-destructive material, and ε is the strain of the non-destructive material. When $D = 0$, the rock–concrete composite material is in an undamaged state and intact; when $D = 1$, the rock–concrete composite material is completely destroyed.

In the loading and unloading tests, if we ignore the single cyclic damage change, the material stress σ can be calculated. Therefore, it can be treated as a known constant in Equation (5). By taking the derivative of ε on both sides of Equation (5), we can obtain the following:

$$\mathrm{d}D = \frac{\sigma}{E\varepsilon^2}\mathrm{d}\varepsilon \tag{6}$$

Assume that ε_0 represents the axial strain at the beginning of the cycle, with the damage variable $D = 0$ at this point. ε_e represents the axial strain at the point of failure after various cycles, with the damage variable $D = 1$ when the specimen fails. Integrating D in Equation (6) from 0 to 1 and ε from ε_0 and ε_e, respectively, the Equation (6) can be expressed as follows:

$$\int_0^D \mathrm{d}D = \int_{\varepsilon_0}^{\varepsilon_e} \frac{\sigma}{E\varepsilon^2}\mathrm{d}\varepsilon \tag{7}$$

$$D = \frac{\sigma}{\mathrm{E}}\left(\frac{1}{\varepsilon_0} - \frac{1}{\varepsilon}\right) + C \tag{8}$$

By substituting $D = 0$ and $\varepsilon = \varepsilon_0$ into Equation (8), $C = 0$ can be obtained. By substituting $D = 1$, $\varepsilon = \varepsilon_e$, and $C = 0$ into Equation (8), the following can be obtained:

$$1 = \frac{\sigma}{E}\left(\frac{1}{\varepsilon_0} - \frac{1}{\varepsilon_e}\right) \tag{9}$$

Therefore, the damage variable D can be expressed as follows:

$$D = \frac{\varepsilon_e}{\varepsilon}\left(\frac{\varepsilon - \varepsilon_0}{\varepsilon_e - \varepsilon_0}\right) \tag{10}$$

According to Equation (10), the damage variable of the composite specimen under a stress amplitudes of 21.33 MPa was calculated using the data of residual strains, and the damage evolution curve of the composite specimen is shown in Figure 15. It can be seen from Figure 15 that the composite specimen's damage variable during the cyclic fatigue process can be divided into three stages. In the initial stage (I), the damage variable decreased rapidly with the increase in cycles. During the steady-state stage (II), the damage variable increased at a constant rate with the increase in cycles. During the acceleration stage (III), as the damage accumulated to a certain extent, the damage variable started to increase rapidly until the composite specimen was destroyed. Compared to other rocks, the composite specimen contained more pores, especially in the concrete part. Under cyclic loading, the micro-cracks from pressure-sealing cracks and new cracks further expanded during the initial cycle, leading to some irreversible deformation. When the damage accumulated to a certain extent, a large number of new cracks began to appear in the composite specimen and gradually connected. Consequently, the damage variable started to increase more rapidly.

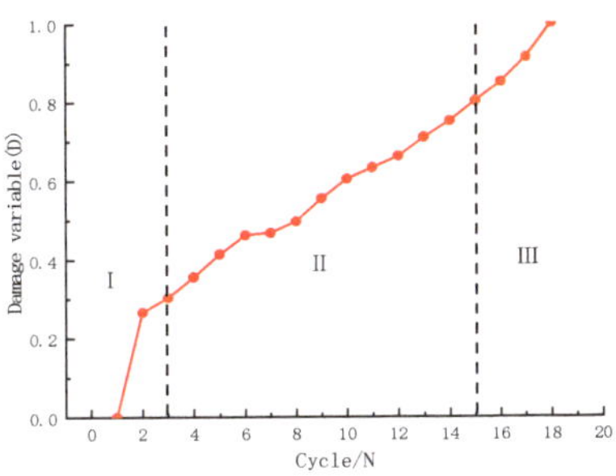

Figure 15. The damage evolution curve of the composite specimen under a stress amplitude of 21.33 MPa.

The damage variable of dissipated energy [26], shown in Equation (11), and the damage variable of residual strain [27], shown in Equation (12), were used for a comparative analysis of the damage evolution of the composite specimen.

$$D_{\mathrm{i}} = \frac{\sum_{\mathrm{i}}^{\mathrm{n}} U_{\mathrm{ip}}}{U_{\mathrm{p}}} \tag{11}$$

where D_{i} is the damage variable of the i-th cycle, $\mathrm{U_{ip}}$ is the cumulative energy dissipation at the i-th cycle, and Up is the total energy dissipation.

$$D_{\mathrm{i}} = \frac{\sum_{\mathrm{i}}^{\mathrm{n}} \varepsilon_{\mathrm{ip}}}{\varepsilon_{\mathrm{p}}} \tag{12}$$

where D_{i} is the damage variable of the i-th cycle, $\varepsilon_{\mathrm{ip}}$ is the cumulative residual strain at the i-th cycle, and ε_{p} is the total residual strain.

Figure 16 illustrates the calculated damage evolution of a composite specimen under a stress amplitude of 21.33 MPa in three different ways. It can be seen from Figure 16 that both the damage variable of cumulative energy dissipation and the damage variable of cumulative residual strain present a straight line with an increase in cycles. However, neither of them accurately represents the trend of rapid propagation, steady development, and accelerated expansion of cracks in rocks. This indicates that the damage calculated by these two methods cannot reflect the damage evolution characteristics of the specimen.

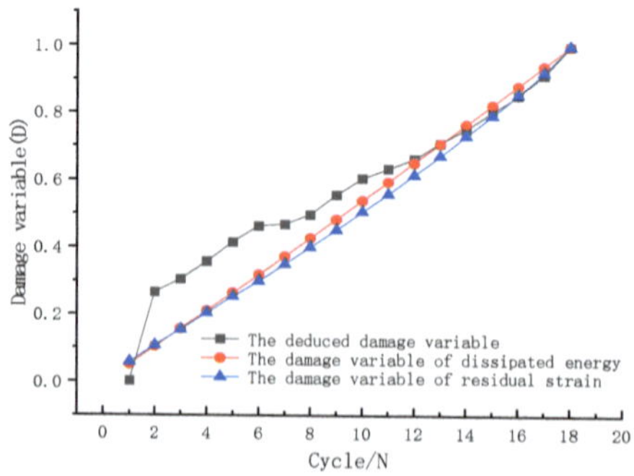

Figure 16. The damage evolution curve of the composite specimen under a stress amplitude of 21.33 MPa.

5. Conclusions

In conclusion, this study investigated the mechanical properties of rock and concrete composite specimens under cyclic loads. In accordance with the experimental results, the deformation characteristics and energy dissipation of composite specimens under cyclic loading were analyzed. The damage variable equation was deduced to evaluate the damage evolution of the specimen. Based on our research results, the following conclusions can be drawn.

Referring to the division method of the rock creep curve, the curve of residual strain and cycles can be divided into two stages for stress amplitudes less than 21.33 MPa: the deceleration accumulation stage and the stable accumulation stage.

The elastic modulus, as well as the dissipated energy, under different stress amplitudes decreased rapidly in the first three cycles. Subsequently, the elastic modulus gradually decreased at a constant rate. With the increase in stress amplitude to 21.33 MPa, the elastic modulus showed a rapid decreasing trend, while the dissipated energy showed an accelerated increase before specimen failure.

A damage variable was deduced, and the damage evolution curve of the composite specimen exhibited three distinct stages: the initial stage, the steady-state stage, and the acceleration stage.

Author Contributions: The first draft of the manuscript was written by H.L. and Z.H.; project administration, B.Z.; data curation, H.M. All authors have read and agreed to the published version of the manuscript.

Funding: This research was funded by the National Natural Science Foundation of China, grant number 41302223, the Scientific and Technological Research Program of Chongqing Municipal Education Commission, grant number KJZD-K202101505, And Tibet Autonomous Region Science and Technology Program XZ202301ZY0037G.

Data Availability Statement: The data underlying this article are available in the article.

Conflicts of Interest: Author Hongjun Li was employed by the company China Anergy Group Third Engineering Bureau Co., Ltd. Author Zhengjun Hou was employed by the company Chengdu Sunvison Intelligent Technology Co., Ltd. The remaining authors declare that the research was conducted in the absence of any commercial or financial relationships that could be construed as a potential conflict of interest.

References

1. Xue, F.; Lin, Z.; Wang, T. Experimental study on effects of cyclic loading paths on cracking behavior and fracture characteristics of granite. *Eng. Fract. Mech.* **2024**, *295*, 109761. [CrossRef]
2. Liu, Y.; Dai, F. A review of experimental and theoretical research on the deformation and failure behavior of rocks subjected to cyclic loading. *J. Rock Mech. Geotech. Eng.* **2021**, *13*, 1203–1230. [CrossRef]
3. Arora, K.; Chakraborty, T.; Rao, K.S. Experimental study on stiffness degradation of rock under uniaxial cyclic sinusoidal compression loading. *Rock Mech. Rock Eng.* **2019**, *52*, 4785–4797. [CrossRef]

4. Meng, Q.; Liu, J.; Pu, H.; Yu, L.; Wu, J.; Wang, C. Mechanical properties of limestone after high-temperature treatment under triaxial cyclic loading and unloading conditions. *Rock Mech. Rock Eng.* **2021**, *54*, 6413–6437. [CrossRef]
5. Xiao, F.; Jiang, D.; Wu, F.; Zou, Q.; Chen, J.; Chen, B.; Sun, Z. Effects of prior cyclic loading damage on failure characteristics of sandstone under true-triaxial unloading conditions. *Int. J. Rock Mech. Min. Sci.* **2020**, *132*, 104379. [CrossRef]
6. Tang, L.X.; Fan, J.Y.; Li, Z.Z.; Chen, J.; Liu, W. A new constitutive model for salt rock under cyclic loadings based on state variables. *Geoenergy Sci. Eng.* **2024**, *233*, 212433. [CrossRef]
7. Guo, J.C.; Li, X.Y.; Lu, C. Modified Model and Simulation Verification of Rock-Fatigue Damage Considering Repeated Discharge Impact. *Processes* **2023**, *11*, 2366. [CrossRef]
8. Ren, C.H.; Yu, J.; Liu, X.Y. Cyclic constitutive equations of rock with coupled damage induced by compaction and cracking. *Int. J. Min. Sci. Technol.* **2022**, *32*, 1153–1165. [CrossRef]
9. Li, Y.Q.; Huang, D. Mechanical Responses and Damage Model of Anchored Jointed Rock Mass under Fatigue Shear Load. *Int. J. Geomech.* **2022**, *23*, 3. [CrossRef]
10. Li, P.; Cai, M. Energy evolution mechanism and failure criteria of jointed surrounding rock under uniaxial compression. *J. Cent. S. Univ.* **2021**, *28*, 1857–1874. [CrossRef]
11. Peng, R.D.; Ju, Y.; Wang, J.G.; Xie, H.P.; Gao, F.; Mao, L.T. Energy Dissipation and Release During Coal Failure Under Conventional Triaxial Compression. *Rock Mech. Rock Eng.* **2015**, *48*, 509–526. [CrossRef]
12. Sun, Y.D.; Yang, Y. Energy and Fatigue Damage Evolution of Sandstone under Different Cyclic Loading Frequencies. *Shock. Vib.* **2021**, *2021*, 5585983. [CrossRef]
13. Meng, Q.B.; Liu, J.F.; Pu, H.; Huang, B.X.; Zhang, Z.Z.; Wu, J.Y. Effects of cyclic loading and unloading rates on the energy evolution of rocks with different lithology. *Geomech. Energy Environ.* **2023**, *34*, 100455. [CrossRef]
14. Graf, O.; Brenner, E. Experiments for investigating the resistance of concrete under often repeated compression loads. *Bull. Dtsch. Aussch. Fur Stahlbeton Berl.* **1934**, *35*, 17–25.
15. Cook, D.J.; Chindaprasirt, P. Influence of loading history upon the compressive properties of concrete. *Mag. Concr. Res.* **1980**, *32*, 89–100. [CrossRef]
16. Long, X.; Mao, M.H.; Su, T.X.; Su, Y.T.; Tian, M.K. Machine learning method to predict dynamic compressive response of concrete-like material at high strain rates. *Def. Technol.* **2023**, *23*, 100–111. [CrossRef]
17. Yang, S.C. Special Issue on Fatigue, Performance, and Damage Assessments of Concrete. *Appl. Sci.* **2024**, *14*, 1845. [CrossRef]
18. Bian, H.B.; Liu, Y.Z.; Guo, Y.D.; Liu, Y.; Shi, W.J. Investigating stress–strain relationship and damage constitutive model of basalt fiber reinforced concrete under uniaxial compression. *J. Build. Eng.* **2023**, *73*, 106789. [CrossRef]
19. Lei, X.; Sun, X.J.; Yu, Z.P.; Guan, Z.X.; Long, A.X.; Lian, H.H.; Lian, Y.J. Experimental study and theoretical analysis on dynamic mechanical properties of basalt fiber reinforced concrete. *J. Build. Eng.* **2022**, *62*, 105334.
20. Li, X.N.; An, H.W.; Chen, Y.L.; Qin, Y.; Li, Y.C.; Zhou, Y.J. Mechanical properties of recycled aggregate concrete under cyclic loading and unloading in alpine areas. *J. Build. Eng.* **2024**, *85*, 108759. [CrossRef]
21. Feng, S.W.; Zhou, Y.; Wang, Y.; Lei, M.D. Experimental research on the dynamic mechanical properties and damage characteristics of lightweight foamed concrete under impact loading. *Int. J. Impact Eng.* **2020**, *140*, 103558. [CrossRef]
22. He, H.Q.; Liang, X.; Shang, J.K.; Long, Y.S. Experimental study on residual mechanical properties after cyclic loading of steel-concrete composite-laminated beams. *Case Stud. Constr. Mater.* **2023**, *19*, 202257. [CrossRef]
23. Zhang, B.; Gao, Y.H.; Zheng, H.Z.; Xu, F.; Zhang, Q.B.; Wang, S.Z. Seismic behaviour and modelling of rectangular FRP-concrete-steel tubular columns under axial compression and cyclic lateral loading. *Structures* **2023**, *48*, 1505–1518. [CrossRef]
24. *DZ/T 0276.19-2015 19*; Regulation for Testing the Physical and Mechanical Properties of Rock-Part 19: Test for Determing the Deformability of Rock in Uniaxiail Compression. Standards Press of China: Beijing, China, 2015. (In Chinese)
25. Li, T.T.; Pei, X.J.; Guo, J.; Meng, M.H.; Huang, R.Q. An Energy-Based Fatigue Damage Model for Sandstone Subjected to Cyclic Loading. *Rock Mech. Rock Eng.* **2020**, *53*, 5069–5079. [CrossRef]
26. LI, S.C.; Xu, J.; Tao, Y.Q.; Tang, X.-J.; Yang, H.-W. Low cycle fatigue damage model and damage cariable expression of rock. *Rock Soil Mech.* **2009**, *30*, 1611–1614+1619. (In Chinese)
27. Liu, Y.; Dai, F.; Dong, L.; Xu, N.; Feng, P. Experimental investigation on the fatigue mechanical properties of intermittently jointed rock models under cyclic uniaxial compression with different loading parameters. *Rock Mech. Rock Eng.* **2018**, *51*, 47–68. [CrossRef]

buildings

MDPI

Article

Theoretical Analysis of the Influence of Bearing Plate Position on the Bearing Performance of Soil around the CEP Antipull Force Double Pile

Yongmei Qian [1,*], Lin Sun [1], Lishuang Ai [2], Ying Zhou [1] and Mingxiao Li [3]

1 College of Civil Engineering, Jilin Jianzhu University, Changchun 130118, China; sunlin@student.jlju.edu.cn (L.S.); joey052zy@163.com (Y.Z.)
2 China Power Engineering Consulting Group, Northeast Electric Power Design Institute Co., Ltd., Changchun 130022, China; ailishuang@nepdi.net
3 Tongyuan Design Group Co., Jinan 250101, China; 18166845289@163.com
* Correspondence: qianyongmei@jlju.edu.cn

Abstract: With the development of large-scale projects such as high-rise buildings, deep-sea platforms, bridges, etc., these construction facilities are affected by many factors such as environment and geological conditions, which put forward higher requirements for pile-bearing capacity. Compared with the straight-hole grouted piles, the CEP (concrete expanded-plate) piles have an increased bearing plate, which has stronger resistance to pullout under the action of axial tension. The location of the bearing plate is the main factor affecting the bearing capacity of a CEP pile. This study simulates and analyzes CEP double piles on ANSYS software (Ansys R19.0 versions) under ideal conditions, designs five types of model piles with different bearing plate positions, and divides them into six groups for simulation. Finally, a complete model of the two-pile system is established. It is obtained that when the bearing plate is in the same position, the longer the pile length above the bearing plate, the greater the ultimate bearing capacity of the CEP double piles; when the bearing plates of a double pile are at different positions, the antipull-force-bearing capacity of the double pile mainly depends on the pile with a smaller pile length above the bearing plate, and determines the calculation mode of a CEP double-pile antipull-force-bearing capacity at different bearing plate positions, so as to provide a theoretical basis for the design and application of CEP pile foundations in large building structures in the future.

Keywords: CEP pile; bearing plate position; ANSYS simulation; double-pile model; soil destruction state; antipull force failure; bearing capacity

1. Introduction

As a new type of variable section cast-in-place pile gradually applied in engineering in recent years, the concrete expanded-plate (CEP) pile has the advantages of high bearing capacity, flexible design, small and even settlement, and simple construction technology; moreover, the new pile generator has the ability to improve the working environment on site [1], achieving the effect of high efficiency, energy conservation, and environmental protection [2–4], as shown in Figure 1. Compared with ordinary concrete straight hole cast piles, CEP piles have increased the bearing plate at the pile body position, and the bearing plate can be set flexibly at the pile body position as well as each parameter of the bearing plate, which changes the force mechanism of the pile body, greatly increases the contact area between the pile and the soil body, and increases the end-bearing force, which greatly improves the bearing capacity and stability of the pile body. At present, the foundation research on single piles of CEP pile has become mature [5–8], but the foundation research on pile groups is mostly based on ordinary concrete straight-hole cast-in-place piles [9,10], and research on CEP pile groups is still in the blank stage [11];

it seriously affects the development and application of CEP piles. When the pile spacing is insufficient, adjacent piles will have a greater mutual influence due to the special pile shaft structure of a CEP pile [12,13]. Thus, the bearing capacity of pile group foundations cannot be simply calculated as an integral multiple of single-pile foundations [14], and the position of the bearing plate is the main factor affecting the bearing capacity of CEP pile groups [15].

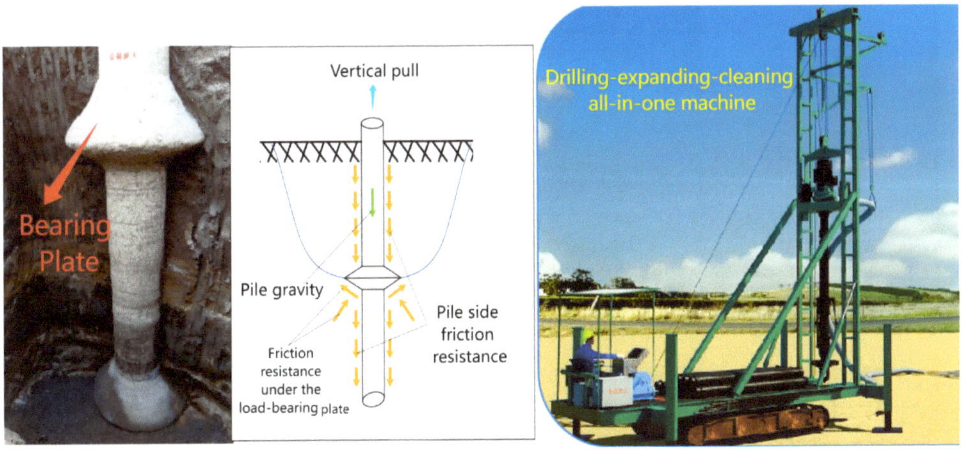

Figure 1. Schematic diagram of the CEP pile.

This study takes CEP double piles as the research object in an ideal condition, where the influence of the change of soil density during the penetration process of the pile [16,17] and the increase of the load during pile driving on the soil mass is neglected [18,19], sets the important parameter of the bearing plate position, and designs five types of model piles to be divided into six groups for finite element simulation of the same plate position and different plate positions. Compared with ABAQUS software (ABAQUS 2021 versions), ANSYS software (Ansys R19.0 versions) can achieve better simulation in static, quasi-static, and dynamic problems and linear analysis [20]; to this end, it is necessary to establish and address the following objectives and tasks through ANSYS software (Ansys R19.0 versions):

(1) Establish the finite element model of CEP double-pile buckling resistance under different pile plate locations, and apply the face load of vertical tension on the top surface of the pile.
(2) The top displacement of the pile, the failure state of soil around the pile, and the shear stress of each node of the pile shaft under the action of vertical tension are recorded. The six groups of results obtained are compared and analyzed, and the corresponding parameters are determined.
(3) The calculation mode of the antipull-force-bearing capacity of CEP double piles with different bearing plates is modified, thereby providing a reliable theoretical basis for the calculation and design application of the bearing capacity of CEP pile groups.
(4) Propose the main design principles of CEP double-pile buckling resistance in actual projects.

2. ANSYS Finite Element Model Construction

Basic assumptions for finite element simulation:

(1) Since the deformation and force state of the soil body around the pile are mainly studied rather than the damage of the CEP pile itself, it is assumed that only the soil body is damaged throughout the simulation analysis, and the CEP double pile always

maintains the elastic working state before reaching the ultimate bearing capacity. The pile body is set as a linear elastic material.

(2) In order to make the goal of simulation analysis more clear and mainly study the influence of plate position, the soil body around the fixed pile adopts a single soil layer; a clayey soil layer is used, and the soil layer is set as elastic–plastic material.

(3) The simulation analysis does not consider the influence of time effect, and the load is continuously applied step by step; the pile and soil are regarded as ideal materials with isotropic and uniform material.

(4) Since the modeled length and width of soil body are much larger than the disc diameter of concrete spreading pile in the simulation, the influence of boundary conditions on the simulation results is not considered.

2.1. Model Size

With the bearing plate position as the only variable, the five model piles are divided into six groups for simulation, namely, MM1 group, MM2 group, MM3 group, MM4 group, MM5 group, and MM6 group. The bearing plates of two model piles in groups MM1–MM4 are the same, whereas the bearing plates of two model piles in groups MM5–MM6 are different. The detailed dimensions of the six groups of model piles are shown in Table 1 and Figures 2 and 3.

Table 1. Pile size parameters for 2D models in ANSYS simulations.

Group	Double-Pile Length above Bearing Plate d_1 (mm)	Pile Length L (mm)	Pile Diameter d (mm)	Pile Spacing S (mm)	Plate Diameter D (mm)	Plate Cantilever Path R_0 (mm)	Plate Slope Angle (°)	Clearance between Plates p (mm)
MM1	3200/3200	8000	500	3200 ($4R_0$)	2100	800	$\alpha = 36°$ $\beta = 21°$	1100 ($1.38R_0$)
MM2	4800/4800							
MM3	5360/5360							
MM4	6400/6400							
MM5	1400/6400							
MM6	3200/4800							

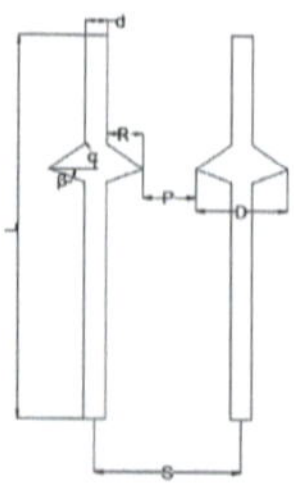

Figure 2. Design details of the simulated pile.

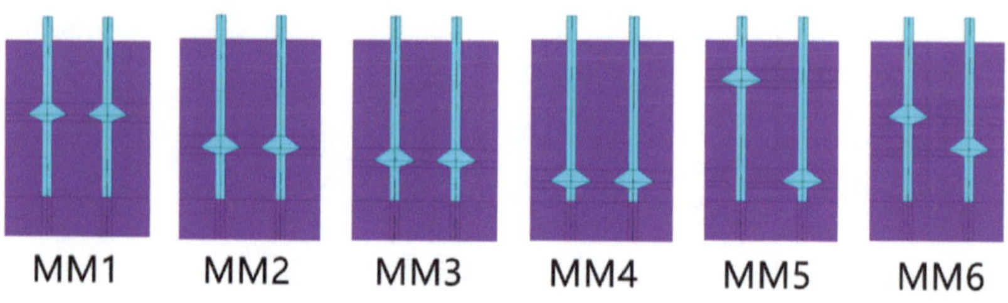

Figure 3. Schematic of the model pile for simulation analysis.

2.2. Material Parameters

The soil parameters in the simulation are consistent with those in the actual geological exploration report to ensure that the simulation analysis is consistent with the actual situation [21,22]. The specific data of soil and pile parameters are shown in Table 2.

Table 2. Pile–soil parameter settings.

Material	Density (t/mm^3)	Elastic Modulus (MPa)	Poisson Ratio	Cohesion (MPa)	Friction Angle (°)	Dilation Angle (°)
Pile	2.25×10^{-9}	3.465×10^4	0.2	–	–	–
Clay	1.488×10^{-9}	25	0.35	0.04355	10.7	10.7

2.3. Establishment of Finite Element Model

In accordance with the material properties of the CEP pile shaft and soil around the pile, SOLID65 element is used for the CEP pile shaft during finite element simulation. SOLID65 element is a commonly used concrete element for ANSYS simulation, with tensile cracking and crushing properties. The soil around the pile adopts SOLID45 element, which is defined by eight nodes and has three degrees of freedom in X, Y, and Z directions.

The establishment process of the CEP double-pile finite element model is basically the same as that of the CEP single-pile finite element model [23,24], except that the soil model is changed from a cylinder to a cube compared to a single pile. The soil model is built from bottom to top, with the length, width, and height of 12,000, 10,000, and 8000 mm, respectively. The model pile is deducted from the soil by Boolean operation, as shown in Figure 4, and then the model pile is combined with the soil. It conforms to the construction process of drilling–expanding plate body–lowering reinforcement cage–pouring concrete in the actual project.

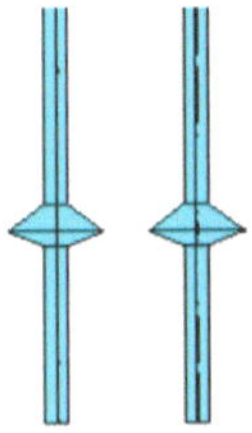

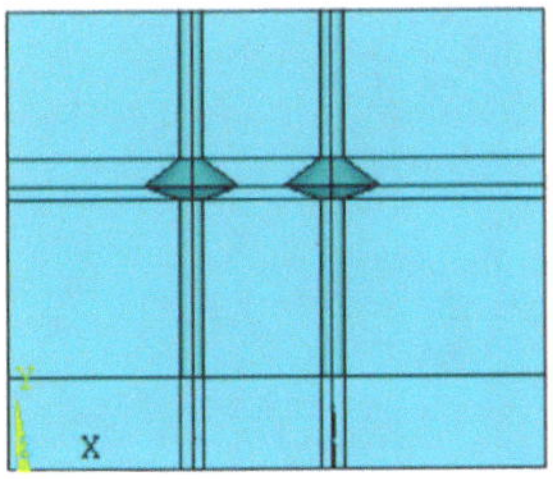

Figure 4. Establishing the pile–soil model.

The model after mapped grid division is shown in Figure 5.

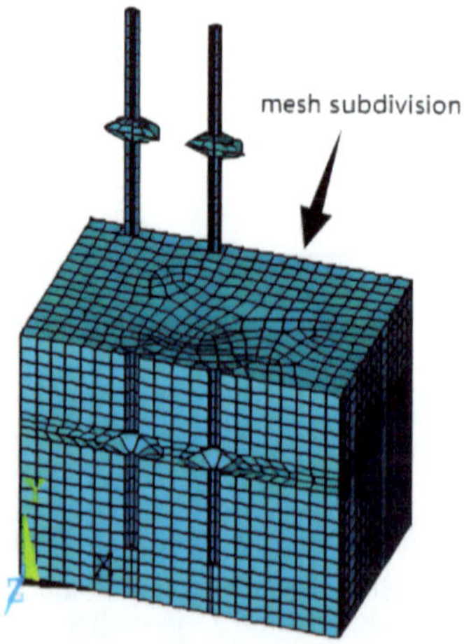

Figure 5. Meshing.

2.4. *Constraints and Load Methods*

2.4.1. Setting of Constraints

In the loading process, the degrees of freedom of each surface of the hexahedron soil must be constrained to prevent large errors caused by soil movement [25]. The front of the hexahedron only constrains the degrees of freedom in the Y direction, and the top face constrains the degrees of freedom in both the X and Y directions. The remaining four faces are constrained in all three directions.

2.4.2. Load Application

In this simulation, we apply the axial tensile load, and adopt the progressive loading mode of long-term loading, with each level of 100 kN. It is applied to the top surface of the pile in the form of surface load. In accordance with the load displacement data extracted from the simulation analysis, when the pile top displacement reaches a sudden change or 100 mm, the load is considered a limit load, terminates the continued loading, and determines the result, as shown in Figure 6.

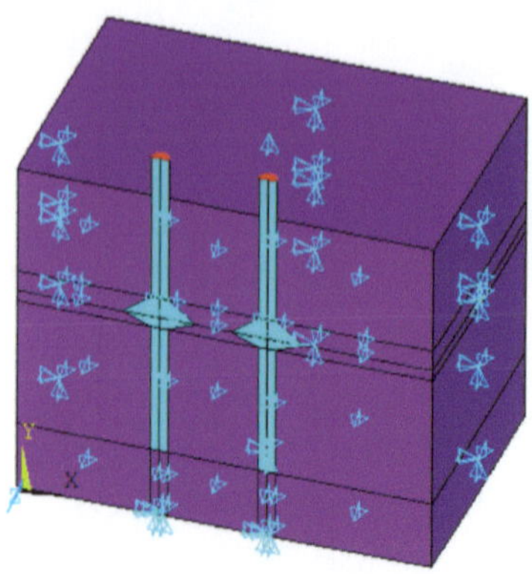

Figure 6. Schematic of restraint and load action position.

3. Displacement Nephogram Analysis of ANSYS Simulation Results

The displacement nephogram of the model pile under various loads is obtained, the analysis data are extracted, and the load displacement curves of the six groups of CEP double piles are drawn through ANSYS finite element analysis. The rules whereby the locations of the same plate and different plates affect the antipull-force-bearing capacity and failure state of CEP double piles are summarized and analyzed.

3.1. Displacement Nephogram Analysis during Loading

3.1.1. The CEP Double-Pile-Bearing Plates at the Same Position

For a more detailed and intuitive observation, the displacement nephogram of CEP double piles and soil around the piles in the simulation analysis is extracted under different loading levels. The MM1 group of pile types at the same position of the bearing plate is selected for detailed description, and the vertical displacement nephogram when the vertical tension is 100, 300, 500, 700, 1100, and 1200 kN is extracted for analysis, as shown in Figure 7.

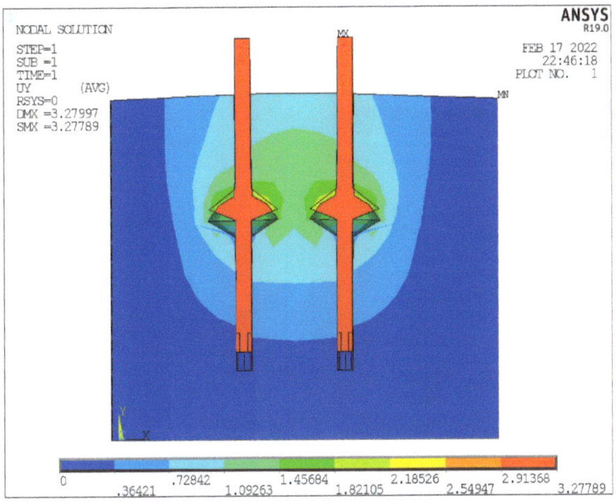

(a)

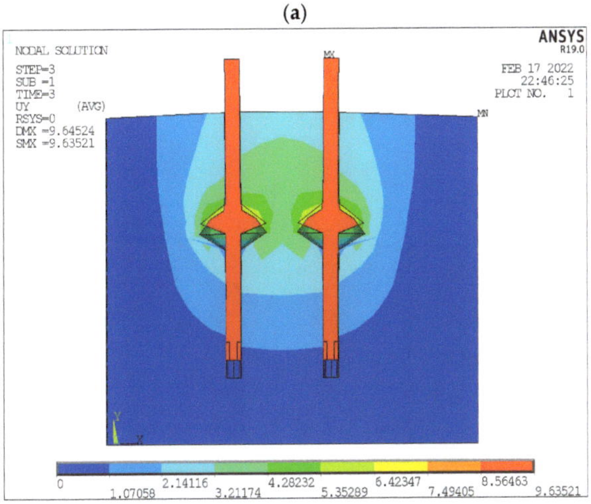

(b)

Figure 7. *Cont.*

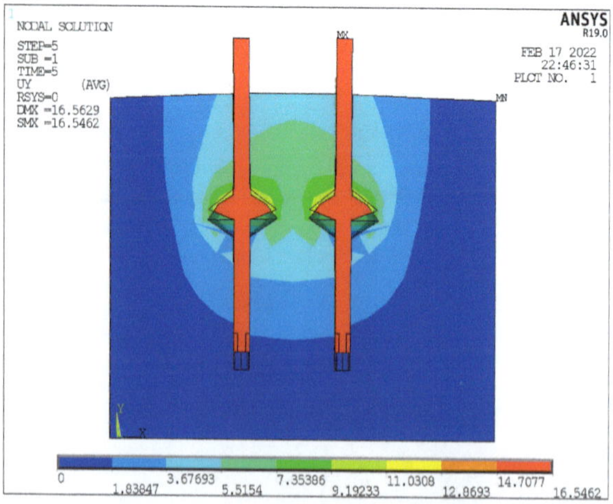

(c)

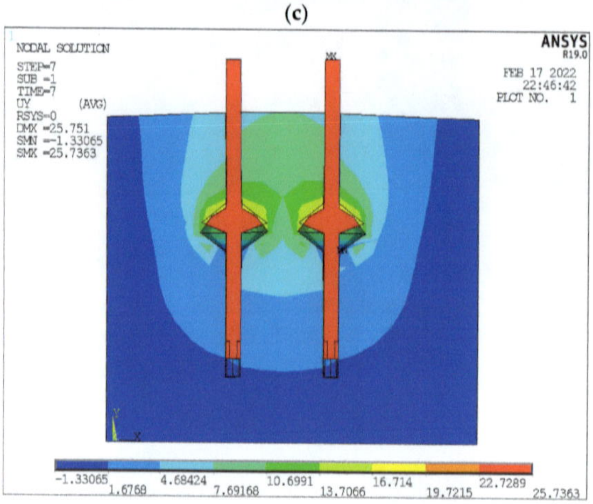

(d)

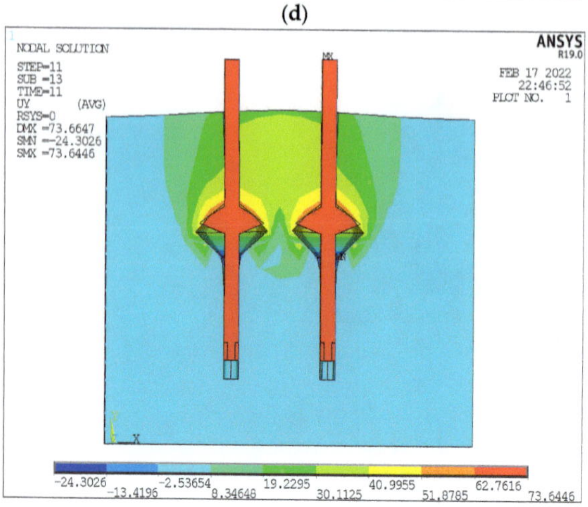

(e)

Figure 7. *Cont.*

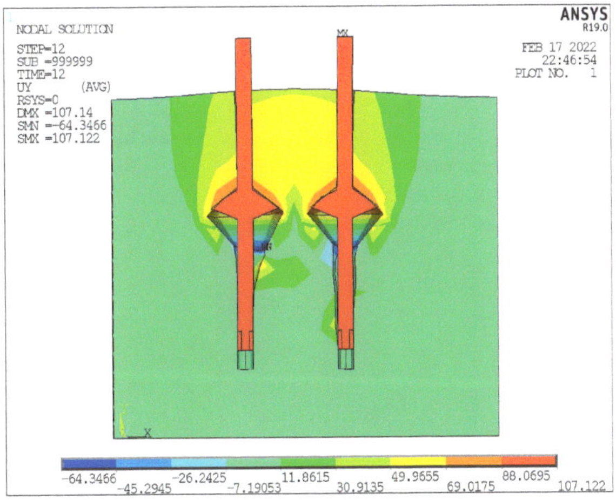

(f)

Figure 7. Displacement cloud diagram of the MM1 group during loading process: (**a**) 100 kN Y direction; (**b**) 300 kN Y direction; (**c**) 500 kN Y direction; (**d**) 700 kN Y direction; (**e**) 1100 kN Y direction; (**f**) 1200 kN Y direction.

When the first stage load of 100 kN is completed, the soil mass above the bearing plate is compressed and deformed by about 1.5 mm. The influence range and deformation of the soil mass around the pile are small, as shown in Figure 7a. When the load is increased to 300 kN, the deformation and influence range of the soil at the upper right of the left pile-bearing plate and the soil at the upper left of the right pile-bearing plate are slightly larger than those at the opposite side due to the interaction between the two piles, but they are still in the linear working stage, as shown in Figure 7b. At 500 kN, the soil at the lower left part of the left pile-bearing plate and the soil at the lower right part of the right pile-bearing plate are pulled apart, resulting in relative displacement, but they have minimal impact on the bearing performance, as shown in Figure 7c. As shown in Figure 7d, the displacement of the pile top is about 25 mm, the relative displacement of the soil mass is connected, the cohesion between the soil particles on and under the bearing plate disappears, shear failure occurs, and the soil mass under the bearing plate stops the work when the load is increased to 700 kN. The soil-bearing capacity mostly disappears after 700 kN due to the increased relative displacement of the soil between the bearing plates of the two piles, as shown in Figure 7e. With the increase in load, the two piles as a whole gradually produce punching failure to the soil on two sides, and the displacement increment increases remarkably. When the load reaches 1200 kN, the ANSYS simulation analysis stops working, reaches failure, and the loading is terminated, as shown in Figure 7f.

3.1.2. The CEP Double-Pile-Bearing Plate at Different Positions

(1) Displacement nephogram analysis of the MM5 group loading process

The vertical distance between the bearing plates of two piles in the MM5 group is $6.25R_0$. The corresponding vertical displacement nephogram when the load is 300, 600, 700, and 900 kN is taken for analysis, as shown in Figure 8.

When the load is 100–300 kN, the displacement of pile top increases linearly with the increase in load, as shown in Figure 8a. When the displacement reaches 25 mm, the load is about 615 kN, as shown in Figure 8b,c, which is less than the load of the first four groups, indicating that the bearing capacity is greatly reduced mainly because the pile length d1 above the left pile-bearing plate is small. A large vertical displacement difference is observed because of the large difference between the positions of the bearing plates of the two piles. The settlement difference between the two piles is 3.95 mm at 615 kN, and the vertical displacement difference between the two piles reaches 25.33 mm at final failure, as shown in Figure 8d. This finding shows that the vertical displacement difference between two piles increases with the increase in load. In the actual project, large additional stress will be caused in the foundation, which should be avoided in the design.

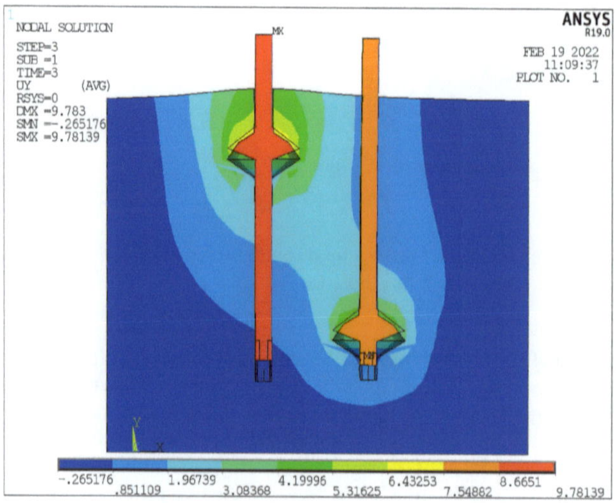

(a)

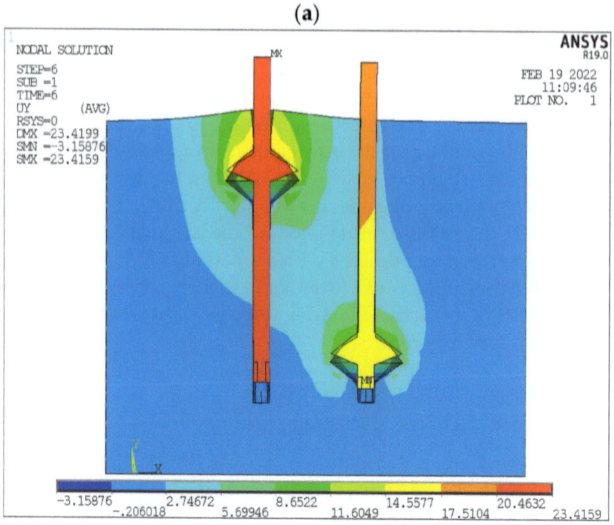

(b)

Figure 8. *Cont.*

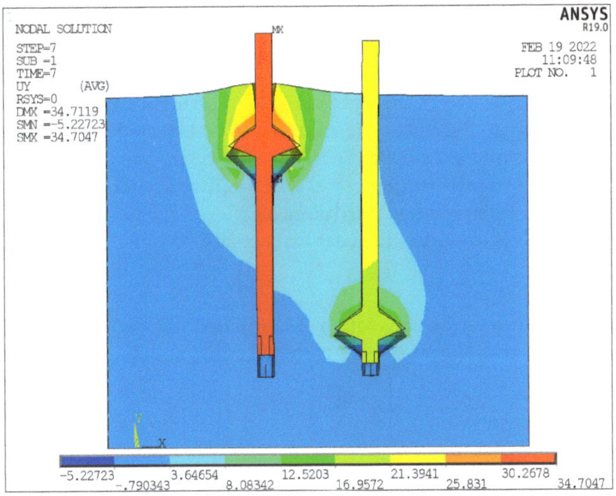

(c)

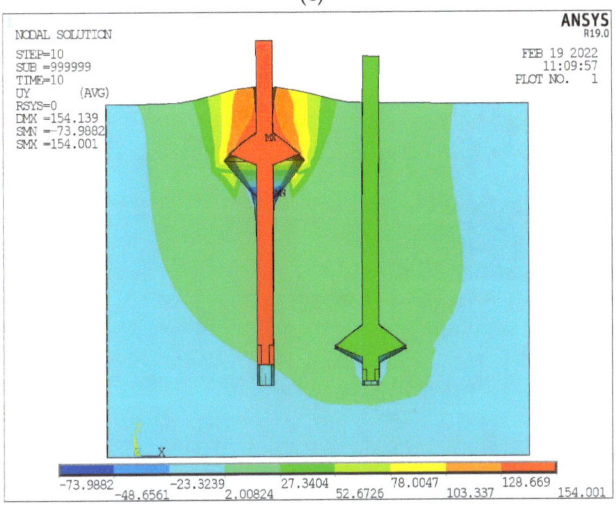

(d)

Figure 8. Displacement cloud diagram of the MM5 group during loading process: (**a**) 300 kN Y direction; (**b**) 600 kN Y direction; (**c**) 700 kN Y direction; (**d**) 900 kN Y direction.

(2) Displacement nephogram analysis of the MM6 group during loading process

The vertical distance between the bearing plates of two piles in the MM6 group is $2R_0$. Different from the first five groups of piles, the two piles have horizontal displacement during the loading process of MM6 group because the soil mass between the two pile plates affects each other, so the vertical displacement and lateral displacement nephograms at 500 and 700 kN are extracted for analysis, as shown in Figure 9.

When the load increases from 0 kN to 500 kN, the displacement of the pile top increases linearly, and the soil around the pile deforms in a "band" shape along the connecting line of the two plates, as shown in Figure 9a. The displacement reaches about 24 mm at 700 kN, as shown in Figure 9b. As shown in Figure 9c,d, the top of the two piles moves away from the direction of the other pile, reaching about 3.5 mm at 700 kN. This condition is because the soil on the upper right side of the left side is subjected to the normal stress inclined to the right, which is transferred to the vicinity of the right pile top, affecting the right pile, and causing the right pile top to move to the right. The left pile top is subjected to the normal stress inclined to the left on the upper left side of the right side, causing the left pile top to move to the left. Therefore, the two pile-bearing plates will produce additional moments of the pile shaft due to mutual influence when the distance between the two plates in the Y direction is $2R_0$, which will adversely affect the bearing capacity.

3.2. *Load Displacement Curve Analysis*

The load displacement values of the six groups of simulated piles are extracted to draw the load displacement curves of the six groups of model piles, as shown in Figure 10.

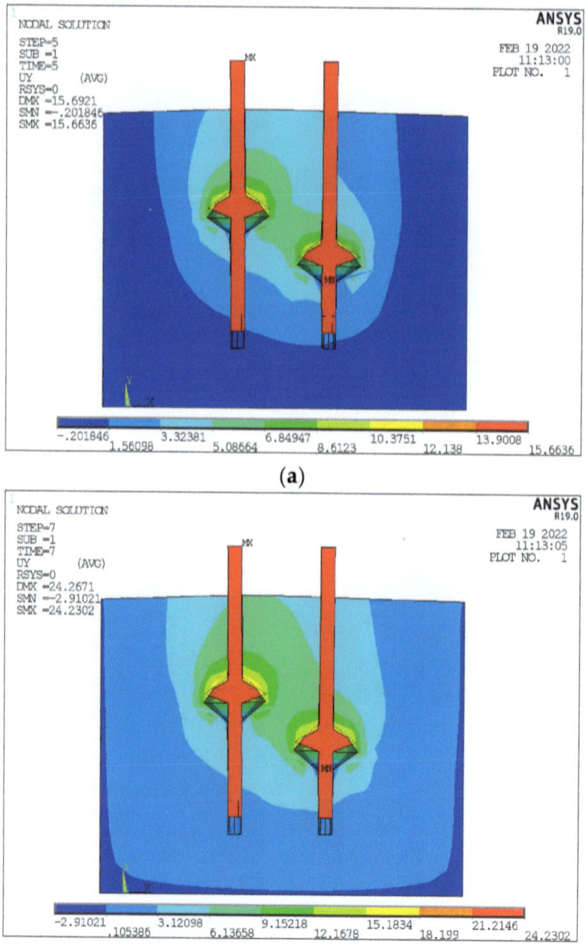

Figure 9. *Cont.*

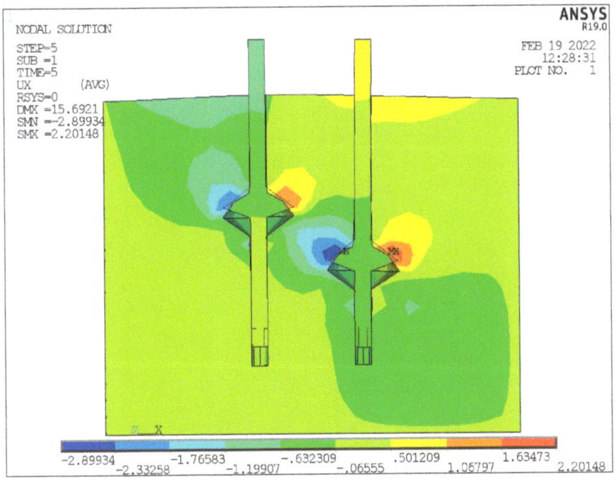

(c)

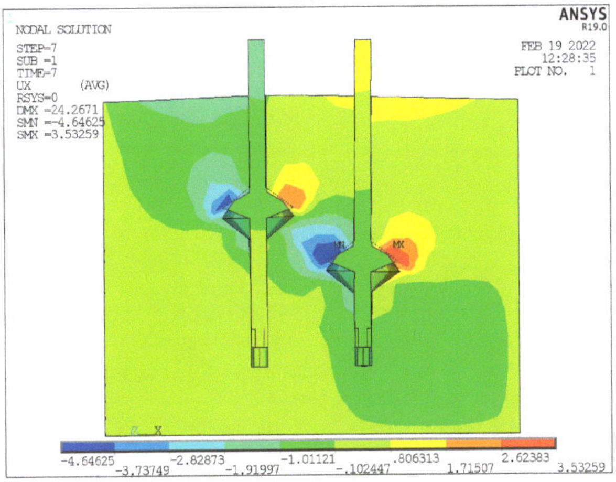

(d)

Figure 9. Displacement cloud diagram of the MM6 group during loading process: (**a**) 500 kN Y direction; (**b**) 700 kN Y direction; (**c**) 500 kN X direction; (**d**) 700 kN X direction.

The observations made from analyzing the load displacement curve in Figure 10 are as follows:

(1) At the initial simulation loading stage of the MM1–MM6 groups of double-pile models, the change trend of load displacement curve is the same, as shown in the red section of Figure 10. At this time, the displacement of the pile top increases linearly with the increase in load, mainly relying on the side friction of double piles and the local soil above the bearing plate to provide the bearing capacity. The bearing plate does not play a role in the bearing capacity of CEP double piles.

(2) In the middle and late stages of loading (outside the red line), the farther the bearing plate is from the soil surface, the smaller the pile top displacement, the greater the load corresponding to failure, and the greater the bearing capacity of the pile with the pile length d1 above the bearing plate under the same load. However, the greater the distance between the bearing plate and the soil surface, the slower the growth rate of its bearing capacity. Under the same load, the pile top displacement of the MM6 group is remarkably smaller than that of the MM5 group. Thus, the antipull-force-bearing

the bearing capacity of the MM6 group is remarkably greater than that of the MM5 group, and the bearing capacity of the bearing plate of the MM5 group is not fully utilized, and the soil has been damaged, so the antipull-force-bearing capacity mainly depends on the pile with a small pile length d1 above the bearing plate.

(3) Under the same load, the antipull-force-bearing capacity of groups M2–M4 is larger, and that of group M4 is the largest. Therefore, the antipull-force-bearing capacity of double piles with the bearing plate at the same position is greater when the pile length d1 above the bearing plate is greater than $4R_0$.

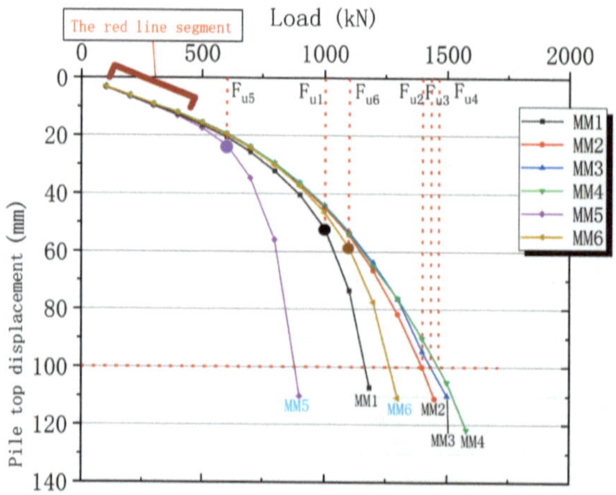

Figure 10. Load–displacement curve of each group of model piles.

For the steep drop load–displacement curve (MM1, MM5, MM6), the ultimate bearing capacity takes the load value corresponding to the starting point where the obvious steep drop occurs; for the slow variable load–displacement curve (MM2, MM3, MM4), the ultimate bearing capacity can be determined according to the pile top displacement, taking the corresponding load value when the pile top displacement is 100 mm.

The ultimate bearing capacity Fu diagram of each group is shown in Figure 11. A large difference is observed between MM1, MM5, MM6, and the other three groups of Fu, where the MM5 group is only 600 kN, and the MM1 and MM6 groups are about 1000 kN. This condition is because the pile length d1 on the bearing plate is extremely short, which is caused by punching failure.

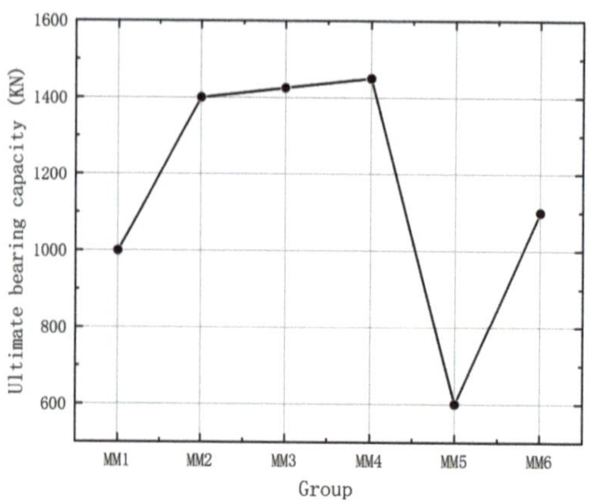

Figure 11. Ultimate bearing capacity of each group of piles.

4. Shear Stress Curve Analysis of ANSYS Simulation Results

For the six groups of model piles, the shear stress values of each node of the pile shaft in the loading process of groups MM1–MM4 with bearing plates at the same position and groups MM5 and MM6 with bearing plates at different positions are analyzed.

4.1. The CEP Double-Pile-Bearing Plates at the Same Position

The right side of the left pile and the left side of the right pile are called the "influence side", as shown by the yellow bold line in Figure 12. The other two sides are called the "normal side", as shown by the pink bold line in Figure 12.

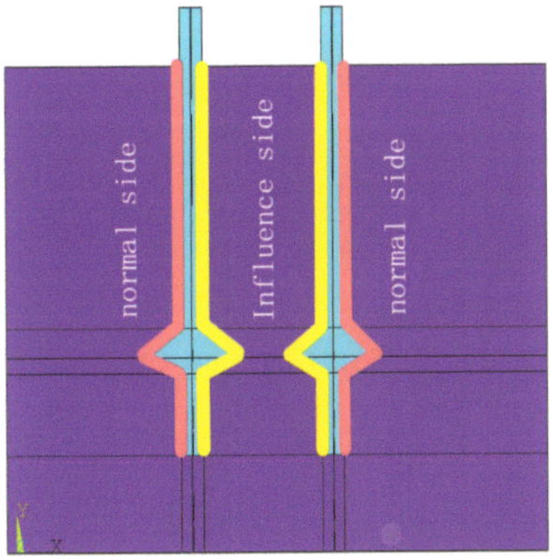

Figure 12. Schematic of the influence side and the normal side.

The MM3 group is selected to study the loading process at the same position of the bearing plate. Sixteen points are evenly selected along the pile shaft on two sides of the left pile, and the loading process of 0–700 kN (at this time, the displacement of the pile top increases linearly with the increase in load) is extracted. The shear stress values of 32 points on the normal side and the influence side of the left pile are shown in Figure 13, and the shear stress change diagrams of each part of the pile shaft on two sides with the loading of MM3 pile are shown in Figures 14 and 15.

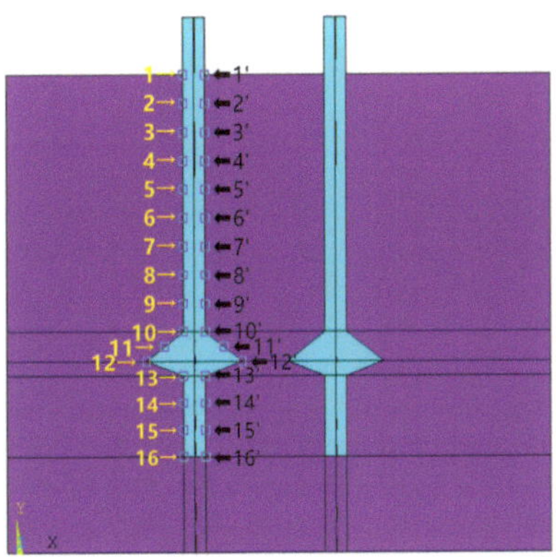

Figure 13. Node distribution diagram of left pile body.

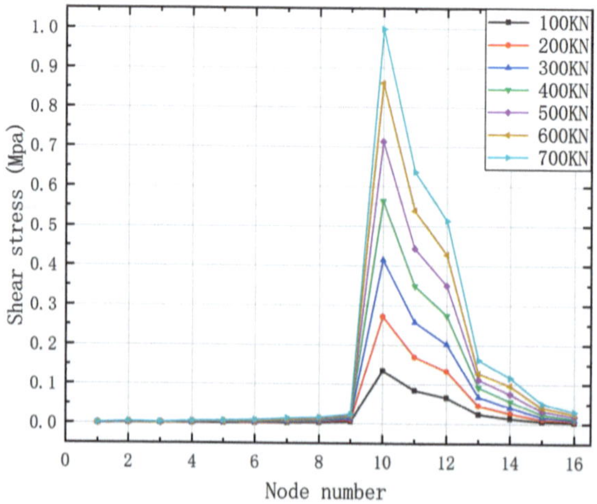

Figure 14. Variation of shear stress on the normal side of the left pile in the MM3 group.

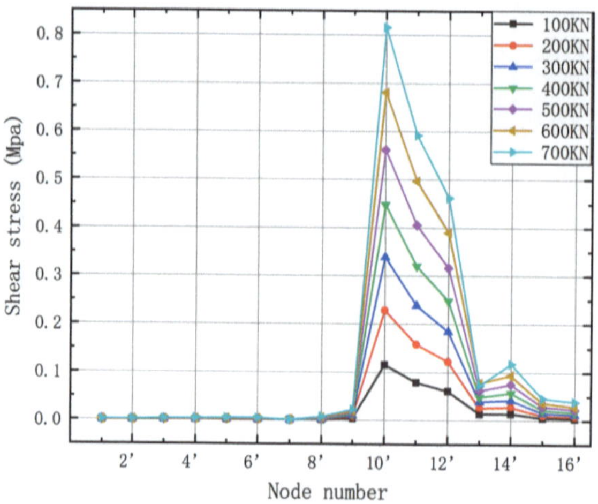

Figure 15. Variation diagram of shear stress on the influence side of the left pile in the MM3 group.

The observations made by analyzing Figures 14 and 15 are as follows:

(1) Compared with the six corresponding shear stress curves when loading on the same side, the position of the bearing plate from the beginning of loading bears approximately more than 80% of the antipull of the pile shaft, and the compressive pile is lower than this value, about 60%. This condition is because the soil mass at the bottom of the compression pile can provide most of the end-bearing capacity and bear down pressure together with the bearing plate. The antipull force pile only has the bearing plate to provide the end-bearing capacity, which causes excessive concentration of the shear stress of the bearing plate and the pile diameter near the bearing plate. Therefore, the antipull force CEP pile should be further densified with steel bars near the bearing plate compared with the compression pile during the design to ensure the safety of the project.

(2) The maximum shear stress occurs at numbers 10 and 10′, that is, the junction of the bearing plate and the upper pile diameter. Extremely large sudden changes in shear stress are observed between points 9 and 10, and between points 9′ and 10′, approximately from zero to the maximum, which is prone to brittle failure. The main

reason for this phenomenon is because the pile top transmits the vertical antipull force to the bearing plate through the upper pile diameter, and the bearing plate bears approximately 80% of the load, so an extremely large shear stress mutation occurs at the junction of the bearing plate and the upper pile diameter. Therefore, this position can be designed as a circular arc during antipull force resistance in actual projects. Although the engagement between the bearing plate and the soil around the plate is weakened, this condition will greatly reduce the shear stress mutation at the junction of the bearing plate and the upper pile diameter, and decrease the risk of damage.

(3) The shear stress of the lower pile diameter is greater than that of the upper pile diameter, and the shear stress of the upper pile diameter is approximately zero when CEP piles are antipull-force-resistant by comparing numbers 1–9, 1′–9′ nodes and numbers 13–16, 13′–16′ nodes. This finding shows that the role of the upper pile diameter is mainly to transfer the antipull force load of the pile top to the bearing plate and then transfer the stress to the soil around the pile by the bearing plate and the lower pile diameter to provide the bearing capacity.

The maximum point of shear stress of the bearing plate is point 10, then point 11, point 12, and point 13 in order. From Figure 16, except for point 13, the ratio of shear stress between two sides of the pile at the remaining three points is between 0.75 and 0.95, and the ratio is relatively concentrated. It shows that the mutual influence between two piles at these three points is not much different, and the shear stress on the influence side is discounted to about 0.9 compared with that on the normal side (taking the average value). The shear stress at point 13 is very small, only 15% of that at point 12, so its ratio of shear stress is negligible. The ratio of shear stresses on both sides is slowly decreasing with the increase of load. It shows that when the load is small, the reduction of shear stress on the influence side is small compared with the normal side, i.e., the degree of mutual influence between the two piles is weak. With the increase of load, the shear stress reduction also becomes larger, and the ratio decreases from 0.9 at 0 kN–700 kN to 0.86 at 1000 kN.

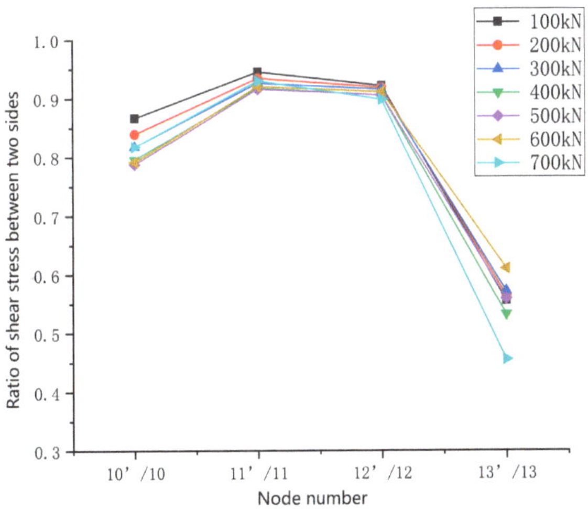

Figure 16. MM3 group left pile-bearing plate near both sides of the stress ratio graph.

4.2. The CEP Double-Pile-Bearing Plate at Different Positions

The vertical distance between the bearing plates of two piles in group MM5 is $6.25R_0$. The two plates have minimal effect in the load range of 0–700 kN because their vertical distance is extremely large, so group MM6 is taken to study the shear stress on two sides of the double piles at different positions.

The shear stresses corresponding to the 700 kN load on the normal side of the left pile, the influence side of the left pile, the influence side of the right pile, and the normal side of the right pile are extracted. Sixteen points along the pile shaft at each side are taken, with a total of 64 points, and the shear stress change curve of the pile shaft under 700 kN load is shown in Figure 17.

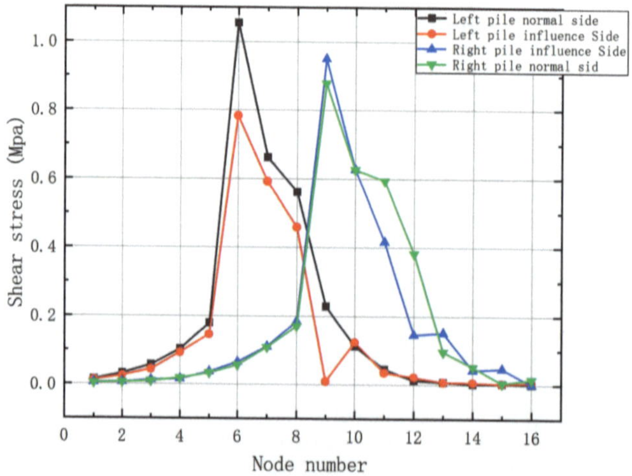

Figure 17. Variation curve of shear stress under 700 kN load of the MM6 group.

As shown in Figure 17, the shear stress at the bearing plate with different positions decreases remarkably, unlike the shear stress change when the bearing plate is at the same position. This condition is due to the dislocation of the bearing plates of two piles, and the disturbed soil cannot provide sufficient shear stress for the bearing plate.

The left pile affects the shear stress of the side bearing plate more seriously than the right pile, which is 74.31% for the left pile and 92.43% for the right pile, with a difference of about 18%. Specifically, the impact on the shear stress of the plate with relatively small burial depth is greater when the two pile-bearing plates are at different positions. However, the total shear stress of the two piles is equal, which indicates that the normal side of the pile with a smaller pile length d1 on the plate will bear a greater shear stress, which is also the reason why the maximum shear stress of the normal side of the left pile is greater than that of the right pile. Therefore, special attention should be paid to the pile with smaller pile length d1 above the bearing plate when it is in different positions in the actual project.

5. Calculation Mode for Antipull-Force-Bearing Capacity of CEP Double Piles

On the basis of the original calculation model of the antipull-force-bearing capacity of CEP single piles, the reduction factor of the antipull-force-bearing capacity of CEP double piles at different plate locations when the distance between two piles is $4R_0$ is α. The antipull-force-bearing capacity calculation mode of CEP double piles is proposed.

5.1. Calculation Mode of CEP Single-Pile-Bearing Capacity

When the soil on the bearing plate is not subject to punching failure under vertical tension, the soil around the pile is subject to sliding failure [26]. On the basis of the virtual work principle and slip line theory, a formula for calculating the bearing capacity of CEP antipull force for a single pile is proposed (1)~(4). The strain field in the Prandtl region of the soil on the bearing disk and the range of variation of the pile lateral friction resistance are shown schematically in Figures 18 and 19.

$$F_{pull\ out} = F_{disk\ end} + F_{pile\ side} + G_{pile} \quad (1)$$

$$F_{disk\ end} = \frac{1}{2}\pi(c\ cot\ \phi\ R_0)\left(e^{2\theta tan\phi} - 1\right) \tag{2}$$

$$F_{pile\ side} = f_{side}\pi dL_0 = f_{side}\pi d(L - H - L'_b + \gamma' L'_a) \tag{3}$$

$$G_{pile} = \frac{\pi\gamma_G}{12}\left[3d^2(L-H) + H\left(D^2 + Dd + d^2\right)\right] \tag{4}$$

[1] D, d, and R_0 = plate diameter, pile diameter, and cantilever diameter {plate cantilever diameter = $\frac{1}{2}$(plate diameter–pile diameter)};

c—cohesive force of soil;

ϕ—internal friction angle of soil;

L—pile length;

H—pile plate height;

γ'—increase factor, 1.1–1.2;

L'_a—increased range of horizontal compressive stress around the main pile on the plate;

L'_b—the horizontal tensile stress range appears around the main pile under the plate.

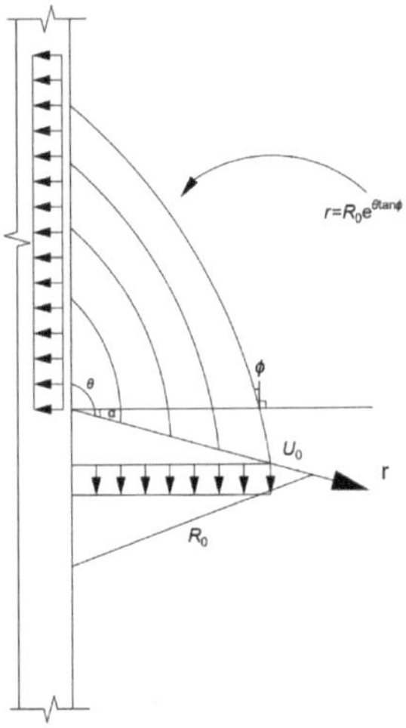

Figure 18. Strain field in the Prandtl region of the soil on the bearing plate.

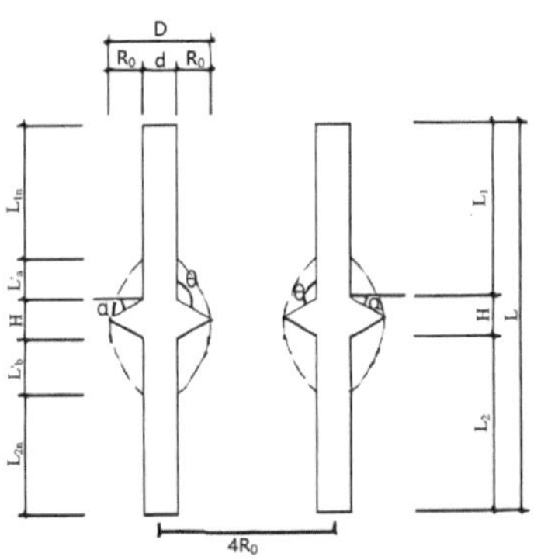

Figure 19. Schematic diagram of the range of variation of pile lateral resistances.

5.2. The Reduction Factor α and the Determination of the Calculation Formula

Antipull force simulation finite element analysis of MEP double piles of the MM1–MM4 groups corresponding to MEP single piles of the M1–M4 groups are conducted, and the pile top load displacement data are extracted. Linear interpolation is used to extract the limit load values of M1–M4 and MM1–MM4 groups, and the ratio of the MM1–MM4 groups of double piles to the M1–M4 groups of single piles is calculated. The results are shown in Table 3. Therefore, the CEP double-pile tensile-bearing capacity is calculated as shown in Equation (5).

$$F_{d-pull\ out} = \alpha\left(F_{disk\ end} + F_{pile\ side}\right) + 2G_{pile}, \quad (5)$$

Table 3. α values of different plate positions.

—	MM1/M1	MM2/M2	MM3/M3	MM4/M4
Pile length above bearing plate (d_1)	$4R_0$	$6R_0$	$6.7R_0$	$8R_0$
α	1.8252	1.8400	1.8462	1.8480

6. Conclusions

(1) In this paper, the simulation study on the pull-out bearing capacity of CEP double piles mainly analyzes the influence of the bearing plate position on the bearing performance of CEP double piles, compared with the bearing performance of single piles. It can be learned that within a certain pile spacing, the bearing capacity of CEP double piles is not twice the bearing capacity of single piles, the interaction between the two bearing plates of CEP double piles results in the folding down of the bearing capacity, which is due to the fact that when CEP double piles are subjected to the vertical tensile force, the load is transferred to the bearing plate through the pile body; the bearing plate and the soil around the pile are in contact with each other, the bearing plate plays the main bearing role, the soil is extruded above the bearing plate, and the soil above the bearing plate generates slip; with the change of different bearing plate positions, the bearing capacity of the double pile is also changing.

(2) When the pile spacing and bearing plate position of CEP double piles are set reasonably, the failure state of soil around each pile is the same as that of the single pile. When the bearing plates of double piles are at the same position, the greater the pile length d1 above the bearing plates, the greater the ultimate bearing capacity of CEP double piles. When the bearing plates of double piles are at different positions, the antipull-force-bearing capacity of double piles mainly depends on the pile with a smaller pile length d1 above the bearing plate. The pile length d1 above the bearing plate should be greater than $4R_0$ as much as possible to avoid punching failure.

(3) The main design principles of CEP double-pile antipull force resistance in actual projects are proposed as follows:

A. The position of the bearing plates of two piles should be the same as much as possible, and the pile length d1 above should be greater than $4R_0$.

B. When the positions of the bearing plates of two piles are different, the pile length d1 above the bearing plates should be greater than $4R_0$, and the vertical distance between the bearing plates of two piles should meet the requirement of no mutual influence and does not need to be larger.

C. At the time of design, tensile CEP piles should be rebar-encrypted near the load-bearing plate compared to antipressure CEP piles.

D. During antipull force, the interface between the bearing plate and the main pile diameter can be designed as a circular arc.

(4) The calculation formula of CEP double-pile pullout bearing capacity was established, which can effectively arrange the number of piles used in the actual project and reduce the waste of concrete.

7. Outlook

In this paper, the influence of tensile capacity of CEP double piles under different pile plate positions is deeply studied by ANSYS finite element simulation software, which fills in the gap of the research on tensile capacity of CEP double piles, and provides a reliable theoretical basis for the calculation of the capacity of CEP group piles and the design and application of CEP group piles, which reasonably arranges the number of piles and reduces the waste of concrete compared with the straight-hole grouted piles. However, there are many other factors affecting the tensile-bearing capacity of CEP double piles in actual projects, such as soil layer properties, number of bearing plates, bearing plate angle, etc. In the next step, it is necessary to consider the influence of many factors on the bearing capacity of CEP double piles, and to improve the theory of the research on CEP piles.

Author Contributions: Conceptualization, Y.Q. and L.S.; methodology, Y.Q.; software, L.S.; validation, L.A., Y.Z. and M.L.; formal analysis, Y.Q.; investigation, Y.Q.; resources, Y.Q.; data curation, Y.Q.; writing—original draft preparation, L.S.; writing—review and editing, L.S.; visualization, Y.Q.; supervision, Y.Q.; project administration, Y.Q.; funding acquisition, Y.Q. All authors have read and agreed to the published version of the manuscript.

Funding: This research was funded by the National Natural Science Foundation of China under Grant number 52078239.

Data Availability Statement: Not applicable.

Conflicts of Interest: The authors declare no conflict of interest.

References

1. Baohan, S. Special Lecture on New Technology of Pile Foundation Construction(thirty-one) Construction method of drilling-expanding-clearing integrated machine and its multi-section drilling and expanding pile. *Eng. Mach. Maint.* **2014**, 106–108+110–114.
2. Qian, Y.; Li, M.; Wang, R.; Jin, Y. Theoretical study on the load carrying capacity of concrete drilled and expanded piles with different stiffness under horizontal load. *IOP Conf. Ser. Earth Environ. Sci.* **2021**, *760*, 012002. [CrossRef]
3. Zhang, M.; Xu, P.; Cui, W.; Gao, Y. Bearing behavior and failure mechanism of squeezed branch piles. *J. Rock Mech. Geotech. Eng.* **2018**, *10*, 935–946. [CrossRef]
4. Ting, L.; Peng, X.; Yang, G. Investigation into bearing performance of concrete expanded-plates piles: Field test and numerical modelling. *Eng. Struct.* **2022**, *271*, 114954.
5. Ma, H.; Wu, Y.; Tong, Y.; Jiang, X. Research on bearing theory of squeezed branch pile. *Adv. Civ. Eng.* **2020**, *2020*, 6637261. [CrossRef]
6. Qian, Y.; Wang, J.; Wang, R. The analysis of the vertical uplift bearing capacity of single CEP pile. *Open Civ. Eng. J.* **2015**, *9*, 598–601. [CrossRef]
7. Chen, Y.; Qian, Y.; Hong, G.; Jin, Y.; Wang, R. Study of undisturbed soil test about slope angle of the expanded-plate affecting the failure mechanism of the NT-CEP pile under vertical tension. *J. Phys. Conf. Ser.* **2022**, *2202*, 012023. [CrossRef]
8. Tian, W.; Qian, Y.; Lang, S.; Wang, R. The Undisturbed-Soil Experimental Analysis of the Destructive-State Influence of the Plate Diameter under the Vertical Tension on the CEP Pile. *Chem. Eng. Trans.* **2018**, *66*, 445–450.
9. Hadi, D.H.; Waheed, M.Q.; Fattah, M.Y. Effect of Pile's Number on the Behavior of Piled Raft Foundation. *Eng. Technol. J.* **2021**, *39*, 1080–1091. [CrossRef]
10. Fattah, M.Y.; Al-Obaydi, M.A.; Al-Jalabi, F.A. Effect of number of piles on load sharing in piled raft foundation system in saturated gypseous soil. *Int. J. Civ. Eng. Technol.* **2018**, *9*, 932–944.
11. Al-Suhaily, A.S.; Abood, A.S.; Fattah, M.Y. Bearing capacity of uplift piles with end gates. In Proceedings of the China-Europe Conference on Geotechnical Engineering, Vienna, Austria, 13–16 August 2016; Springer International Publishing: Cham, Switzerland, 2018; Volume 2, pp. 893–897.
12. Liang, F.; Yu, F.; Han, J. A simplified analytical method for response of an axially loaded pile group subjected to lateral soil movement. *KSCE J. Civ. Eng.* **2013**, *17*, 368–376. [CrossRef]
13. Zhang, H.; Li, C.; Yao, W.; Long, J. A novel approach for determining pile spacing considering interactions among multilayered sliding masses in colluvial landslides. *KSCE J. Civ. Eng.* **2019**, *23*, 3935–3950. [CrossRef]
14. Fan, Y.; Wang, J.; Feng, S. Effect of spudcan penetration on laterally loaded pile groups. *Ocean. Eng.* **2021**, *221*, 108505. [CrossRef]
15. Qian, Y.; Chao, S.; Wang, R. Theoretical Analysis of Plate Location Effect on Soil Failure State under Horizontal Force of a Concrete Plate-Expanded Pile Used for Oceanographic Engineering. *J. Coast. Res.* **2020**, *106*, 655–659. [CrossRef]
16. Dobrzycki, P.; Kongar-Syuryun, C.; Khairutdinov, A. Vibration reduction techniques for Rapid Impulse Compaction (RIC). *J. Phys. Conf. Ser.* **2019**, *1425*, 012202. [CrossRef]
17. Herbut, A.; Khairutdinov, M.; Kongar-Syuryun, C.; Rybak, J. The surface wave attenuation as the effect of vibratory compaction of building embankments. *IOP Conf. Ser. Earth Environ. Sci.* **2019**, *362*, 012131. [CrossRef]
18. Khayrutdinov, M.; Kongar-Syuryun, B.; Khayrutdinov, M.; Tyulyaeva, Y.S. Improving Safety when Extracting Water-soluble Ores by Optimizing the Parameters of the Backfill Mass. *Occup. Saf. Ind.* **2021**, *1*, 53–59. [CrossRef]
19. Golik, I.; Kongar-Syuryun, B.; Michałek, A.; Pires, P. Ground transmitted vibrations in course of innovative vinyl sheet piles driving. *J. Phys. Conf. Ser.* **2021**, *1921*, 012083. [CrossRef]

20. Baca, M. Numerical Modeling of Pile Installation Influence on Surrounding Soil. In Proceedings of the 17th Internationa Multidisciplinary Scientific GeoConference, SGEM 2017, Albena, Bulgaria, 27 June–6 July 2017.
21. Han, W.F.; Feng, B.B.; Zhou, J.; Lu, C.Y. The Study on the Engineering Properties of Squeezed Branch Piles under Combined Loa *Appl. Mech. Mater.* **2014**, *580–583*, 371–375. [CrossRef]
22. Yin, L.; Fan, X.; Wang, S. A study on application of squeezed branch pile in clay soil foundation. *IOP Conf. Ser. Earth Environ. Sc* **2017**, *61*, 012091. [CrossRef]
23. Qian, Y.; Yang, Y.; Hong, G.; Tian, W.; Song, Y. Theoretical Analysis of the Influence of a Sectional Form of a Flexible NT-CEP Pil Disc on the Anti-tilting Loading Performance. *KSCE J. Civ. Eng.* **2022**, *26*, 4709–4716. [CrossRef]
24. Baca, M.; Rybak, J. Pile Base and Shaft Capacity under Various Types of Loading. *Appl. Sci.* **2021**, *11*, 3396. [CrossRef]
25. Kahyaoglu, M.R.; Imançli, G.; Önal, O.; Kayalar, A.S. Numerical analyses of piles subjected to lateral soil movement. *KSCE J. Ci Eng.* **2012**, *16*, 562–570. [CrossRef]
26. Qian, Y. *Study on Damage Mechanism of Soil around Concrete Spread Plate Pile and Vertical Bearing Capacity of Single Pile*; Chin Building Industry Press: Beijing, China, 2018; Volume 11.

MDPI AG
Grosspeteranlage 5
4052 Basel
Switzerland
Tel.: +41 61 683 77 34

Buildings Editorial Office
E-mail: buildings@mdpi.com
www.mdpi.com/journal/buildings